Cindy Majewski
152-60-2296-5

Physical Chemistry Laboratory:
Principles and Experiments

Physical Chemistry Laboratory

PRINCIPLES AND EXPERIMENTS

Hugh W. Salzberg
Jack I. Morrow
Stephen R. Cohen
Michael E. Green

MACMILLAN PUBLISHING CO., INC.
New York
COLLIER MACMILLAN PUBLISHERS
London

COPYRIGHT © 1978, MACMILLAN PUBLISHING CO., INC.

PRINTED IN THE UNITED STATES OF AMERICA

All rights reserved. No part of this book may be reproduced or transmitted in any form or by any means, electronic or mechanical, including photocopying, recording, or any information storage and retrieval system, without permission in writing from the Publisher.

Earlier edition entitled *Laboratory Course in Physical Chemistry* copyright © 1966 by Academic Press, Inc. Earlier edition entitled *Physical Chemistry: A Modern Laboratory Course* copyright © 1969 by Academic Press, Inc.

MACMILLAN PUBLISHING CO., INC.
866 Third Avenue, New York, New York 10022

COLLIER MACMILLAN CANADA, LTD.

Library of Congress Cataloging in Publication Data

Main entry under title:

Physical chemistry laboratory.

   Published in 1966 under title: Laboratory course in physical chemistry, by H. W. Salzberg, J. I. Morrow, and S. R. Cohen.
   Includes bibliographies and index.
   1. Chemistry, Physical and theoretical—Laboratory manuals. I. Salzberg, Hugh W. II. Salzberg, Hugh W. Laboratory course in physical chemistry.
QD457.P49   541'.3'028   77-3883
ISBN 0-02-405350-3

Printing: 1 2 3 4 5 6 7 8   Year: 8 9 0 1 2 3 4

Preface

Most chemists earn their living in laboratories, either doing or directing laboratory work. Laboratory physical chemistry is therefore a discipline in its own right. This text is intended to train undergraduates for the physicochemical experimentation they will be doing in their future careers when, without the guidance of an instructor, they will decide on methods, select appropriate apparatus, take measurements, process data, and write reports.

The first half of the book contains discussions of most of the basic measurements and equipment used in the field, together with chapters on general laboratory practice and safety, data processing, drawing graphs, and writing reports. This information is the mental equipment that trained chemists should have at their disposal no matter what set of experiments they have performed during their coursework.

The second part consists of a selection of experiments designed to teach basic laboratory procedures and at the same time to reinforce basic theory. The measurements are used to calculate important quantities. Most of these experiments were in the previously published versions of the text and have been performed successfully for years. New experiments have been added and some of the originals have been modified. All have been use-tested with relatively large numbers of students and instructors. We have not included experiments involving a few methods, such as nmr, which we have not been able to use-test. We have also omitted experiments involving only analytical techniques, such as equilibrium and kinetics experiments that require just sampling and titrating. In most cases, the experiments in the book can be done with standard commercially available apparatus so that students may become familiar with the type of equipment they will encounter in the field. There are, however, some special ad hoc setups for unusual experiments, such as that on electrokinetic phenomena.

We want students to realize that each apparatus has several uses and that

PREFACE

each measurement can be made in several different ways. The directions for using the apparatus are written in general terms, to help the student to understand the principle of the measurement, no matter what the make or model of the apparatus. They are intended to supplement the manufacturer's instruction booklet, since no one set of detailed directions could be applied to all the available models. For apparatus that is not commercial, the directions are more specific.

The format of the experiments has been devised so that the student must study beforehand. The experiment cannot be done while reading line by line. Warnings and suggestions are provided under the heading "Notes," but to make use of these, the student must have read the assigned text material and the experiment.

We have tried to include as much background information as possible without writing either an encyclopedia or a textbook of theory. There is included, however, a chapter on thermodynamics of irreversible processes, on the grounds that it is both necessary for the performance of the experiment on electrokinetic phenomena and so important in its own right that the student should have some contact with it. We have included in the chapter on electrical measurements some AC theory and some basic electronics, in the belief that most students have very little knowledge of or experience in this important area. The appendixes contain tables of necessary data and some important practical information that is not readily available to students.

We wish to thank Professors Francis Condon, Joseph Rennert, Henri Rosano, Horst Schulz, Amos Turk, Michael Weiner, and Arthur Woodward and Dr. Hilton Evans for their help, suggestions, and criticism. Our special thanks go to our editor, Elisabeth Belfer, for her professional expertise and her labors at keeping this work in the English language.

H. W. S.
J. I. M.
S. R. C.
M. E. G.

Contents

Part One
PRINCIPLES
of Laboratory Practice

1.	Introduction: General Instructions to the Student	3
2.	Data Processing	8
3.	Drawing Graphs from Experimental Data	23
4.	Selected Numerical and Graphical Methods	25
5.	Recording and Reporting Data	42
6.	Temperature Measurement and Control	50
7.	Calorimetry	63
8.	Density Measurement	75
9.	Pressure Measurements, Vacuum Systems, Handling Gases	79
10.	Vapor Pressures and Boiling Points of Liquids and Solutions	99
11.	Thermal Methods of Analysis	105
12.	Viscosity	113
13.	Surfaces and Membranes	124
14.	Electrical Measurements and Circuits	139
15.	Electrical and Electronic Devices	168
16.	Electrochemistry	186

17. Polarimetry 217
18. Refractometry 221
19. Absorption and Emission of Radiation 226
20. X-Ray Diffraction 251
21. Magnetochemistry 269
22. Mass Spectrometry 289
23. Thermodynamics of Irreversible Processes 302
24. Techniques of Chemical Kinetics 316
25. Introduction to Computers 324

Part Two
EXPERIMENTS:
Applications of Laboratory Techniques

1. Viscosity of a Gas / *Pressure–Time Measurement* 345
2. Velocity of Gas Molecules / *Speed-of-Sound Method* 347
3. Viscosity of a Liquid / *Ostwald Viscometry* 351
4. Vapor Pressure Versus Temperature / *Ramsay–Young, Isoteniscope, or Internal Manometer Method* 354
5. X-Ray Diffraction Photographs of a Cubic Solid / *Debye–Hull–Scherrer Method* 356
6. Repeat Distance in a Fiber / *Rotating-Crystal X-Ray Method* 360
7. Repeat Distance in a Single Crystal / *Rotating-Crystal X-Ray Method* 362
8. Molecular Weight of a Solid / *Cryoscopy* 364
9. Cryoscopic Constant / *Beckmann Method* 366
10. Activity of a Solvent / *Vapor Pressure by Ramsay–Young Method* 369
11. Mean Ionic Activity Coefficient / *EMF Measurement* 372
12. Transference, Ionic Conductance, and Ionic Mobility / *Conductance, Moving-Boundary Method* 376

13. Energies and Heats of Combustion and Formation / *Constant Volume Calorimetry* 379
14. Resonance Stabilization Energy of *o*-Phthalic Anhydride / *Constant Volume Calorimetry* 381
15. Heat of Neutralization and Dilution / *Constant Pressure Calorimetry* 383
16. Standard Free Energy, Enthalpy, and Entropy of Reaction / *EMF Measurement* 386
17. Partial Molar Volume / *Pycnometry* 389
18. Liquid–Solid Equilibrium / *Thermal Analysis* 392
19. Phase Diagram of a Binary Liquid–Vapor System / *Choppin–Cottrell Method* 395
20. Vapor and Sublimation Pressure / *Internal Manometer Method* 400
21. pK_a of an Indicator / *Spectrophotometry* 402
22. Ionization Constant of a Weak Acid / *Kohlrausch Conductance Method* 406
23. Formation Constant of a Complex Ion / *Polarography* 409
24. Solubility of an Insoluble Salt / *EMF Method* 411
25. Decomposition Pressure of Ammonium Carbamate / *Mercury Isoteniscope Method* 413
26. Dimerization of Acetic Acid Vapor / *Mercury Isoteniscope Method* 415
27. Kinetics of the Iodide-Catalyzed Decomposition of H_2O_2 / *Gas Volume Measurement* 418
28. Rate of the Acid-Catalyzed Hydrolysis of Sucrose / *Polarimetry* 421
29. Effect of Ionic Strength on the Rate of a Reaction / *Spectrophotometry* 424
30. Kinetics of a Fast Reaction, by a Relaxation Technique / *Concentration-Jump Method* 427
31. Kinetics of the Formation of Peroxochromic Acid / *Stopped-Flow Method* 431
32. Cross Sectional Area of Molecules in a Soluble Monolayer / *Surface Tension Measurement* 435

33. Insoluble Monolayers on a Liquid Surface / *Surface Pressure Measurement* — 437
34. Surface Area of a Powder / *BET Gas Adsorption Method* — 440
35. Dipole Moment of a Polar Liquid / *Dielectric Constant Measurement and Refractometry* — 445
36. Bond Moments, Bond Angles, and Dipole Moments / *Dielectric Constant Measurement and Refractometry* — 451
37. Dissociation Energy of Halogen Gases / *Ultraviolet Spectrophotometry* — 453
38. Molecular Weight and Shape of Dissolved Polymer Molecules / *Viscometry* — 456
39. Electronic Structure of an Ion / *Magnetic Susceptibility Method* — 458
40. Isotopic Composition of an Element / *Mass Spectrometry* — 461
41. Appearance Potentials of Gaseous Ions / *Mass Spectrometry* — 463
42. The Hydrogen Emission Spectrum / *Emission Spectrography* — 465
43. Vibrational-Rotational Spectra of Gases / *Infrared Spectrophotometry* — 468
44. Effect of Isotopic Mass Change on Molecular Vibration / *Infrared Spectrophotometry* — 471
45. The ESR Spectrum of a Free Radical — 473
46. Quantum Efficiency of a Photochemical Reaction / *Chemical Actinometry* — 475
47. Equilibrium Constant of a System in an Excited State / *Fluorimetry* — 479
48. Electrokinetic Phenomena — 483
49. Correlation of Polarographic Reduction Potentials and Electron Affinity / *Polarography* — 488

APPENDIXES

1. Atomic Weights of the Elements — 491
2. Units and Important Physical Constants — 494

3. Relative Density and Volume of Water	498
4. Temperature Correction for Barometric Readings	500
5. Density of Mercury	501
6. Glassworking	502
7. Electrical Work	508
8. Color Code for Electronic Components	511
9. Soldering	512
Index	515

Part One

PRINCIPLES
of Laboratory Practice

Chapter 1

Introduction: General Instructions to the Student

Preparation

You should know what you are doing and why you are doing it at every step. Proper preparation before starting the experimental work will enable you to accomplish more with less effort, to avoid unnecessary repetitions and loss of time, and to avoid errors in principle and method that may invalidate the entire experiment.

Before starting the experiment:

1. Read the assigned experiment and study the appropriate sections in your text. Read the references for additional information. This is particularly important if the experiment must be performed before the theory has been discussed in class.
2. Study the reading assignment and learn the principles of the experimental method and how to use the apparatus.
3. Make a rough estimate of the value of each quantity to be measured and of the final calculated results. Look up published data, if necessary, and make some sample calculations for practice.
4. Plan the organization of the data in your notebook.
5. Write the preliminary report.

Laboratory Work

Work carefully, in an organized fashion. Repeat each type of measurement at least once, to get an estimate of your precision.

If possible, graph or compute your results as you proceed to see how the experiment is progressing and which measurements, if any, are suspect and should be repeated.

Measure each quantity to the number of significant figures needed to obtain the desired precision in the final results. If you can obtain more precise results without additional time or effort, do so, but do not spend time and effort on unnecessary precision.

Before putting your equipment away, check your data and make some rough calculations of the final result so that, if necessary, additional measurements may be made.

Reporting the Results

Write the report as soon as possible after doing the experiment. Do not let a week or more elapse, for you may forget what you did, or omit some observation.

Do not let several reports accumulate and then try to do them simultaneously. Not only will the reports be badly written, but you will confuse the concepts that the experiments were intended to illustrate.

Safety Precautions

Be safety conscious. Industrial chemists pay the same insurance premiums as construction workers.

Never work alone in the laboratory unless there is someone nearby who can respond to a call for help.

Become familiar with the location and operation of the fire extinguishers, the safety shower, and other safety devices and be prepared to use them, if necessary, without asking permission or discussing it with your neighbor. Check to see that the extinguishers are full.

Never perform an unauthorized experiment. If you desire to change a procedure, check with the instructor beforehand to get permission.

Wear eye shields or eyeglasses, even if your experiment does not require them. Protect yourself against neighbors who might be careless.

Have an instructor check all electrical equipment that does not seem to function properly. In the event that you must make adjustments or repairs, disconnect the power to the apparatus and short out all capacitors before making any tests inside the chassis. Work with only one hand to avoid a shock from arm to arm through your heart. Be sure that the floor and laboratory desk are nonconductors. Some cement floors have a reinforcing wire grid that can act as a conductor.

Always read the instruction manual for an instrument before using it and *always* go through the manual before attempting a repair.

Do not eat or smoke in the laboratory.

Keep all bottles of flammable, poisonous, and corrosive materials closed. Use these materials in the fume hood and with adequate traps.

Never pipet poisonous materials without interposing a trap or using a mechanical pipetting device.

Before starting an experiment, lay out your materials so that contact with vapors will be minimized by optimum use of available hoods and ventilators. Provide standby equipment for the removal of spills. Above all, look up the properties of all the materials that are to be used. Suggested references are *Dangerous Properties of Industrial Materials* [Sax], *The Merck Index*, and *Handbook of Laboratory Safety* [Steere].

If you are developing a new procedure, try beforehand to think of things that might go wrong and prepare for them. Unfortunately, often it is the unexpected that happens, but at least you will be prepared for the obvious hazards. It is always wise to assume that anything that can go wrong will go wrong.

Health Hazards

Safety hazards such as fire, explosion, and electric shock have results that show themselves rapidly. Health hazards, on the other hand, may cause injuries that will not become apparent for years.

Health hazards should always be in the forefront of the chemist's thought. Recent work on occupational safety shows that industrial workers have suffered from exposure to many materials that were simply assumed to be harmless. Chemists, by the nature of their work, are also exposed to such materials, and the incidence of cancer among chemists is significantly higher than the average. Unlike many workers, industrial chemists do have some control over their environment and can set up operations with adequate traps, safety valves, ventilators, and other devices.

In general, toxic compounds enter the body as inhaled vapors or particles. Secondary routes of entry include the mouth and digestive tract (recall the ban on eating in the laboratory) and the skin, particularly for fat-soluble materials. Most ingested materials, if completely insoluble, do no harm and are passed through the intestinal tract. To be harmful they must be sufficiently soluble in water or in fat to pass through the gut wall and enter the circulatory system. Exceptions are radioactive materials and certain finely divided materials such as asbestos fibers. Inhaled materials need not be soluble to do damage. Insoluble gases such as NO or NO_2 that get into the lungs cause severe irritation and pulmonary edema. Insoluble solids produce pneumoconioses, such as

Black Lung, the Coal Worker's Pneumoconiosis. The effect of asbestos, which can cause lung cancer, seems to be aggravated by cigarette smoke. Asbestos workers who smoke have cancer rates 45 times higher than those who do not smoke and 90 times higher than people who neither smoke nor work with asbestos.

The following health hazards associated with some common materials illustrate the need for caution.

1. Heavy Metals / Many heavy metals are neurotoxic or affect the liver and kidney. Except for mercury, these metals have low vapor pressures in ordinary circumstances, so heavy metals are inhaled only as powders or volatile compounds (lead acetate powder, tetraethyllead, etc.). If the metal is in the form of a soluble salt, skin contact should be avoided, especially if there are open cuts. Some substances, such as chromium compounds, cause long-lasting skin ulcers that may become cancerous.

In the case of mercury, the damage appears to be irreversible and cumulative. The vapor pressure is appreciable at room temperature. Mercury vapor is often a permanent feature in laboratories because of spillage from broken instruments such as thermometers and manometers. Spilled mercury should be removed as completely as possible; suction devices will pick up visible droplets. Unfortunately, there are always small cracks in floors and benches where small amounts of mercury remain unobserved. This mercury should be neutralized as much as possible with sulfur.

2. Aromatic Compounds / Benzene causes liver and kidney damage as well as various blood diseases, including leukemia. It attacks the bone marrow, causing anemia, a reduction in red cells and platelets.

Toluene has narcotic effects, but its overall effect on blood is less than that of benzene, so toluene should be substituted for benzene wherever possible. It is itself far from harmless, however.

Polycyclic aromatics are, in many cases, carcinogens. Naphthalene does not appear to be but it, at least, is a powerful eye irritant which should not be used without good ventilation.

3. Hydrogen Sulfide / Hydrogen sulfide, H_2S, has been used so heavily in undergraduate analytical laboratories that familiarity has caused contempt. The gas is absorbed rapidly into the bloodstream, and if its concentration there reaches 700 parts per million, it can no longer be detoxified by the body mechanisms. The gas reaches the brain and stops the impulses that produce breathing. Even low-level exposures that do not cause unconsciousness or death can have cumulative effects.

4. Chlorinated Hydrocarbons / Chloroform and carbon tetrachloride are severely toxic to the liver and kidneys. (One of the authors of this text has suffered chloroform-induced hepatitis.) An insidious feature of chloroform and carbon tetrachloride vapors is that they are toxic at levels too low to smell. Trichloroethylene and tetrachloroethylene are narcotic and cause kidney and severe liver damage; they may also be carcinogenic.

Halogenated ethers, which have long been used as anesthetics, recently have been found to have a markedly harmful effect on women, increasing the incidence of involuntary abortion and stillbirth. They affect not only operating room nurses and female physicians but also the wives of male operating room personnel. Apparently, the vapors adhere to the skin, hair, and undergarments and are brought home.

Vinyl chloride, though not a common chemical, illustrates the need for constant wariness. For years this compound was thought to be nontoxic because no effect had been observed. Then, suddenly, it was discovered that exposure to air containing only a few parts per million of vinyl chloride vapor causes an otherwise rare form of liver cancer.

5. *Aromatic Amines and Nitro Compounds* / Some of these, such as aniline, can be absorbed through the unbroken skin. Nitrobenzene and aniline combine with hemoglobin and produce symptoms similar to those of carbon monoxide poisoning. Benzidine and several other compounds in this group cause bladder cancer in a very high proportion of exposed workers.

Obviously, this sketchy listing cannot cover the hundreds of common chemicals on which we have definite warning data, let alone the hundreds of thousands of materials about which nothing is known. Furthermore, chemicals sometimes work together to produce effects that are not always expected.

References

Braker, W., and A. L. Mossman, *Effects of Exposure to Toxic Gases, First Aid and Medical Treatment*. Matheson Gas Products, East Rutherford, N.J., 1970.

Christensen, H. E., T. T. Luginbyhl, and B. S. Carroll, *Toxic Substances List*. U.S. Department of Health, Education and Welfare, National Institutes of Health, Rockville, Md.

Green, M. E., and A. Turk, *Health and Safety in the Chemistry Laboratory*. Macmillan, New York, 1978.

Manufacturing Chemists Association, *Guide for Safety in the Chemical Laboratory*. Van Nostrand Reinhold, New York, 1972.

Merck & Co., Inc., *The Merck Index*, 8th ed. Merck & Co., Inc., Rahway, N.J., 1968.

Sax, N. I., *Dangerous Properties of Industrial Materials*, 3rd ed. Van Nostrand Reinhold, New York, 1968.

Steere, N. V., *Handbook of Laboratory Safety*, 2nd ed. Chemical Rubber Co., Cleveland, 1971.

Stellman, J., and S. Daum, *Work Is Dangerous to Your Health*. Pantheon, New York, 1973.

Chapter 2

Data Processing

In laboratory physical chemistry the experimenter usually wants to evaluate quantities that cannot be measured directly. Instead, the values are calculated from the results of measurements made on such quantities as temperature, pressure, and time. In addition, the experimenter not only wants the value of the calculated quantity, but also needs an estimate of the reliability of the calculated result before reporting to the boss, publishing a paper, or building a plant.

First the experimenter calculates the average and estimates the precision and reliability of the individual measurements, perhaps discarding some data that are probably in error. Next, he or she calculates the precision of the average. These steps are taken for each of the experimental quantities that go into the final calculation of his result. Then the experimenter calculates the final quantity from the averages of the measured quantities and finally estimates the reliability of the calculated value from the precision of the separate averages.

If a mathematical relationship rather than a numerical value is wanted, an equation or a graph is fitted to the data. Estimates of the "goodness of fit" may be made, and sometimes the equation or graph is integrated or differentiated.

It is of vital importance for a scientist or engineer to be familiar with and to understand the common procedures used to process data. He or she must be able to make calculations and report results intelligibly and to evaluate the results of others.

Experimental Numbers

There are three kinds of numbers encountered in experimental work:

Counted numbers, such as the number of persons or cells, must be integers.

Mathematical numbers that appear in formulas and mathematical relationships are defined and are known to any desired accuracy, whether they are rational or irrational. For example, to the mathematician the number 5 may be 5.0 or 5.000—to as many zeros as desired. The number 2 in the formula $A = \pi r^2$ is 2.0 . . . , with an infinite number of zeros, and the quantity π is worked out to as many decimal places as desired. So are other irrational numbers, such as the square root of 2.

Measured numbers are the results of experimental measurement of such quantities as length and mass. There is always a limit to the certainty with which they are known. On a trip scale the mass of an object may be 2.1 g. On a balance the measured mass may be 2.094 g. With a better balance the measured mass may be 2.09442 g, or some such number. Ultimately, no matter how good the apparatus, a limit is reached to the accuracy of the measurement. For one thing, the experimental conditions will not be constant. Dust may settle on or be blown off the balance pan; the temperature may change, causing the arm of the balance to expand or contract; or other factors may change. The sum of all these random changes in conditions will change the measured result by an unknowable and uncontrollable amount. Consequently, all measurements have an inherent uncertainty, or *random error*, the magnitude of which depends upon the method, the apparatus, and the skill of the experimenter.

Reliability and Precision

Since there is always random error associated with a measured quantity, methods are needed for estimating and for expressing the reliability of measurements. There are two expressions of reliability. *Accuracy* is the agreement of the experimental measurement with the true or accepted value of the quantity. *Precision* is the agreement of measurements with each other. Usually, the true value of a quantity is not known except by definition or agreement. For example, the atomic weights listed in Appendix 1 are accepted, by international agreement, as the true values. In many cases of interest in physical chemistry the true value is not known. Consequently, precision rather than accuracy must be used as an estimate of the reliability. Usually, we assume that in the absence of any consistent error, such as an incorrect zero, the more precise the results, the greater the probability of their being accurate.

Expressions of Precision

The limit of error or "precision index" of a number must always be reported whenever it is necessary to estimate the precision of quantities derived from that number. It is appended to the number. For example, 2.500 ± 0.002 shows that the value lies between 2.498 and 2.502. Sometimes two limits are used, as in

$16.24^{+0.02}_{-0.01}$. As a matter of common sense, the limit of error should not be reported to more than two significant figures (see below) and usually to not more than one. A precision index of 0.428 cm means that there is an uncertainty of 0.4 cm plus an additional uncertainty of 0.02 cm and another, additional uncertainty of 0.008; but, obviously, an uncertainty of 0.008 is meaningless in the presence of an uncertainty of 0.4. (Even the limit of error is ambiguous unless it has been defined as the standard deviation, etc.)

If only a rough indication of precision is needed, it is usually shown with *significant figures*. These are figures all of whose digits are known with certainty. For example, 23 cm is known to be 23 cm, not 22 and not 24. We generally assume that the 23 is actually 23 ± 0.5, with an uncertainty of half a unit in the last indicated place. The four-significant-figure number 64.86 is therefore 64.86 ± 0.005. Note that the number of significant figures is not affected by the position of the decimal point. A volume of 0.005 ml is known to only one significant figure. Another way to express this is to say that the number of significant figures is independent of the choice of units; thus 0.01 m is the same as 1 cm. To avoid confusion when changing units, retain the same number of significant figures and use exponential notation to locate the decimal point. For example, write 1.5 cm as 1.5×10^8 Å, not as 150,000,000 Å. Note that significant figures give only a rough indication of precision. In significant figure notation the number 23.0 ± 0.3 must be written either as 23.0, which implies too great precision, or as 23, which implies too little.

Operations with Significant Figures

The precision of a computed result is limited by that of the least precise quantity in the computation. In addition and subtraction the number of significant figures in the result depends on the term with the largest *absolute uncertainty*. Consider the following sum:

Value	Absolute Uncertainty
+ 0.3442	0.00005
+ 1207.6	0.05
+ 1.33	0.005
− 5.002	0.0005
+ 1204.3	0.06

The precision in this example is limited by the number 1207.6 because it has the greatest absolute uncertainty, or error (error is used interchangeably with uncertainty, although the two words actually have somewhat different meanings). The error in a sum must be at least equal to the largest of the individual errors. If one of the component numbers has an error of 0.05, the sum must have an error of at least 0.05. Note that the errors are added, although the computation itself may involve subtraction.

In multiplication and division, the number of figures in the computed quantity depends upon the term with the largest relative uncertainty. *Relative uncertainty*, or *relative error*, is the absolute error divided by the value of the quantity. For the number 23.4 ± 0.2, the absolute error is 0.2 and the relative error is $0.2/23.4 \approx 1\%$. For example,

$$\frac{47.61 \times 0.0024}{2.83} = 0.040 = 4.0 \times 10^{-2}$$

The calculated result has the same number of significant figures as the term with the fewest number of significant figures. In this example, the precision is limited by the value of 0.0024 because it has the greatest relative error, roughly 2%. The result must therefore have a relative error of 2% and must have the same number of significant figures as the number 0.0024.

However, in some cases the answer may have a greater number of significant figures than that of the limiting quantity. For example, $9.0 \times 1.20032 = 10.8$ rather than 11. The answer is expressed to three figures rather than to two because the relative error in the number 9.0 is 0.05/9, about the same as 0.05/11, or roughly 0.5%. On the other hand, 0.5/11 is about 5%. There are certain errors that can result from mechanically applying rules for using significant figures but these may be avoided by remembering and applying the principle behind the rules.

In operating with significant figures, round off the numbers to one more than the correct number of significant figures. Then after performing the computation, round off the result to the correct number of figures. Note that in the summation example, if the numbers had been rounded off to the first decimal place, the answer would have been 1204.2, which is 0.1 too low. In rounding off numbers ending in 5, either add or subtract 5 to obtain an even number: 21.5 becomes 22, while 20.5 becomes 20.

Calculations of Best Values, Precision, and Reliability

In using the measured data, we run into some basic problems because each experimental value includes a random error that can never be known. These random errors contribute to an error in the final calculated result.

We need to know the most accurate or "best" value for each experimental quantity and its precision, or reliability. We want to know when we can reject a measurement that might be incorrect. We need to know the best value of the final quantity calculated from the measured data, and we want to know the reliability of this quantity, i.e., how much confidence we can put in our answer.

Best Value of an Experimental Number

There are two types of measured quantities. Quantities such as mass, length, resistance, and voltage are single-valued under constant conditions and do not change. The results of several measurements will be different because of random errors, but there really is a single *true value*. On the other hand, the quantity of material weighed out or poured out during a run is not fixed. It varies from run to run. There is no single true value, and repeating the procedure does not really repeat the measurement.

In measuring fixed single-valued quantities, if the errors are truly random, the best or most probably correct value is the mean of the measurements, $\bar{x}$.

Unless there is a bias or *constant error*, each experimental value, x_i, differs from the true value x_0 by an unknown or residual error, r.e.$_i$, which may be either positive or negative.

$$x_i = x_0 + \text{r.e.}_i$$

$x_0 =$ true value
$x_i =$ measured value
r.e.$_i =$ error

Since x_0 is unknown, the problem is to obtain from the data some best value of the quantity that will most closely approximate the true value. Consider the difference between the experimental value x_i and some arbitrarily assumed best value x:

$$\Delta_i \equiv x_i - x$$

Statistical theory assumes that if the sum of the squares of these differences is at a minimum, the probability is greatest that x will be the true value x_0. The sum of the squares of the differences is given by

(2–1) $$\sum \Delta_i^2 = \sum (x_i - x)^2$$

When equation (2–1) is differentiated with respect to x and the derivative is set equal to zero,

$$\frac{d(\sum \Delta_i^2)}{dx} = \frac{d \sum (x_i - x)^2}{dx} = 0$$

On solving the resulting equation one obtains as the best value

(2–2) $$x = \frac{1}{n} \sum x_i = \bar{x}$$

where n is the number of measurements going into the mean. Note that under these conditions the sum of the experimental residual errors is zero, as would be expected for random errors. This is the statistical justification for using $\bar{x}$ as the best approach to x_0.

The value of $\bar{x}$ will approach x_0 closely but will not necessarily equal it. For an infinite number of measurements $\bar{x}$ will equal x_0, but for a finite number there may be an unknown random difference. The larger the number of measurements and the greater the care taken with each, the closer the approximation will be.

Measuring quantities that have no single true value, such as the poured-out amount of a liquid, poses a somewhat more difficult problem than measuring fixed-value quantities. The amount of material used must be known accurately, but if the liquid is measured with a pipet or buret, there is no way to tell how much liquid is delivered unless it is run into a receiver and weighed. Although at each temperature the volume of a pipet is fixed and definite, the quantity of material actually delivered will be different each time. The volume delivered depends on such irreproducible factors as the location and shape of the meniscus, the speed at which the liquid runs out, the time the experimenter waits for the liquid to run out, and the angle at which the pipet is held.

The experimenter is not interested in the volume contained by the pipet but only in the volume the pipet actually delivers when it is used. Unfortunately, there is no way to know this, no way to predict the exact amount that will be delivered. All that can be said is that, depending on the skill of the experimenter, the volume delivered on any run will probably be somewhere near the mean of a number of runs.

To estimate how much a pipet or other volumetric apparatus will deliver on any single run, a series of deliveries is made and the volume delivered determined for each. The mean $\bar{x}$ and the standard deviation s (see the next section) of the raw data are calculated. The volume delivered any one time is

$$\bar{x} \pm ks$$

where k is a number obtained from statistical tables. The value of k depends on two factors: the desired confidence and the number of measurements used to find the mean.

First, the larger the value of k, the more confident the experimenter can be that the volume delivered will fall between the limits of $\bar{x} \pm ks$. For example, if a 5 ml pipet on a series of calibration runs delivers an average volume of 4.98 ml, then on any one run the probability is much greater that the pipet will deliver 4.98 ± 0.1 ml than that it will deliver 4.98 ± 0.01 ml. The experimenter who wants to have a specified confidence in the limits of the amount delivered selects the appropriate k. The greater the confidence desired, the greater the value of k. (A 95% confidence means that 95 times out of 100, or 19 times out of 20, the result will be within the stated limits; that is, the odds are 19 : 1 in favor of the experimenter.)

TABLE 2-1 / Representative Values of k

Number of Runs	Confidence Limits (%)			
	99	95	90	50
2	77	16	7.6	1.2
3	10	4.7	3.2	0.90
4	6.4	3.5	2.6	0.84
5	5.1	3.1	2.3	0.82
6	4.4	2.9	2.2	0.80
8	3.9	2.7	2.1	0.79
10	3.4	2.4	1.9	0.74
∞	2.6	2.0	1.6	0.67

Second, the greater the number of runs used to find $\bar{x}$, the smaller the value of k needed for a given confidence level. Obviously, one can have more confidence in the results of 100 runs than of 10 runs.

In Table 2-1 are some representative values of k rounded off to two significant figures. If a pipet has been calibrated on the basis of 10 runs and the experimenter wishes a 90% confidence that the pipet will deliver a volume within stated limits, the standard deviation is multiplied by 1.9. That is, on 9 runs out of 10, the pipet will deliver a volume of $\bar{x} \pm 1.9s$.

Analogous considerations apply to the calibration of other apparatus and to the estimation of the precision of other nonrepeatable quantities.

Reliability of the Experimental Number

The reliability of the mean depends on the reliability of the data that went into the calculation of the mean and on the number of measurements made. The greater the number of measurements, the more valid are the statistical assumptions upon which the choice of the sample mean is based and the less effect that any one random error can have on the mean. Also, the more closely the individual measurements agree with each other, the greater the confidence that can be placed in the calculated value.

There are four common measures of the precision of an individual measurement: the average deviation, the probable deviation, the root mean square deviation, and the standard deviation.

The *average deviation*, a.d., is the sum, without regard to sign, of the deviations of the individual data from the mean divided by n, the number of experimental measurements:

(2-3) $\qquad \text{a.d.} = \frac{1}{n} \sum |\delta_i|, \qquad \delta_i = x_i - \bar{x}$

The average deviation is the easiest precision index to calculate without a computer or a calculator.

The *standard deviation*, s, is the square root of the sum of the squares of the individual deviations divided by $n - 1$:

$$(2\text{-}4) \qquad s = \sqrt{\frac{\sum \delta_i^2}{n - 1}}$$

The denominator is $n - 1$ rather than n because there are only $n - 1$ independent values of δ_i, since by definition

$$\sum \delta_i \equiv 0$$

With a good desktop or pocket calculator, the standard deviation is as easy to calculate as the average deviation and much more useful. The standard deviation is the fundamental precision index in error theory. If the number of measurements is large enough, 32% of the measurements will differ from the mean by at least the standard deviation, s. Therefore, there is a 32% probability that any individual measurement will differ from the mean by s or more. There is only a 5% probability of the measurement differing by 2s from the mean, and the probability of the deviation being at least 3s from the mean is only 0.3%; that is, only 3 measurements out of 1000 are that far from the mean. These probabilities form the basis for the rejection of individual measurements.

For the normal, or Gaussian, distribution of random errors, which is commonly assumed to apply to experimental measurements, the standard deviation is 1.25 times the average deviation, and so it is more probable that the measurements will fall within the range $\bar{x} \pm s$ than within the range $\bar{x} \pm$ a.d.

The *root mean square deviation*, r.m.d., is the square root of the average of the squares of the deviations:

$$(2\text{-}5) \qquad \text{r.m.d.} = \sqrt{\frac{\sum \delta_i^2}{n}}$$

For large values of n the standard and the root mean square deviations are the same, and even for small values they may be considered interchangeable. The *probable deviation*, p.d., is 0.68s. It has the special significance that half of the deviations are smaller and half of the deviations are larger than the probable deviation. Most physical chemists use either the more fundamental standard deviation or the more easily computed average deviation.

There are four precision indices for the mean, which correspond to the respective precision indices of the individual measurements. These are calculated from the indices for individual measurement by dividing by the square root of the number of measurements:

1. The average deviation of the mean, A.D., is

$$\text{A.D.} = \frac{\text{a.d.}}{\sqrt{n}}$$

2. The standard deviation of the mean, S.D., is

$$\text{S.D.} = \frac{s}{\sqrt{n}}$$

3. The root mean square deviation of the mean, R.M.D., is

$$\text{R.M.D.} = \frac{\text{r.m.d.}}{\sqrt{n}}$$

4. The probable deviation of the mean, P.D., is

$$\text{P.D.} = \frac{\text{p.d.}}{\sqrt{n}}$$

Relative precision indices are often used in addition to or instead of the above *absolute* precision indices. The relative precision index of a quantity is the corresponding absolute precision index divided by the absolute value of the quantity. The relative standard deviation of the mean is S.D./$\bar{x}$, and the other relative indices are analogous.

The square root dependence of the precision indices places a practical limit on the increase in precision obtainable from repeated measurements. Whereas the precision of a result is doubled by making four times the number of runs, to increase the precision by a factor of 10 requires 100 times the number of runs. After 10 runs the increase in the precision obtained rapidly becomes outweighed by the increased effort required.

Precision indices obtained for individual measurements and for the mean should be used with common sense, based on the experimenter's experience. If the results of a few runs agree with each other exactly, the precision index should not be taken to be zero, since this indicates no experimental error. The agreement could have occurred by chance, or perhaps the apparatus was not sensitive enough to detect fluctuations. A thermometer calibrated to 0.1°C is not sensitive enough to detect fluctuations in a thermostat that is constant to ± 0.001°C, and according to the thermometer the temperature would appear to be absolutely constant. If all the measurements agree with each other, the precision measure should be taken as the smallest change to which the instrument is sensitive, or to which it can be read, namely, 0.1 of a scale division.

Another error to be avoided is the indiscriminate use of *precision* to mean *accuracy*. For example, an ammeter might give a reading reproducible to ± 0.01 A, but this is not necessarily its "accuracy." If the ammeter is certified by the manufacturer or calibrated by the experimenter as having an accuracy of $\pm 2\%$, a current of 1 A should be taken to be 1.00 ± 0.02 A.

Confidence Limits

It is often more meaningful to indicate the reliability of research data by confidence limits rather than simply giving one of the common precision indices. A *confidence limit* is a range, on both sides of the mean, within which one may have the specified confidence that the "true value" will be found. (The true value here is the mean of an infinite number of repeated measurements.) A 90% confidence means that there is a 90% probability that the true value is within the range $\bar{x} \pm$ limit, and also implies that there is a 5% chance that the true value is above the range and another 5% chance that it is below the range. For normally distributed errors (in the absence of a constant error) the confidence limits are found by multiplying the standard deviation of the mean, S.D., by a factor t, which is selected from tables on the basis of the desired confidence and the number of measurements that went into the computation of the mean. The desired confidence and the number of runs must be decided by the experimenter on the basis of his own judgment, not by any statistical method. Some values of t are listed in Table 2–2.

Suppose that an experimenter has made five measurements of a quantity from which he calculates the mean, and that he wants to be 95% confident that the true value lies within a given range computed from his measurements. He computes the S.D. and multiplies it by 2.8, the factor selected for 95% confidence in the result of five runs. The odds will then be 19 : 1 or 95%, that the true value will be within the range $\bar{x} \pm 2.8$ S.D.

On inspection, Table 2–2 shows two interesting properties. First, it shows the overwhelming importance of repeated measurements, especially when a high confidence in an experimental value is desired. For 90% confidence t decreases from 6.3 to 2.4, thus decreasing the range by a factor of 2.6, when the number of runs is increased from two to four, while for 99% confidence t decreases from 63 to 5.8, a factor of 11, when going from two to four runs. Conversely, even with the same standard deviation of the mean, a mean of a few measurements is much less reliable than one of many measurements. These

TABLE 2–2 / Some Values of t

Number of Runs	Desired Confidence (%)		
	99	95	90
2	63	13	6.3
3	9.9	4.3	2.9
4	3.8	3.2	2.4
5	4.6	2.8	2.1
10	3.2	2.3	1.8
∞	2.6	2.0	1.6

features reflect the influence of the occasional large error in small sample statistics. The table can be used to help an investigator decide beforehand how many runs to make and what confidence can reasonably be expected.

Some tables give t values not in terms of the confidence that the true value is within the interval, but in terms of the probability that it lies outside. For this alternative convention the t values would be listed under headings beginning with 0.01 (or 1%) probability. The convention used in any particular table can be found by looking at the specified confidences.

Treatment of Doubtful Values

One of the most difficult decisions to make in handling data is when to accept or reject a value that differs appreciably from the others. With a large number of repeated measurements the problem is well defined statistically and it is comparatively easy to set up rigid standards, but in such a case the decision is rarely important, since one bad result will not greatly affect the mean of a large number of measurements. With only a few measurements, not only is the problem much more difficult, because the statistical distribution is not well defined, but the effect of the decision becomes more important.

One of the difficulties of using small numbers of measurements is that any procedure that will reject most of the bad results will also reject a large fraction of the valid measurements. Conversely, any procedure that will accept most of the valid measurements will also accept many invalid results. As a result, there is no widely accepted procedure to be applied. It becomes a matter of personal preference, with as many methods as there are textbooks.

One widely used procedure, which has the merit of simplicity, is the following. First, average the results without the suspect value and compute the average deviation. Compare the doubtful value to the average. Discard it if it deviates from the mean by at least four times the average deviation. Do not, however, discard more than one value for each five values, and do not discard a value if it agrees with one or more other values. (See, for example, the accompanying tabulation.)

Values	Deviations
27.3	0.2
27.1	0.0
27.8[a]	0.7[a]
26.9	0.2
27.0	0.1
$\bar{x} = 27.1$	a.d. = 0.1

[a] Value is doubtful or suspect.

The $\bar{x}$ and the a.d. computed without the suspect value are 27.1 and 0.1, respectively. The deviation of the suspect value from 27.1 is 0.7, which is more than four times 0.1. The value is therefore discarded. If the value were 27.4 or 26.8, it could not be discarded, and a new mean would have to be calculated using the questioned datum.

Another procedure is to apply the Q test. The difference between the suspect result and its nearest neighbor is divided by the difference between the smallest and the largest values. If the ratio, Q, exceeds a critical value, the number is rejected. There are standard values of Q_{crit} for each combination of desired confidence and number of measurements [Dean and Dixon].

Another method, used with a minimum of five replicate measurements, is to discard the highest and the lowest value.

Precision of a Calculated Result

The best value of a calculated result is that obtained from the averages of the measured quantities after each of these has been corrected by rejection of all invalid data. Clearly, since the most reliable values of each experimental quantity are used for the final calculation, the result is the best that can be obtained from the experimental data.

The precision of the final result is calculated from the precision indices of the experimental measurements. The details of the calculation depend upon the type of the experimental function being investigated and upon whether the most probable error is desired or the maximum error. The procedure is based upon the calculus.

Let Y, the quantity calculated, be some function of a, b, c, ..., the measured quantities:

$$Y = f(a, b, c, \ldots)$$

The change in Y resulting from change in a is

(2-6) $$\delta Y_a = \left| \left(\frac{\partial Y}{\partial a} \right)_{b, c, \ldots} \delta a \right|$$

From statistical theory the change in Y caused by all the changes in $a, b, c, \ldots$ is

(2-7) $$\delta Y = \sqrt{(\delta Y_a)^2 + (\delta Y_b)^2 + \cdots}$$
$$= \sqrt{\left[\left(\frac{\partial Y}{\partial a} \right)_{b, c, \ldots} \delta a \right]^2 + \left[\left(\frac{\partial Y}{\partial b} \right)_{a, c, \ldots} \delta b \right]^2 + \cdots}$$

where δY is the most probable error.

Two special cases are worth considering. If the operations are only addition and subtraction, such as in the formula

$$Y = a + b - c$$

then

(2-8) $$\delta Y = \sqrt{(\delta a)^2 + (\delta b)^2 + (\delta c)^2}$$

If the operations are multiplication and division, without powers or roots, such as in the formula

$$Y = \frac{ab}{c}$$

then

(2-9) $$\frac{\delta Y}{Y} = \sqrt{\left(\frac{\delta a}{a}\right)^2 + \left(\frac{\delta b}{b}\right)^2 + \left(\frac{\delta c}{c}\right)^2}$$

with the $S.D._a$, $S.D._b$, ... or other precision indices substituted for δa and δb, equations (2-8) and (2-9) are then used to calculate the absolute and relative errors in Y. These two equations also provide the basis for the rules for handling significant figures in computations. It is clear by inspection that large *uncertainties* greatly outweigh small ones. For example, if one uncertainty is 10 times another, its contribution is 100 times as great.

If maximum error is desired rather than most probable error, the formula used is

$$\delta Y = |\delta Y_a| + |\delta Y_b| + |\delta Y_c|$$

This formula is much more convenient, but with three or more experimental quantities it overstates the error of the computed quantity.

To find the confidence limits of a computed quantity from the confidence limits of the experimental data, replace the

$$\delta a, \delta b, \text{etc.}$$

by

$$t_a S.D._a, t_b S.D._b, \text{etc.}$$

The t's are obtained from tables of t values, such as Table 2-2, using the number of runs and the confidence appropriate to each quantity. Note that the confidence limit of the answer is the same as that of a, b, etc. There cannot be 99% confidence in Y if t_a, t_b, etc., are selected from 95% confidence tables. To apply these formulas, the equations for Y as a function of a, b, c, ... must be known.

The preceding discussion applies to data that can be averaged so that the desired quantity can be calculated from the means of the measured quantities.

Sometimes, however, it is not possible to average data because the individual measurements of duplicate experiments are not comparable. One example is the determination of molecular weight by cryoscopy. The repeated runs are made using different amounts of solute and solvent, and there are also changes in room temperature and differences in the extent of supercooling. Clearly, corresponding experimental measurements such as the freezing point depression cannot be averaged. Instead, the data from each experimental run are used to calculate the molecular weight; the calculated molecular weights are then averaged; the precision of the individual values of the molecular weight is calculated and results discarded if necessary. The desired quantity is the mean, and the precision is the standard, average, or other deviation value of the mean.

Summary of Calculation Procedures

For data for which corresponding quantities may be averaged:
  Average the measurements of each quantity.
  Determine the desired precision index and use it to determine which, if any, of the data should be discarded. If necessary, compute the arithmetic mean again.
  Calculate the precision of each mean. Report each $\bar{x}$ value together with its desired precision measure or confidence limit.
  Use the $\bar{x}$ of each experimental quantity to calculate the quantity desired.
  Calculate the precision of the calculated quantity from the precision indices of the means of the measured quantities and the desired confidence limits.

For data for which corresponding measurements cannot be averaged:
  Calculate the desired quantity for each separate run.
  Calculate the mean of the value of the desired quantity from the individual runs.
  Reject any values that the data indicate are not reliable.
  Compute the mean again, if necessary, and compute the desired precision index of this mean.

References

Blaedel, W. J., V. W. Meloche, and J. A. Ramsay, A comparison of criteria for the rejection of measurements. *J. Chem. Educ.* **28**, 643 (1951).

Calder, A. B., The statistical approach in analytical chemistry. *Anal. Chem.* **36**, 25A–34A (1964).

Dean, R. B., and W. J. Dixon, Simplified statistics for small numbers of observations. *Anal. Chem.* **23**, 636 (1951).

Margenau, H., and G. M. Murphy, *The Mathematics of Physics and Chemistry*, 2nd ed., ch. 13. Van Nostrand, Princeton, N.J., 1956.

Mellor, J. W., *Higher Mathematics for Students of Chemistry and Physics*, 4th ed., ch. IX. Dover, New York, 1955.

Wilson, E. B., *An Introduction to Scientific Research*. McGraw-Hill, New York, 1952.

Worthing, A. G., and J. Geffner, *Treatment of Experimental Data*. Wiley, New York, 1943.

Youden, W. J., *Statistical Methods for Chemists*. Wiley, New York, 1951.

Young, G., *Statistical Treatment of Experimental Data*. McGraw-Hill, New York, 1962.

Chapter 3

Drawing Graphs from Experimental Data

In reporting data, a good graph is worth several hundred words; a bad graph is worth nothing. A graph can quickly show trends of the data, minima and maxima, and regions where the data are either contradictory or insufficient. It is a good idea to plot a rough graph as soon as the data are in hand, so that if additional measurements are required, they can be made immediately with the same apparatus and materials. The following common sense rules will make it easier to plot and interpret graphs.

Use good quality graph paper with sufficient lines to the centimeter or inch. For most purposes millimeter-ruled paper is adequate. Special papers with logarithmic and semilogarithmic axes are readily available.

Caption each graph and choose a set of coordinates that are easy to read and to use for calculations. Use the ordinate for the dependent variable. The major divisions should be integers multiplied by the appropriate power of 10, such as 10^2 or 10^{-3}, not numbers such as 0.17, 0.19, 0.35. The two axes may have different scales. Each axis must be labeled fully, with the quantity plotted and the scale values and units shown.

Draw circles around the individual points if there are uncertainties in the values of both coordinates. Draw vertical lines if the uncertainty is only in the ordinate. The radius of the circle or the half-height of the vertical line should equal the precision measure of the data, that is, the average deviation or the standard deviation of the mean, if it is known or can be estimated. If the precision is not known, indicate the data point only by a symbol, such as + or ×.

Select a scale that places all the data, rather than just part of the data, on the graph. If necessary, use a second graph or an insert to show an expanded portion of the curve. Try to have the curve fill the entire sheet, but do not expand the scale so much that the circles or points become too large. Remember that if you expand the scale in order to expand the distance between points, you will also expand the points themselves. Not only do scales that are too large give a false impression of the data, but they make it harder to draw a smooth curve defined by the points.

If absolute values are to be emphasized, put the zero or the reference value on the scale. If changes are to be shown, start the scale just below the lowest value to be graphed. (More often than not, the zero or reference mark is omitted.) Whatever the ranges covered by the ordinate and the abscissa, the purpose of the graph is to convey meaning. When more than one curve is plotted on a single sheet of paper for purposes of comparison, label each curve clearly and shape the data points differently. Standard symbols are open and filled circles, squares, triangles, inverted triangles, ×'s, plus signs, and daggers. If necessary, dotted and dashed lines of various types can be used to distinguish among curves.

Each point represents the result of an experiment with an experimental error. It is extremely unlikely that all the points, or that even any one point, will lie exactly on the curve. Do not connect each point to the next with a straight line, but draw the best smooth curve among the points. (One important exception is a calibration graph, for example, true buret volume plotted against scale reading; such graphs are drawn with straight lines connecting successive experimental points.) Try to have the data scatter randomly about the curve so that for any appreciable segment of the curve the total scatter above and below the curve is roughly equal.

Do not attempt to draw graphs freehand. Use a straightedge, or if the line is curved, use a spline or a French curve. To obtain the best straight line, a handy tool is a transparent ruler with a black line running through it. Place the ruler on the graph with the black line making the best straight line to the eye among the points. Note the position of the line but do not draw the line. Remove the ruler and place it again on the paper to obtain another equation for the straight line. Rotate the paper 90–180° to remove any consistent bias and repeat the process. The result will be a group of straight lines giving an average or best straight line and a scatter for the precision of the slope and the intercept. Then draw the best straight line.

In drawing a graph of a mathematical or theoretical relationship, do not show individual data points and do not show circles indicating any experimental error, since, in fact, there are no experimental points.

Reference

Worthing, A. G., and J. Geffner, *Treatment of Experimental Data*, ch. 2. Wiley, New York, 1943.

Chapter 4

Selected Numerical and Graphical Methods

Fitting Equations to Experimental Data

It is very often necessary to find an equation that best fits the data. There are two parts to the problem: first, selecting a suitable function; and second, finding the best values for the parameters in the function. In many cases the proper function is known in advance. For example, in the study of the vapor pressure of a liquid as a function of temperature, the Clausius–Clapeyron equation is known to apply. In other cases it may have been decided to express the data as a polynomial for convenience in mathematical handling. Under these circumstances the first part of the problem is more or less solved.

If the proper function is not known, the first step is usually to graph the data, as discussed in Chapter 3, keeping in mind any conditions imposed by the problem, such as no negative absolute temperatures. Once a smooth curve has been drawn, a fairly good guess at the proper function can often be made, based on the judgment and intuition of the experimenter. Common sense must of course be used to decide when a particular deviation is functional and when it is simply due to random scatter of the data. If the data are widely scattered, there may be no point in even attempting to find an equation. In general, the greater the number of maxima, minima, and inflection points in the smoothed curve, the greater the number of adjustable parameters necessary and the more laborious the curve fitting.

Methods for fitting a straight line to data are simpler and easier to use than methods for fitting any other function. Consequently, various mathematical functions of the data are plotted in the hope of finding one that is linear. Inspection of a graph of the data will usually suggest a suitable function to be tried. Some of the more important of these are shown in Table 4-1, together with their linear forms. The last two columns give the slope and the intercept of the straight line expressed in terms of the constants of the nonlinear form of the

TABLE 4–1 / Linear Forms of Common Equations

Equation	Linear Form	Linear Coordinates	Slope	Intercept
$y = ae^{bx}$	$\ln y = \ln a + bx$	$\ln y$ vs. x	b	$\ln a$
$y = ab^x$	$\log y = \log a + x \log b$	$\log y$ vs. x	$\log b$	$\log a$
$y = ax^b$	$\log y = \log a + b \log x$	$\log y$ vs. $\log x$	b	$\log a$
$y = a + bx^2$	—	y vs. x^2	b	a
$y = a \log x + b$	—	y vs. $\log x$	a	b
$y = \dfrac{a}{b+x}$	$\dfrac{1}{y} = \dfrac{b}{a} + \dfrac{x}{a}$	$\dfrac{1}{y}$ vs. x	$\dfrac{1}{a}$	$\dfrac{b}{a}$
$y = \dfrac{ax}{1+bx}$	$\dfrac{1}{x} = \dfrac{a}{y} - b$	$\dfrac{1}{x}$ vs. $\dfrac{1}{y}$	a	$-b$
	$\dfrac{1}{y} = \dfrac{1}{ax} + \dfrac{b}{a}$	$\dfrac{1}{y}$ vs. $\dfrac{1}{x}$	$\dfrac{1}{a}$	$\dfrac{b}{a}$

equation. Unfortunately, not all functions can be reduced to linear form. For example, the very important relation

$$y = a(1 - e^{bx})$$

has no linear form. Special methods are required in such cases. If the trial function does put the data in linear form, it may be accepted as a suitable function, and the constants in the equation are then computed from the slope and intercept of the straight line. An equation may have several linear forms, each giving somewhat different best values for the constants. This is seldom a serious problem, but one should be aware of it. Where there are serious discrepancies between the constants computed from different linear forms, one must decide which values, if any, are the proper values to represent the data [Dowd and Riggs]. After constants have been found from some linear form, the original nonlinear form should be drawn, using the calculated constants, to check its fit to the experimental data.

The three methods commonly used to determine the constants of a straight line are plotting by eye, the method of averages, and the method of least squares. The easiest way to fit the straight line is by eye (page 24). This method is perfectly satisfactory for many purposes, and the resulting line agrees surprisingly well with that calculated by either of the mathematical procedures.

METHOD OF AVERAGES

The data are used to determine two averaged points, through which the straight line is to be drawn. To obtain the two averaged points, the data are arranged in order of increasing magnitude of x (or y) and then split into two equal groups, one containing the first half of the data points, and the other containing the remainder of the data points. If there are an odd number of points, the central data point can be arbitrarily assigned to either group or divided between them. Once the two groups of points have been separated, the x and y coordinates of the points in each group are averaged separately to determine the two averaged points, (X_1, Y_1) and (X_2, Y_2).

The straight line can either be drawn directly through the two points, or it may be found algebraically by solving the two simultaneous equations $Y_1 = mX_1 + b$ and $Y_2 = mX_2 + b$. (The second procedure is equivalent to combining the data into two simultaneous equations, which have the formula $\sum y = m \sum x + nb$.) A better algebraic method is to compute the slope of the line from

$$m = \frac{Y_2 - Y_1}{X_2 - X_1}$$

and to substitute this slope, together with one averaged point, into the equation

$$Y = mX + b$$

and solve for b.

Occasionally, some experimental data appear to split into two separate groups when plotted. If so, it is sometimes preferable to use these natural groupings. In all cases the greater the separation of the two averaged points, the greater the precision of the straight line. Therefore, the two groups of data must not overlap, which is why the points are arranged sequentially before being grouped. (It is not permissible to split the data into two groups of odd and even points as is sometimes seen in the older literature.)

Tables 4–2 and Figure 4–1 show the method of averages. Note that because of the increase in significant figures, due to the summation, there is one more significant figure in m and in b than there is in the experimental data. Note also that 8.11/4.5 is taken here to be 1.802, rather than 1.80, because the relative error in 1.802 is much closer to that of 8.11/4.5 than is the relative error in 1.80.

The procedure is modified slightly if either the slope or the intercept is already known or is imposed by the conditions of the study. If the slope is known, the points are not grouped. Instead, the entire set of data is combined into one average point. The coordinates of this point and the known value of the slope are inserted into the equation for a straight line, which is then solved for the intercept.

When one point is known in advance but the slope is unknown, the problem is more complex. If the known point is far from the measured data, the x and y coordinates of the measured points are averaged. The desired straight line is the one through this average point and the known point. However, if the average point is close to the known point, a small displacement in either point will produce a large change in the calculated slope. In addition,

TABLE 4–2 / Equation of a Straight Line by the Method of Averages

Data		Group 1		Group 2	
x	y	x	y	x	y
0.03	−3.01	0.03	−3.01		
0.95	−0.97	0.95	−0.97		
2.04	0.96	2.04	0.96		
3.11	3.08	3.11	3.08		
3.96	4.86	3.96/2	4.86/2	3.96/2	4.86/2
5.03	7.11			5.03	7.11
5.99	9.03			5.99	9.03
7.01	10.93			7.01	10.93
8.10	13.28			8.10	13.28
		8.11	2.49	28.11	42.78
Average = sum/4.5		1.802	0.553	6.247	9.507

$$m = (9.507 - 0.553)/(6.247 - 1.802) = 2.014$$
$$b = 0.553 - 1.802 \times 2.014 = -3.076$$
$$y = 2.014x - 3.076$$

FIGURE 4–1 / Method of averages. The dots are the data points of Table 4–2; the crosses are the two averaged data points. The straight line is that obtained by solving two simultaneous equations.

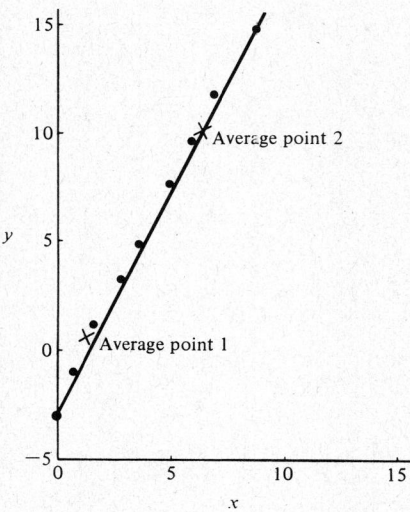

the assumptions underlying the method of averages are no longer valid, and it is better to use the method of least squares. If for some reason the method of least squares cannot be applied, obtain two averaged data points, as in the general case, and calculate the slope. Then draw a straight line with this slope through the known point.

METHOD OF LEAST SQUARES

The basic assumption of the method of least squares is that the best straight line is the one for which the sum of the squares of the deviations of the data points from the calculated line is a minimum. Usually, for mathematical convenience it is assumed that the error lies solely in the dependent variable y, and that all data points are equally trustworthy.

For the ith point, the residual, or random, error r.e.$_i$ is

$$\text{r.e.}_i = y_i - \bar{y}_i = y_i - mx_i - b$$

where $\bar{y}_i$ is the true value of the variable and y_i is the measured value. The sum of the squares of the residuals is

(4-1) $$\sum \text{r.e.}_i{}^2 = \sum (y_i - mx_i - b)^2$$

This sum is a function of the values of each of the measured data points and of the two parameters m and b. The parameters define a family of straight lines with different values of m and b which can be drawn through the data. The value of the sum of the squares of the residuals will vary with the choice of the line, that is, with the values of m and b. To select the values of m and b for which the sum of the squares of the residuals will be at a minimum, equation (4-1) is differentiated with respect to m and b, the derivatives are set equal to zero, and the two resulting equations are solved. The expressions for the slope and the intercept for n data points then become

(4-2) $$m = \frac{n \sum x_i y_i - \sum x_i \sum y_i}{n \sum x_i{}^2 - (\sum x_i)^2}$$

(4-3) $$b = \frac{\sum y_i \sum x_i{}^2 - \sum x_i \sum x_i y_i}{n \sum x_i{}^2 - (\sum x_i)^2}$$

To use the method of least squares, tabulate the data as shown in Table 4-3 and insert the sums at the bottom of the columns into equations (4-2) and (4-3) to obtain the values of m and b.

There is a considerable amount of tedious arithmetic involved, and at any stage the results may be invalidated by an arithmetical error. Furthermore,

TABLE 4–3 / Equation of a Straight Line by the Method of Least Squares

x	y	x^2	xy
0.03	−3.01	0.0009	−0.0903
0.95	−0.97	0.9025	−0.9215
2.04	0.96	4.1616	1.9584
3.11	3.08	9.6721	9.5788
3.96	4.86	15.6816	19.2456
5.03	7.11	25.3009	35.7633
5.99	9.03	35.8801	54.0897
7.01	10.93	49.1401	76.6193
8.10	13.28	65.6100	107.5680
36.22	45.27	206.3498	303.8113

$$m = \frac{n \sum x_i y_i - \sum x_i \sum y_i}{n \sum x_i^2 - (\sum x_i)^2} = \frac{9(303.8113) - (36.22)(45.27)}{9(206.3498) - (36.22)^2}$$
$$= 2.008$$

$$b = \frac{\sum y_i \sum x_i^2 - \sum x_i \sum x_i y_i}{n \sum x_i^2 - (\sum x_i)^2} = \frac{(45.27)(206.3498) - (36.22)(303.8113)}{9(206.3498) - (36.22)^2}$$
$$= -3.049$$

since the separate terms in the numerator and denominator are of the same order of magnitude, the usual rules of significant figures cannot be applied, and there is no rounding off. The computations therefore become almost impossible without a desktop or pocket calculator.

One short cut in the least squares computation employs smaller and more manageable numbers to compute corrections to an arbitrary line through the data. The corrections are then added to the arbitrary line, giving the parameters of the best straight line through the data. The equation whose constants are desired is

$$y = mx + b$$

The equation of the arbitrary line is

$$y' = m'x + b'$$

The equation of the difference is

(4-4)
$$\Delta y = (y - y')$$
$$= (m - m')x + (b - b') = \Delta m x + \Delta b$$

For each data point, the difference Δy_i between the measured value and the arbitrary value y'_i is determined. The least squares procedure is then

Chapter 4. Selected Numerical and Graphical Methods

TABLE 4–4 / Modified Least Squares Calculation

x_i	y_i	$y_i - y' = \Delta y$	x_i^2	$x_i \Delta y_i$
0.03	−3.01	−0.07	0.0009	−0.0021
0.95	−0.97	+0.13	0.9025	+0.1235
2.04	0.96	−0.12	4.1616	−0.2448
3.11	3.08	−0.14	9.6721	−0.4354
3.96	4.86	−0.06	15.6816	−0.2376
5.03	7.11	+0.05	25.3009	+0.2515
5.99	9.03	+0.05	35.8801	+0.2995
7.01	10.93	−0.09	49.1401	−0.6309
8.10	13.28	+0.08	65.6100	+0.6480
36.22		−0.17	206.3498	−0.2283

employed to find the best value of Δm and Δb, and these are added to the arbitrary m' and b'. If the arbitrary line lies reasonably close to the data and if m' and b' are reasonably simple, the labor of the computations is greatly reduced. The method is illustrated in Table 4–4, with the data of Table 4–3 already used in demonstrating the method of least squares. The arbitrary equation is $y' = 2x - 3$. We have

$$\Delta m = \frac{n \sum x_i \Delta y_i - \sum x_i \sum \Delta y_i}{n \sum x_i^2 - (\sum x_i)^2}$$

$$= \frac{9(-0.2283) - (36.22)(-0.17)}{9(206.3498) - (36.22)^2} = +0.008$$

$$\Delta b = \frac{\sum x_i^2 \sum \Delta y_i - \sum x_i \sum x_i \Delta y_i}{n \sum x_i^2 - (\sum x_i)^2}$$

$$= \frac{(206.3498)(-0.17) - (36.22)(-0.2283)}{9(206.3498) - (36.22)^2} = -0.049$$

$$m = 2 + 0.008 = 2.008$$
$$b = -3 - 0.049 = -3.049$$

The best straight line is therefore $y = 2.008x - 3.049$, which is the same as that obtained from the method of least squares applied directly to the experimental data. The labor saved by doing a least squares computation with two-digit numbers amply justifies the work of obtaining the Δy's from the data and the arbitrary line. Also, if the arbitrary equation reproduces the data fairly well, large values of the correction terms Δm and Δb will indicate an arithmetical error in the computation.

Another method of saving labor is to shift the axes in order to use smaller numbers in the computation.

If either the slope or the intercept of the line is known in advance, the computations are much reduced. When b is known,

$$(4\text{-}5) \qquad m = \frac{\sum x_i y_i - b \sum x_i}{\sum x_i^2}$$

and when m is known,

$$(4\text{-}6) \qquad b = \frac{\sum y_i - m \sum x_i}{n}$$

The preceding equations have been derived on the assumptions that all of the experimental error is in y. There is actually no reason, in most cases, for making this assumption. The error, in fact, may lie partly or wholly in x. If it is assumed that the error lies entirely in x, one procedure for finding the "least squares on x" is to reverse the roles of the variables. An equation is set up of the form

$$x = \alpha y + \beta$$

$$\alpha = 1/m$$
$$\beta = -b/m$$

and a least squares equation is employed to find α and β. The data of Table 4–3, if treated by the method of least squares on x, give

$$\alpha = 0.4975 \qquad \beta = 1.520$$

from which

$$m = 2.010 \qquad b = -3.055$$

Note that the same data gave three different best straight lines when three different commonly used procedures were used:

Method of averages: $y = 2.014x - 3.076$
Least squares on x: $y = 2.010x - 3.055$
Least squares on y: $y = 2.008x - 3.049$

Within experimental error, all three equations are equally correct. That is, the differences between them are less than the experimental error in determining y and less than the difference that would be produced by making different assumptions about the distributions of errors. In practice, therefore, unless the data scatter greatly, the easiest method, given the calculating equipment at hand, is the best.

One caution is necessary in any work involving the fitting of parameters to a straight line. The computed line should always be drawn through the data to show how the points scatter. If the scatter is random, without any definite trend, the data are best represented by the straight line. If, however, there is a definite trend—for example, if higher and lower points are consistently above while intermediate points are below—the data will be better represented by a curve.

ESTIMATION OF ERRORS IN m AND b

In the method of averages the simplest procedure is to assume that the error will reside solely in y and that the values of y all come from the same population and therefore have the same deviation.

From the definition of Y_1 and Y_2, the y values of the two averaged points, we have

$$Y_1 = \frac{1}{n/2} \sum_{i=1}^{n/2} y_i$$

From the definition of the standard deviation, σ, we have

$$\sigma_{Y_1}^2 = \frac{4}{n^2} \sum_{i=1}^{n/2} \sigma_{y_i}^2$$

Since all y's are from the same population, $\sigma_{y_1}^2 = \sigma_{y_2}^2 = \cdots = \sigma_y^2$, and therefore

(4-7) $$\sigma_y^2 = \frac{1}{n-2} \sum (y_i - mx_i - b)^2$$

This is computed and substituted into

(4-8) $$\sigma_m^2 = \frac{4\sigma_y^2}{n(X_2 - X_1)^2}$$

(4-9) $$\sigma_b^2 = \frac{2\sigma_y^2 (X_2^2 + X_1^2)}{n(X_2 - X_1)^2}$$

to compute the standard deviation in m and in b.

In calculating the errors in the slope and intercept of a straight line obtained by the method of least squares, the same two assumptions are made, that is, that the values of y are all taken from the same population and that all the experimental errors are found in y rather than in x. After the appropriate partial differentiations and substitutions, we get

(4-10) $$\sigma_m^2 = \frac{n\sigma_y^2}{n \sum x_i^2 - (\sum x_i)^2}$$

(4-11) $$\sigma_b^2 = \frac{\sigma_y^2 \sum x_i^2}{n \sum x_i^2 - (\sum x_i)^2}$$

where σ_y^2 is defined by equation (4-7).

Graphical and Numerical Integration

By definition, the integral of a function between two limits is the area under the curve between the two limits. Even if the function does not fit the experimental data very closely, the area under the curve will usually be a good approximation of the integral because, unless the curve is close to the base line, the largest contribution to the integral will be made by elements of area well removed from the curve. Figure 4–3 shows this. Curve *a* is the curve whose integral is to be evaluated. Curves *b* and *c* are two trial functions whose values do not correspond particularly closely with those of curve *a*. By inspection it is obvious that the integrals of all three curves correspond fairly closely. Most of the numerical methods of integration used are approximation methods which rely for their accuracy upon the fact that reasonably similar functions will give almost identical definite integrals.

METHOD OF COUNTING

In the *method of counting*, the best smooth curve is drawn among the data points and all the squares under the curve are counted. Each fractional square is counted as one half. This method is laborious but, with care, gives results accurate to a few tenths of a percent. In a variation of the method of counting, the area under the graph is cut out with a pair of scissors and weighed. The weight is then compared with that of a piece of the same paper of known area. The method depends upon having uniform graph paper, but it is accurate if performed carefully. To save the data, use a Xerox copy of the graph. This also gives a better result, since Xerox paper is somewhat more uniform than most graph paper.

AREA BY PLANIMETRY

The *method of planimetry* utilizes the planimeter, a mechanical device for measuring the geometrical area bounded by a closed curve. It consists of a stationary base, a movable index or pointer which is used to trace the boundaries of the area, a wheel connected to the base and to the pointer by a mechanical linkage system, and a counter which shows the number of revolutions of the wheel. The linkage is so designed that, as the curve is traced, the wheel partly rolls and partly slides in such a way that the number of revolutions is proportional to the enclosed area. In principle, the accuracy of a planimeter is limited by the skill of the operator in tracing the curve.

MATHEMATICAL INTEGRATION

In *mathematical integration*, an integrable equation is fitted to the data and integrated. The more closely the trial function approximates the data, the better the accuracy of the integral. As previously mentioned, even an approximate equation will give accurate values of the integral provided that the curve passes reasonably close to each data point.

TRAPEZOIDAL METHOD

In the *trapezoidal method*, the area under the curve is approximated by a set of trapezoids formed by connecting successive data points with straight lines and then drawing a vertical line from each data point to the abscissa, as shown in Figure 4–2 using the data of Table 4–5. The area of a trapezoid is half the product of the base and the sum of the two sides. The approximate value of the integral is the sum of the areas of the individual trapezoids. The greater the number of data points, the closer the approximation. In the special case where the base of each trapezoid is the same, that is, where the data have been taken at equal increments of x, the equation for the integral becomes

(4–12) $$\int y\, dx = \Delta x (\tfrac{1}{2} y_0 + y_1 + y_2 + y_3 + \cdots + y_{n-1} + \tfrac{1}{2} y_n)$$

FIGURE 4–2 / Trapezoidal method. The solid line is the curve for the Bessel function $y = J_0(x)$. The dashed lines are the straight lines connecting the data points.

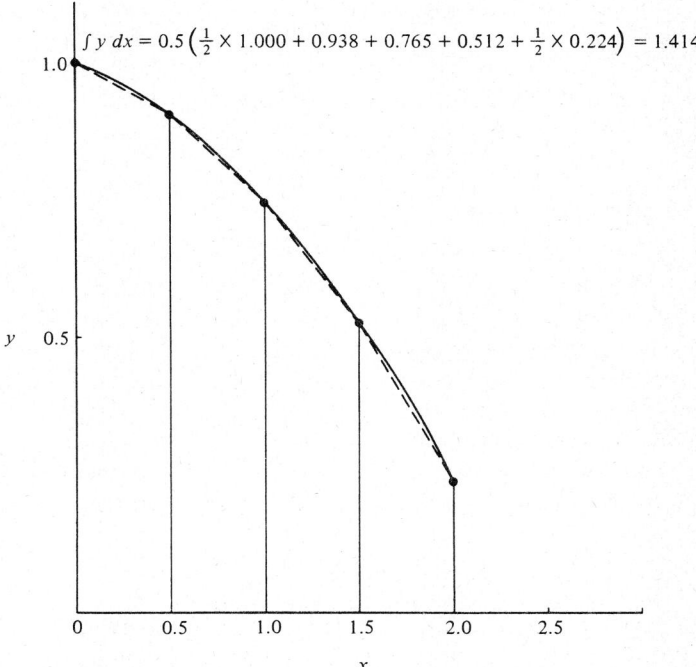

$\int y\, dx = 0.5 \left(\tfrac{1}{2} \times 1.000 + 0.938 + 0.765 + 0.512 + \tfrac{1}{2} \times 0.224\right) = 1.414$

A variation of this method, which can be applied to a graph but not to tabulated data, is to approximate the curve with a set of straight lines drawn so that as far as possible the area lost by this polygon is equal to the area gained. The polygon is broken up into triangles and trapezoids, and its area is then computed by elementary geometry.

SIMPSON'S METHOD

Simpson's method may only be applied to data taken at equal increments of x. The area under the curve is approximated by the area under the figure formed by drawing a parabola through each three successive points. The equation for the integral is

$$(4\text{-}13) \quad \int y\, dx = \tfrac{1}{3} \Delta x [y_0 + 4(y_1 + y_3 + y_5 + \cdots + y_{n-1}) \\ + 2(y_2 + y_4 + \cdots + y_{n-2}) + y_n] \\ = \tfrac{1}{3} \Delta x (y_0 + y_n + 4 \sum y_{\text{odd}} + 2 \sum y_{\text{even}})$$

Simpson's method is the most accurate of the common methods of numerical integration. To use Simpson's method, if there is an odd number of intervals in the data, find the area of either of the end intervals by the trapezoidal method and then add its value to the integral of the rest as determined by Simpson's method.

Figure 4–3 and Table 4–5 show the approximation of the definite integral

FIGURE 4–3 / Approximation of the definite integral between 0 and 2 for the Bessel function $J_0(x)$. Here a is the function $J_0(x)$, b is the quadratic, and c is a straight line.

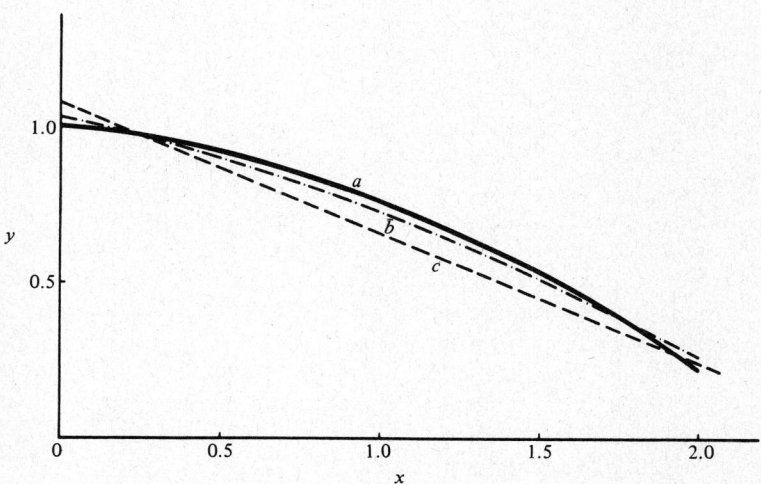

TABLE 4–5 / Approximation of the Integral of $y = J_0(x)$ in the Interval 0 to 2

$y = J_0(x)$	x	Straight Line $y = 1.125 - 0.448x$	Quadratic $y = 1.037 - 0.186x - 0.104x^2$
1.000	0.0	1.125	1.037
0.938	0.5	0.901	0.918
0.765	1.0	0.677	0.747
0.512	1.5	0.453	0.524
0.224	2.0	0.229	0.249

between 0 and 2 for the Bessel function $J_0(x)$. (This function was selected because the integral and derivatives are known.) The straight line has been fitted to the value of $J_0(x)$ by the method of averages, although the curve is obviously not a straight line. The quadratic equation has been fitted by the method of selected points (not discussed here) and is a somewhat better approximation. The definite integral between $x = 0$ and $x = 2$ has been evaluated by integrating each equation, by using the trapezoidal rule, and by using Simpson's rule. The results are

Straight line	1.354
Quadratic	1.425
Trapezoidal rule	1.414
Simpson's rule	1.426
Accepted value	1.426

Even the very ill-fitting linear equation gives an integral within 5% of the accepted value, and the somewhat better approximate quadratic equation gives a value within 0.1% of the accepted value.

Differentiation

The derivative of an experimental function, in contrast to the integral, is very much dependent on how well the selected curve fits the experimental data. Even if the empirical function differs somewhat from the true function, the value of the integral is changed only slightly. However, the slope of the empirical curve may differ greatly from the true value at the experimental point. It is therefore not correct to fit an empirical function to data and then to obtain the slope by differentiating the function. One must approximate the best slope at each particular point from the data.

An approximate method is to obtain, *carefully*, the desired best curve and then to draw the tangent to the curve at the desired point. There are several visual methods of finding tangents to the curve, and inexpensive commercial devices are available that give good results.

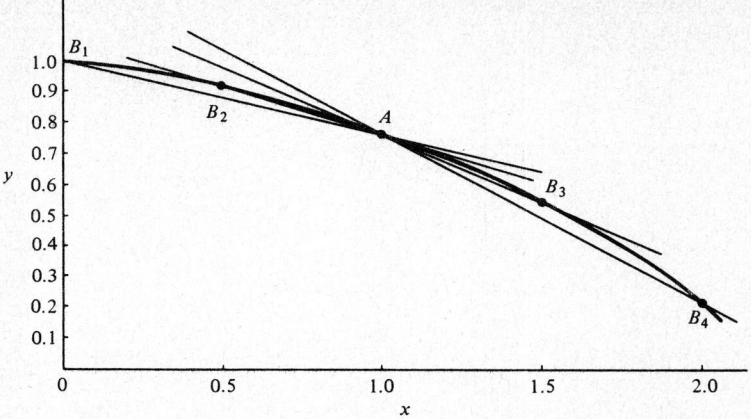

FIGURE 4–4 / Secants through point A on function $y = J_0(x)$.

The *limiting secant method* is probably the most accurate common method. It is more tedious to use than the visual methods, but more accurate. We apply it here to the function $y = J_0(x)$ at $x = 1$. We start at point A on Figure 4–4, the point at which the value of the tangent is to be determined, and draw a series of secants to B_1, B_2, etc., at both higher and lower values of x. The slope of each secant is computed, as in Table 4–6. The slopes are then plotted against the abscissas at the points of intersection. The slope at the point where the intercept of the secant coincides with point A is the slope of the

TABLE 4–6 / Secants Through Point A of the Function $y = J_0(x)$

Point	Coordinates at Point of Intersection		Slope of Secant
	y	x	
B_1	1.000	0	$\dfrac{0.765 - 1.000}{1.0 - 0} = -0.235$
B_2	0.938	0.5	$\dfrac{0.765 - 0.938}{1.0 - 0.5} = -0.346$
B_3	0.512	1.5	$\dfrac{0.512 - 0.765}{1.5 - 1.0} = -0.506$
B_4	0.224	2.0	$\dfrac{0.224 - 0.765}{2.0 - 1.0} = -0.541$

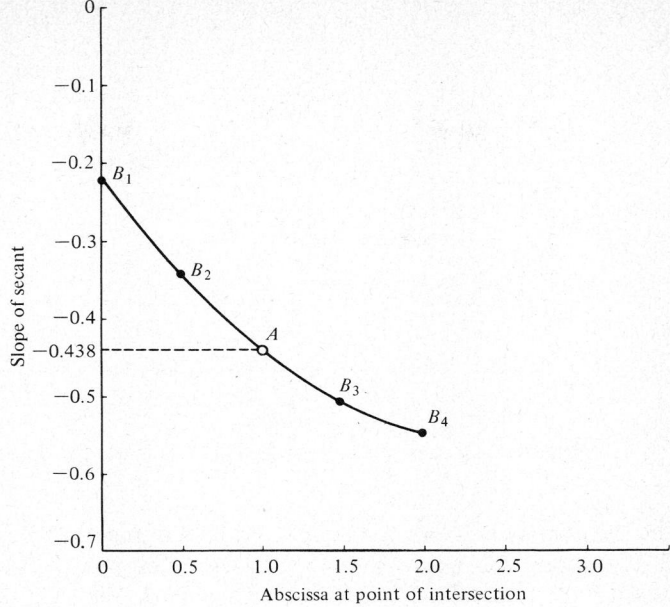

FIGURE 4–5 / Slope of secants through point A on function $y = J_0(x)$.

limiting secant, that is, the tangent at point A. Note that if point A is at a cusp, the derivative does not exist, and the graph of the slope of the secant will consist of two curves, at higher and lower values of B, which do not intersect. From Figure 4–5 the slope of the limiting secant at point $x = 1$ is -0.438. The accepted value is -0.440. Thus, the error is 0.50%. With a little more labor—that is, by plotting additional secants—the error would have been even less.

Inflection Points, Maxima and Minima

Certain quantities—for example, the endpoint of a pH titration—are defined by inflection points on a graph. Often, as in Figure 4–6, the inflection point is not easily ascertained by inspection. For any function the inflection point occurs where dy/dx is a maximum or minimum. Graphical methods for finding dy/dx from a set of data points are frequently cumbersome and inaccurate. Fortunately, it is not necessary to plot a graph of dy/dx to find the inflection point; a graph of the slopes $\Delta y/\Delta x$ can be used instead. For each pair of

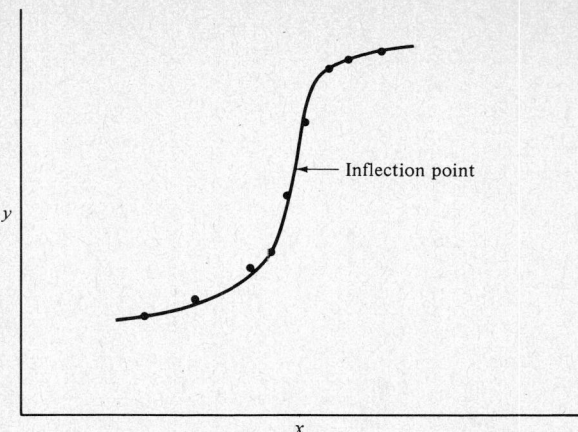

FIGURE 4–6 / Function with an inflection point.

successive data points—that is, for the points n and $n + 1$, $n + 1$ and $n + 2$, $n + 2$ and $n + 3$, and so on—calculate $\Delta y/\Delta x = (y_{n+1} - y_n)/(x_{n+1} - x_n)$. By the theorem of the mean, this is equal to dy/dx at some point within the interval x_n to x_{n+1}. Since this point is not known, assume it to be the midpoint of the interval, $x = \frac{1}{2}(x_{n+1} - x_n)$. Plot these slopes as a function of their corresponding midpoints (Figure 4–7), draw a smooth curve through them (smoothing by eye is sufficient), and determine the maximum or minimum of this difference

FIGURE 4–7 / Difference curve—graph of successive slopes from Figure 4–6.

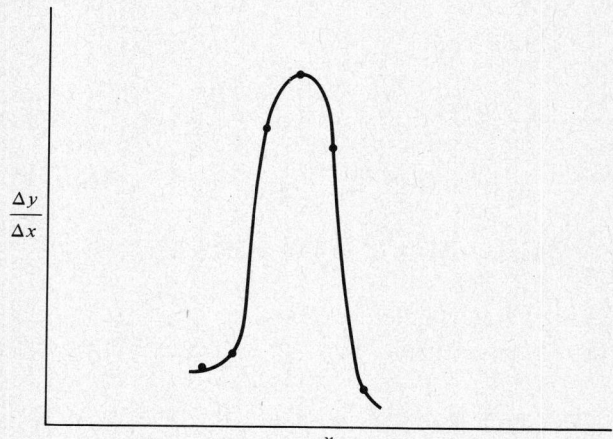

curve. The value of x at this maximum or minimum is the value of x at the inflection point. If the data are too scattered to give a reasonably well defined curve when successive points are used to calculate $\Delta y/\Delta x$, every other point may be skipped and $\Delta y/\Delta x$ calculated from the pairs n and $n + 2$, $n + 1$ and $n + 3$, $n + 2$ and $n + 4$, and so on. To determine the maximum or minimum more accurately than can be done by eye, draw several horizontal chords, through the ascending and descending portions of the difference curve (Figure 4–7). Find the midpoints of these chords. Draw a smooth curve through the midpoints of these chords, and extend it until it intersects the difference curve. This intersection gives the value of x at the inflection point. (When this procedure is used to find the density of a fluid at its critical point, it is known as the *law of the rectilinear diameter*.) When the ascending and descending portions of the difference curve appear to become steeper as the maximum or minimum is approached, a somewhat different procedure, which was first proposed to find the endpoint of a pH titration [Cohen], can be used. Plot the difference curve, draw chords, and find their midpoints as above. Plot the x coordinate of the midpoint as a function of the length L of the chord. Plot the midpoints as a function of L, draw a smooth curve through the points, and extend it to where $L = 0$ to find the value of x at the inflection point. The methods given to find the maximum or minimum of a difference curve can obviously be used to find maxima or minima of any curve. Alternatively, plot a difference curve using points near the desired maximum or minimum and draw a smooth curve through these secants. The desired value of x occurs where $\Delta y/\Delta x = 0$.

References

Cohen, S. R., A simple graphical method for locating the end point of a pH or a potentiometric titration. *Anal. Chem.* **38**, 158 (1966).

Dowd, J. E., and D. S. Riggs, Comparison of estimates of Michaelis–Menton kinetic constants from various linear transformations. *J. Biol. Chem.* **240**, 863 (1965).

Guest, P. G., *Numerical Methods of Curve Fitting*. Cambridge Univ. Press, Cambridge, 1961.

Hamilton, W. C., *Statistics in Physical Science: Estimation, Hypothesis Testing and Least Squares*. Ronald Press, New York, 1964.

Hammes, G. G. (ed.), *Investigation of Rates and Mechanisms of Reactions* (Vol. VI of *Techniques of Chemistry*, edited by A. Weissberger), 3rd ed. Wiley-Interscience, New York, 1974.

Young, H. D., *Statistical Treatment of Experimental Data*. McGraw-Hill, New York, 1962.

Chapter 5

Recording and Reporting Data

All data should be recorded for future reference. Unrecorded results are soon forgotten or, what is worse, remembered incorrectly. Data should either be entered in a permanent laboratory notebook or made part of a laboratory report, or both. Data put into a report instead of into the permanent notebook are usually concerned with short-term studies designed to provide specific answers to specific questions. The data included in such a report might be those obtained in calibrating an instrument or determining a physical property that is to be used in designing an apparatus. For example, the vapor pressures of a series of greases might be studied to select one for a high-vacuum apparatus designed for some long-term study. The laboratory notebook, in this case, would be concerned with the results of the high-vacuum investigation and not with the selection of the grease. The study of vacuum greases would be put into a separate report and kept in the files for use, from time to time, in the selection of greases for other apparatus.

The following general instructions for keeping a notebook and for writing reports are not rigid; omissions, additions, and rearrangements may be made when desirable. As always in laboratory work, common sense and judgment should be applied to each case. The main requirements are completeness, clarity, conciseness, and quick, easy accessibility of necessary information.

The Notebook

Records of laboratory work should be so complete as to be meaningful to the reader even years after completion of the work. If the research is part of a continuing project, as is usually the case, the data should be recorded in a permanent (bound) notebook to which reference may be made at any time. Data that cannot be included in a notebook, such as photographs, microscope slides, and recorder charts, should be labeled and stored in a readily accessible permanent container to which reference is made in the notebook.

All significant raw data should be included in the record. Many times old data are reused in computations based on newer theory. Often apparently inconsequential information turns out to be of great value when past work is reviewed. (Common sense should be used, however. Do not fill up the notebook with trivial information such as the size of the beakers from which solutions were poured.) Not only the data, but also the experimental method and incidental observations of unexpected or inexplicable phenomena should be recorded. All entries in research notebooks should be dated because the priority of publications and patents is often based on entries in permanent laboratory notebooks. Duplicate notebooks, which provide a detachable carbon copy of all entries, are frequently used. With such notebooks one copy of the data is placed in an accessible central file, while the investigator keeps the other copy.

In keeping a notebook, follow these general instructions:

1. Use a bound notebook with numbered pages. Provide space at the front for a table of contents.
2. Write your name, address, and telephone number in several places in the book, including the front and back pages.
3. Enter all pertinent information in detail, including all data taken and the method of obtaining the data. If in doubt, make the entry. Excess information may be deleted, but unrecorded observations are lost.
4. Record all data as soon as possible, and use ink.
5. Date each page. In industry the pages are often signed and the signatures witnessed, but this is not necessary in student notebooks.

Figure 5-1 is an example of a typical page in a student notebook. Note the following:

It is good policy to head each page with the title of the project in case the notebook is to be used for several projects.

Reference to the experimental procedure should be included in the data.

The data are included from which the humidity was calculated. This permits the calculations to be rechecked for mistakes.

The brand of the chemical is not really necessary in this experiment. However, the source could be important in research that is sensitive to the presence of trace impurities, which may vary widely from supplier to supplier.

The two weighings from which the weight of vapor is found have been recorded in case an arithmetic mistake was made.

Discarded data are crossed out in such a way that they can be read if necessary.

2/15/77　　　　　　　　　**Molecular Weight of a Vapor by the Dumas Method**

References: ———
Humidity: Dry bulb 25°C, wet bulb 18°C, humidity 52% (from chart)
Barometric pressure: 754 mm, 24°C, at start, corrected to 751 mm at 0°C
　　　　　　　　　754 mm, 26°C, at end, corrected to 751 mm at 0°C
Liquid: Mallinckrodt reagent grade acetone, 10 ml.

	I	II
Weight of bulb and vapor	32.2485	31.5974
Weight of bulb	32.0684	31.4792
Weight of vapor	0.1801	0.1182
Weight of bulb and water	158.5	156.8
Weight of bulb	32.1	31.5
Weight of water	126.4	125.3
Temperature of water bath	98.1°C	100.2°C
Temperature of water in bulb	25°C ↑	26°C

　　　　　　　　　　　*run discarded because
　　　　　　　　　　　bulb filled only part
　　　　　　　　　　　way when opened
　　　　　　　　　　　under water*

FIGURE 5–1 / Specimen page of a student laboratory notebook.

The Report

Write the report from the point of view of an employee on a continuing project. You have been assigned a task and the report of what you did and what you found is to be made part of the record and consulted from time to time by those who need the information. Your report not only presents the information, but also represents you to whoever reads it.

The report should be brief, well-organized, to the point, neat, and written in correct and literate English. Reports should be typed, double-spaced, on one side of uniform, good quality paper. If you cannot type, write legibly. Graphs should conform to the specifications in Chapter 3. Follow the format suggested below, at least until you are experienced enough to develop your own style. Remember, the purpose of a report is to communicate.

　I. Introductory material
　　A. Title page
　　B. Abstract
　　C. Summary of theory

II. The experiment
   A. Experimental method
   B. Data, graphs, results, and precision analysis
   C. Sample calculations
III. Conclusions
   A. Discussion of results and comparison with accepted data
   B. Error analysis
   C. Suggested alternative methods
   D. Answers to study questions

The title page, the abstract, and the sample calculations should be included in all reports as distinct, separate sections. The other sections may be combined, subdivided, modified, or omitted if in your judgment the report is thereby improved. However, any modification should be clearly for the purpose of improving the report, not to cover up an error or to avoid work.

I-A. The *title page* should contain your name, your partner's name, the date of the experiment, the date of the report, the title of the report, and the course and section.

I-B. The *abstract* is a brief summary of the contents. It should be written last, after all of the results have been calculated and the conclusions drawn. The abstract should be specific and should state the major findings or conclusions, the experimental method, if important, and any other significant information. The following specimen is an acceptable abstract.

> The viscosities of water and benzene were measured with an Ostwald viscometer. For water the viscosity decreases from 1.76 cP at 0°C to 0.671 cP at 40°C. For benzene it decreases from 0.831 cP at 6°C to 0.507 cP at 40°C. The activation energies for fluid flow are 4.1 and 2.5 kcal/mole, respectively. The viscosities of mixtures of benzene and cyclohexane were measured at 25°C and found to be higher than predicted by either of two equations for the viscosities of mixtures. Viscosities were reproducible to 0.4% and were within 3% of accepted values.

I-C. The *summary of theory* should be written as though to refresh the memory of the reader who may be familiar with the material but a bit rusty. There should be no lengthy explanations. If the reader needs to learn the material, he can read a text. The summary should be written in the student's own words. Any text used should be paraphrased. The material must *not* be copied word for word.

II-A. The discussion of experimental method contains a *summary of the method*, a description of the apparatus, and an outline of the procedure. The summary is brief and gives, if necessary, the theory of the method, as distinct from the theory underlying the experiment. Necessary equations should be stated, not derived.

The description of the apparatus and the outline of the procedure should be detailed enough for the reader to duplicate the experiment. If commercial instruments are used, the make and model should be stated. Any special pieces of apparatus should be described, giving critical dimensions or novel design features. Neat, inked diagrams of important apparatus, arrangements, circuits, and so on, should be included. However, ordinary apparatus such as beakers and burners should be diagrammed or described only when essential to the understanding of the experiment. In outlining the procedure, avoid giving cookbook instructions.

This material may be written in one or more sections, depending upon the particular experiment and the report. Whatever the arrangement, the reader should be able to evaluate the method and, if necessary, to repeat the experiment.

II-B. The *data*, *graphs*, and *results* should follow the description of the experimental method and should be organized to permit the reader to follow the experiment from the measured values to the final computed results easily and clearly, with a minimum of turning back and forth from page to page.

Data are best presented in a table, together with the experimental precision and the results. Columnar data paper is preferred. The individual deviation or other precision measure should be reported adjacent to the value to which it refers. The column heading is the name of the parameter if the experimental measurements are averaged, and the final result is calculated from the averages of the experimental parameters. Within each column the successive measured results are placed one underneath the other (see Example I).

If the data cannot be averaged, as is the case when separate charges are run, the results must be computed for each run. In this case there should be a separate column for each run, with the measured quantities arranged vertically in each column (see Example II).

Sometimes several tables will be required for the different portions of the experiment.

The *precision analysis* is presented together with the data and results for

EXAMPLE I / Horizontal Table for Averaged Data:
Viscosity by the Ostwald Method

	Flow Time				Density			
Run	Water	δ	Liquid	δ	Water	δ	Liquid	δ
1								
2								
3								
4								
	$\bar{x}$	a.d. or s	$\bar{x}$	a.d. or s	$\bar{x}$	a.d. or s	$\bar{x}$	a.d. or s

EXAMPLE II / Vertical Table for Data that Cannot be Averaged: Measurement of Heat of Combustion

	Run 1	Run 2	Run 3
Weight of charge			
Final temperature			
Initial temperature	___	___	___
Temperature difference			
Length of unburned wire			

purposes of comparison. First the sample mean of the *experimental measurements* is used to calculate the reliability of the individual measurements. Then the arithmetic mean computed from the *reliable*, or accepted, data is used to compute the final result. Finally, the precision measure of the result is calculated from the appropriate precision measures of the experimental quantities. If the experimental data cannot be averaged, the arithmetic mean of the computed quantity is calculated from the results of the individual runs. This mean is then used to eliminate unreliable runs and, if necessary, a new, better arithmetic mean is obtained from the data of the reliable runs. Finally, the precision index of the result is calculated from the arithmetic mean and the individual values of the separate runs. Suspect, or doubtful, data should be clearly indicated as such, and data omitted from the calculation should be clearly marked or labeled so that no mistakes are made by the reader.

The results should be reported on the data page, together with the data from which they are obtained. However, if putting the calculations, precision, and results all on the same sheet of paper would make the data page cluttered and difficult to read and understand, it is better to use two or three sheets. Correct dimensions and units must be used. The precision measure of the result must be reported wherever the result is reported.

Graphs normally follow either the results or the sample calculations. What is graphed usually depends on the writer's judgment. Graphs should be included if they were used for calculations or calibrations or if they demonstrate a theoretical relationship.

II-C. *Sample calculations* must be shown for each type of calculation. Only one sample calculation is needed. Correct dimensions and units must be specified, but the arithmetic should be omitted. The data used in the sample calculation should be clearly indicated. The sample calculations should follow the results. They may be put on the data page if there is sufficient room.

III-A. The *discussion and interpretation of results* and *comparison with accepted data* should be more than just an assemblage of numbers. Usually, the results will have some points of interest to be discussed. For example, the vapor pressure of a substance may be unusually low, or high, and this might deserve comment. If the purpose of the experiment is to test theory, there should be some discussion of how the results support or contradict the theory, together with a comment on the validity of the results as a test of theory. Excess verbiage and unsupported or trivial conjectures should be avoided, but one purpose of any report is to present and interpret the experimental results.

III-B. The *error analysis* is a reasoned discussion of the precision and accuracy to be expected of the experimental results. Three categories of errors should be discussed.

1. *Inherent errors* generally result from the use of an approximate theory or an approximate mathematical method. They should be noted and their effect estimated. Some sources of inherent errors are the use of the ideal gas law for a vapor, the use of concentrations instead of thermodynamic activities, and the assumption that real solutions obey Raoult's law.

2. *Instrument errors* are of two types: errors caused by a faulty instrument and errors caused by misreading the instrument. Better instruments either are calibrated or have their *limiting accuracy* specified by the manufacturer. For example, a good meter might be accurate to $\pm 0.25\%$ of full scale and would have a temperature calibration chart. If the limiting accuracy is not known and cannot be determined, the instrument is assumed to be accurate and the *limiting precision* of reading is taken as the instrument error. The limiting precision of reading the instrument is determined by making several readings of the same value (e.g., three or four readings of the resistance of a particular thermistor) and calculating the average deviation or other precision measure. However, if there is no way to determine the precision of reading the intrument, the limiting precision of reading is taken as the smallest interval normally estimated.

3. *Random errors* include the scatter produced both by the inevitable small differences between repeated measurements and by such uncontrolled, or uncontrollable, factors as impurities, thermostat temperature fluctuations, thermal noise, and human reaction time. Unlike instrument errors and inherent errors, random errors can only be estimated roughly beforehand. They are best evaluated from the scatter of repeated measurements. They should be listed, and methods of minimizing them should be suggested.

III-C. The *alternative method* of obtaining the information may be either a method found in the literature or a method devised by the student. If it is a literature method, the description should be *paraphrased* and the source stated. If it is an original method, the student should list the required apparatus. The possible limitations, advantages, and disadvantages of the method for the specific purpose of the experiment should be stated.

The Preliminary Report

Before entering the laboratory the student should have prepared a preliminary report for inspection by his instructor. This should consist of the introductory material, a table complete with headings and ready for the entering of the experimental data, and, as far as possible, the *discussion of error*. After the preliminary report has been initialed by the instructor, the student should enter the data, as taken, and then complete the report. The completed report is due the week after completion of the experiment, unless the instructor wishes to allow 2 weeks. It is far easier to write the report immediately after the work has been done, while the procedure and results are still fresh in your mind.

References

Fieser, L. F., and M. Fieser, *Style Guide for Chemists*. Reinhold, New York, 1960.

Gensler, W. J., and K. D. Gensler, *Writing Guide for Chemists*. McGraw-Hill, New York, 1961.

Trelease, S. F., *How to Write Scientific and Technical Papers*. Williams & Wilkins, Baltimore, 1961.

Chapter 6

Temperature Measurement and Control

In theory, *temperature* is completely and unambiguously defined by either the ideal gas law or the concept of thermodynamic efficiency. In practice, however, ideal gases and reversible Carnot engines do not exist. Consequently, a conventional temperature scale has been defined in terms of a number of reproducible systems, such as the triple point of water. The temperatures of these systems have been measured as closely as possible, and by agreement among international organizations of scientists the "best" values have been selected and defined as the temperature of the particular systems. In each temperature range a definite reproducible method for interpolating and measuring temperatures between the defined temperatures has also been agreed upon. The conventional temperature scale arrived at in this fashion is revised from time to time as additional information is obtained.

Temperature measurements depend upon heat transfer. If two objects at different temperatures are in contact with each other, the object at the higher temperature will transfer heat to the object at the lower temperature. The heat transfer will take place even if the object at the lower temperature has more enthalpy than the high-temperature object. For example, a drop of hot water will transfer heat to an iceberg. The heat transfer will continue until thermal equilibrium is reached and both objects are at the same temperature.

When a thermometer is placed in contact with an object at a different temperature, heat will flow into or out of the thermometer, changing its temperature and any properties dependent upon temperature, until a steady state is reached. If the temperature dependence of some property is known, the value the property reaches at the steady state indicates the temperature of the thermometer, and therefore of the object. In using a thermometer, the thermometer and the object must be kept at the same temperature, since the thermometer indicates only its own temperature. Heat transfer by the thermometer from the object to the surroundings must be prevented because even if this heat conduction does not change the temperature of the object, it may produce a steady state in which the thermometer temperature is not that of the object. One extreme case is found in the measurement of the temperature of hot gases in ducts. If a thermometer without radiation shields is placed in the gas stream, the thermometer will "see" the walls of the duct rather than the gas and may indicate a temperature as much as several hundred degrees below the true gas temperature. If the object has a thermal gradient, the thermometer sensing element must be small compared to the gradient in order to measure the temperature at only one point instead of an average temperature of a region.

Theoretically, any property that depends upon temperature may be used to measure temperature. However, the practical thermometric properties are the volume of a liquid, the length of a solid, the pressure of a gas, the electrical resistance of a metal or a semiconductor, the emf developed by a thermocouple, and the brightness or color of a hot object.

Constant Volume Ideal Gas Thermometers

At sufficiently low densities, gases obey the *ideal gas law*:

$$pv = nRT$$

T = absolute or Kelvin temperature
  = 273.16 + Celsius temperature
p = pressure of gas
v = volume of gas
n = number of moles of gas
R = gas constant

If the volume and mass of the gas are kept constant, the relationship between temperature and pressure becomes

(6-1) $$T = 273.16 \frac{p}{p_0}$$

p = pressure at temperature T
p_0 = pressure at the triple point of water, defined as 273.16 K (0.01°C)

The gas thermometer is a bulb filled with an "ideal" gas attached to a mercury manometer, as shown in Figure 6-1. The ideal gas is helium, hydrogen, or nitrogen at pressures low enough for the gas to obey the ideal gas law at the lowest temperature reached. For very precise work, corrections are used to compensate for the nonideality of the gas. Helium may be used at any temperature down to 3 K. Hydrogen is usually used only for temperatures up to 400°C (673 K); nitrogen may be used above 400°C.

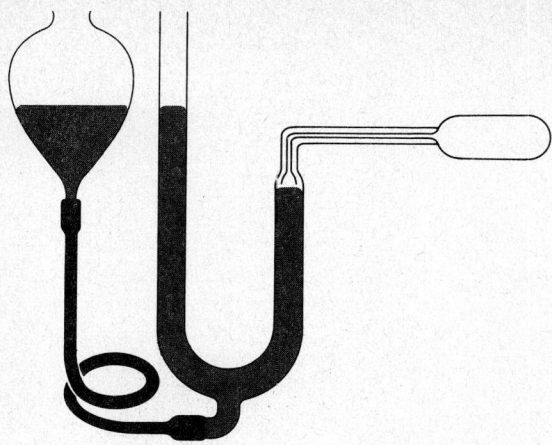

FIGURE 6–1 / Gas thermometer.

When the gas thermometer is in operation, the volume of gas is held constant by raising or lowering the mercury reservoir. The pressure is determined from the difference in the heights of the mercury columns, and the temperature is then calculated from equation (6–1). To make the thermometer more sensitive, the mercury may be replaced by a less dense liquid, provided that its vapor pressure is negligible in the desired temperature range.

Sources of error include differences in temperature between the gas in the bulb and the gas in the "dead" space between the mercury and the capillary; changes in glass volume with changing temperature; a pressure gradient between the bulb and the manometer, if the capillary diameter is about equal to the length of the mean free path of the gas molecules (i.e., at low densities) and if the bulb and the mercury are at different temperatures; adsorption of the gas on the walls of the bulb at low temperature; and compressibility and thermal expansion of the mercury. As a result of these errors and the further limitations imposed by fragility, slowness of response, and the large volume of the sensing element, the gas thermometer is used principally as a primary standard for the calibration of other thermometers rather than as a working laboratory thermometer.

Mercury-in-Glass Thermometers

As the temperature of the thermometer increases, the mercury rises in the capillary because mercury expands more than glass. It is customary to speak of the increase in length of the mercury thread rather than the relative increase in the mercury volume compared to that of the glass. The difference in length of the mercury thread at two different temperatures is proportional to the temperature difference:

(6-2a) $$\frac{t - t_i}{t_s - t_i} = \frac{L - L_i}{L_s - L_i}$$

The subscripts s and i refer to the steam point and the ice point, respectively. Since by definition $t_i \equiv 0°C$ and $t_s \equiv 100°C$ at a pressure of 1 atm, the Celsius temperature, as read on a mercury-in-glass thermometer, is

(6-2b) $$t = \frac{100(L - L_i)}{L_s - L_i}$$

If the capillary bore is uniform, the distance between the mercury lengths at 0 and 100°C may be marked off into 100 intervals, each of which is a "degree" Celsius. A degree Celsius based on the mercury-in-glass thermometer is therefore the change in temperature that causes a change in the length of the mercury column equal to one one-hundredth of the change in length between the ice and the steam points. Defining the degree Celsius with the mercury thermometer in this manner is equivalent to assuming that there is a constant difference between the coefficients of cubical expansion of glass and of mercury. Fortunately, the error resulting from this assumption is less than 5 parts per 1000 between -30 and $+150°C$ and is negligible, except for highly accurate work that requires calibration of the thermometer at each temperature used. Some special purpose thermometers contain liquids other than mercury, such as alcohol or pentane. However, mercury is preferred, for, by coincidence, the mercury-in-glass scale is almost identical with the thermodynamic or Kelvin scale over its useful temperature range.

There are two ways of placing the scale on a mercury-in-glass thermometer. Solid stem thermometers have the scale engraved on the outside of the stem. More expensive thermometers, such as the Beckmann thermometer, have a scale attached to the capillary, with both capillary and scale enclosed in a glass envelope. Since the position of the enclosed scale may shift, these thermometers are generally used to measure temperature differences rather than absolute values of temperature.

Unless the thermometer is immersed up to the level of the mercury meniscus, there is a temperature gradient between the bulb and the meniscus. For accurate work a correction must be added to or subtracted from the observed reading to obtain the true bulb temperature. At 200°C, for example, this correction may amount to 2 or 3°C, depending on the particular thermometer, the length of the exposed stem, and the difference in temperature between the bulb and the meniscus. Modern thermometers usually specify the depth of immersion and are calibrated to compensate for the exposed-stem temperature gradient, provided that the exposed stem is at 20°C. Even so, there remain small exposed-stem corrections. (In enclosed-stem thermometers, these corrections are still smaller.)

The advantages of the mercury-in-glass thermometer include simplicity, reliability and freedom from mechanical failure, ease of handling, cheapness, ease of reading, and direct reading. Disadvantages are fragility—the enclosed scale very often breaks loose from the capillary, and on the solid stem thermometers each calibration mark is a weakening scratch; hysteresis—if a mercury thermometer is kept at 100°C for a few minutes and then put into an ice-water bath, it will read below 0°C for several hours (this effect is due to the slow rate of contraction of glass); pressure sensitivity—an increase in external pressure squeezes the bulb and produces higher readings; variations in capillary bore; separation of the mercury column; need for exposed-stem temperature corrections; changes in readings as the thermometer ages; and slowness of response.

Mercury-in-glass thermometers are not used below $-40°C$, where mercury freezes, or above 360°C. However, some mercury thermometers made of special glass may be used at temperatures as high as 500°C. These are filled with gas under pressure to prevent the mercury in the bulb and stem from boiling and condensing at the top of the thermometer.

For measuring temperature differences, as in cryoscopy, ebullioscopy, and calorimetry, differential thermometers are often used. These have very fine uniform capillaries and scales calibrated for temperature intervals as small as 0.01°C per engraved scale division. In routine, moderate precision combustion calorimetry (page 72), reactions are usually run at about room temperature, and the thermometers commonly have solid stems with scales reading between 19 and 35°C engraved in 0.02°C scale divisions, which (with a lens) can be read to $\pm 0.002°C$. For constant pressure calorimetry and cryoscopy the thermometer often has an enclosed scale calibrated over a 6°C range with scale divisions of 0.01°C, which can be read to $\pm 0.001°C$. For vapor pressure–temperature measurements and for liquid–vapor phase studies, thermometers usually have an enclosed scale with a 50°C range and scale divisions of 0.1°C.

The Beckmann Thermometer

The Beckmann thermometer (Figure 6–2) is the most versatile of the differential mercury thermometers, since it can be adjusted for use at any temperature from -20 to $+140°C$. It has a fine capillary about a foot long attached to a mercury bulb and a mercury reservoir. The capillary and scale are enclosed in a glass envelope. The volume of mercury in the bulb is so large that a 5° change in temperature is sufficient to move the mercury level through the entire length of the capillary. To adjust the thermometer to read at any desired temperature, mercury is added to or removed from the bulb until the mercury level is on

Chapter 6. Temperature Measurement and Control

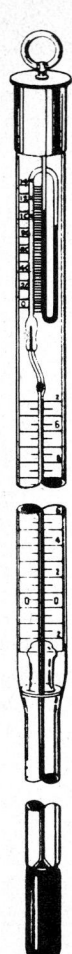

FIGURE 6–2 / Beckmann thermometer.

scale in the desired range. However, if the quantity of mercury in the bulb is changed, a change in temperature will move the mercury level to a different point on the scale. Consequently, the scale reading must be multiplied by a setting factor or range factor to compensate for any change in mercury volume. Equations for setting factors may easily be derived on the basis of the coefficients of cubical expansion of glass and mercury.

The Beckmann thermometer has now been largely replaced, in research laboratories, by thermocouples and thermistor thermometers because it is fragile and has a slow response time. It is still commonly used, however, especially in instructional laboratories.

OPERATING INSTRUCTIONS

Adjust the mercury level so that it is on scale in the desired temperature range. For an approximate adjustment, immerse the bulb in water at a temperature about 1°C higher than the maximum test temperature. After several minutes, invert the thermometer and tap it gently to remove the excess mercury. On cooling, the mercury should contract until, at the highest temperature required, the meniscus is at the top of the scale. For a more exact adjustment, warm the bulb carefully until droplets are forced out the capillary tip. The number of degrees per droplet is marked on the thermometer scale.

If necessary, add mercury to the bulb either by warming the bulb until contact has been made between the mercury in the stem and the mercury to be added or by tilting the thermometer until some of the mercury in the reservoir siphons down into the capillary and bulb. Then turn the thermometer upside down and shake it or tap it sharply, but carefully, to break the mercury thread without breaking the thermometer. Reinvert the thermometer and return the excess mercury to the reservoir.

NOTES / **Tap the stem gently before each reading to prevent the mercury from sticking to the capillary.**

If the thermometer is set for temperatures well below room temperature, do not let it warm up between runs. Mercury may escape from the capillary, changing the setting.

To avoid hysteresis, keep the thermometer at or near the temperature of the measurement. Do not use a thermometer for room-temperature measurements for at least 24 hr after using it at high temperatures.

EXERCISES

1. Derive an equation for the setting factor.
2. Suggest temperature standards for the ranges 20–25, 40–45, and 70–75°C. These standards should be known to within 0.002 C and should be convenient to work with and easily reproducible.

Resistance Thermometers

There are two types of resistance thermometers: *metallic conductors*, in which resistance increases with increasing temperature, and *semiconductors* (thermistors), in which resistance decreases with increasing temperature. The temperature dependence of semiconductors is negative and exponential; that of metallic conductors is positive and almost linear. Whereas electrons in metals move about freely in the conduction band, with low resistance, electrons in semiconductors must pass over energy barriers to get into the conduction band; therefore they do not move as freely and have rather high resistance. Increasing temperature disturbs the crystal structure in metals and so increases resistance, but in semiconductors increasing temperature produces more electrons with enough kinetic energy to get into the conduction band and so decreases resistance.

A metallic resistance thermometer consists of a length of carefully purified and annealed wire, usually platinum or nickel, wound around a thin frame of mica or fused silica in such a way as to avoid strains and increased tension when the wire contracts on cooling. The wire is contained in a thin (for increased sensitivity) pyrex glass or metal sheath in dry air at atmospheric pressure. The resistance under constant pressure and tension is given by the empirical *power series formula*:

(6–3) $$R = R_0(1 + at + bt^2 + ct^3)$$

R = resistance
R_0 = resistance at 0°C
t = temperature, °C
a, b, c = empirical constants whose values are characteristic of the particular wire

Commerical platinum resistance thermometers usually have resistances either about 25.5 ohms at 0°C, with a temperature coefficient of 0.1 ohm/°C, or about 2.55 ohms at 0°C, with a temperature coefficient of 0.01 ohm/°C.

The resistance at any temperature may be measured with either a resistance bridge (page 151) or a potentiometer circuit. The constants in the equation are determined from resistance measurements at standard temperatures. With care, a platinum resistance thermometer can be used to obtain temperatures within ± 0.001°C.

The platinum resistance thermometer is the standard for temperatures between -190 and $+600$°C. It can be used to make accurate measurements up to 1200°C, but above 600°C it is no longer the thermometric standard. The chief drawbacks are the high cost, the large time lag due to the mass of the thermometer, the necessity for controlling the temperature of the bridge in high precision work, and, in some cases, errors from the conduction of heat along the lead wires.

Thermistors, the second type, are semiconductors. They are used primarily for temperature control, but are also valuable for measuring small temperature changes and differences. The equation for the variation of the resistance of a thermistor with temperature is

(6–4) $$R = be^{a/T}$$

R = resistance
T = Kelvin temperature
a, b = constants characteristic of the particular thermistor

The thermistor is a bead of mixed metallic oxides fused to two wires on a ceramic support and enclosed in a glass guard tube. It may be used in the same way as a wire resistance thermometer is used. The great advantage of a thermistor is the very large temperature coefficient of resistance, which enables the experimenter to ignore the changes in the resistance of the lead wires and to eliminate expensive resistance bridges. Other advantages are cheapness, durability, small size, low heat capacity, and rapidity of response. Disadvantages are possible irreversible change on prolonged use, nonreproducibility from one thermistor to another that necessitates recalibration when thermistors are changed, and very high resistances at very low temperatures.

Thermistor thermometers are now being used in hospitals. They are attached to ohmmeters with digital readout calibrated directly in degrees. One such device emits a whistling sound when the temperature reaches equilibrium and stays constant for several seconds.

Thermocouple Thermometers

If a loop is made of two different metals and the junctions are maintained at different temperatures, a potential will develop, causing an electric current to flow in the circuit. The magnitude of the potential depends upon the two metals and on the temperatures of the junctions. The observed electromotive force (emf), first described by Seebeck, is the sum of the contact potentials at the junctions and the potential gradients along the wires. The contact potentials result from the difference in chemical potential between the electrons in the two metals. Potential gradients represent the difference in emf between the hot and cold ends of a metal that results from the greater kinetic energy (and therefore higher chemical potential and lower concentration of the electrons) at the hot end. (See Chapter 23 for a theoretical discussion.) Figure 6–3 illustrates the use of the thermocouple thermometer.

If one junction of the loop is kept at a fixed temperature, the observed emf is a direct function of the temperature of the other junction. The empirical equation is

(6–5a) $\quad E = a + bt + ct^2$

E = observed emf
t = temperature, °C
a = emf when the test junction is at the ice point, 0°C
b, c = constants characteristic of the two metals and the reference temperature

When the reference junction is kept at the ice point,

(6–5b) $\quad E = b_0 t + c_0 t^2$

Chapter 6. Temperature Measurement and Control

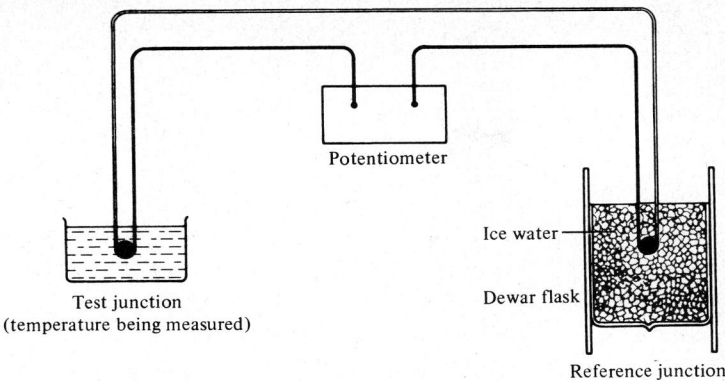

FIGURE 6–3 / Thermocouple thermometer.

The most widely used thermocouples are copper versus constantan, iron versus constantan, chromel versus alumel, and platinum versus platinum–rhodium alloy. (Constantan is an alloy of copper and nickel; chromel and alumel are alloys of nickel with chromium and aluminum, respectively.) At room temperature a copper versus constantan thermocouple has a temperature coefficient of about 0.043 mV/°C. The temperature coefficients of iron versus constantan, chromel versus alumel, and platinum versus platinum–10% rhodium at room temperature are about 0.054, 0.041, and 0.005 mV/°C, respectively.

To make a thermocouple, the ends must either be brazed together or welded together in a very hot flame or electric arc. The wires must be insulated from each other with fiberglass, plastic, cloth, or ceramic, depending upon the temperature at which the thermocouple is to be used.

The emf developed in the thermocouple is measured with a thermocouple potentiometer or a recording potentiometer. The former is a high sensitivity potentiometer that is operated manually over two ranges, between 0 and 16 mV and between 16 and 64 mV. For measurement of temperatures that change over a prolonged period of time, an automatic recording potentiometer is used. For small temperature changes multiple junction thermocouples are used. These consist of several small thermocouples in series. Thermocouples are used in inexpensive, rugged furnace pyrometers. The potential is measured with a millivoltmeter and the cold junction is at room temperature.

The great advantages of the thermocouple are ease of construction, cheapness, ruggedness and durability, quickness of response, reproducibility (the variation between different thermocouples composed of the same metals is negligible), the very large temperature range, and independence of room temperature. The temperatures of the leads and of the potentiometer do not affect the reading, but heat conduction by the leads may cause a significant error.

NOTES / Measure the emf either with a recorder or with a manually operated potentiometer.

Use a reference temperature. An ice–water slush is convenient, but it must be stirred from time to time.

The junctions should be protected with guard tubes.

The thermocouple should be calibrated at two or more temperatures. Some convenient temperatures are those of boiling water, boiling sulfur, melting tin, melting lead, and the sodium sulfate–sodium sulfate decahydrate transition temperature.

Do not keep the potentiometer or recorder too close to the heat source. The emf of the standard cell varies with temperature.

EXERCISE

1. Explain why the emf of the thermocouple is independent of any junction potential across the junctions between the leads and the potentiometer binding posts.

Temperature Control

The simplest thermostats maintain temperature by phase changes, such as a liquid changing into vapor at the boiling point or a solid changing into liquid at the melting point. As long as both phases are present, the temperature of the bath should not change. The only piece of equipment needed is a stirrer. However, the stirring must be very efficient to maintain equilibrium. It is necessary to replenish the phase being used up. It is also often quite difficult to find a system operating at the particular temperature desired by the experimenter.

All laboratories are equipped with adjustable thermostats that can be set at any temperature desired within very broad limits. These adjustable thermostats consist of a bath, bath liquid, stirrer, heater, regulator, and sometimes a cooling device.

Usually, in the 0–100°C range a water bath is used. Oil may be used up to its boiling point or decomposition point. Organic oils decompose above 180°C or so. Silicones may be used up to 250°C. Above 250–300°C ovens and fused-salt baths are used. Above room temperature the thermostat is cooled by air or by cold water circulating through cooling coils. Depending on the season, cold, circulating tap water can maintain temperatures at 10–15°C. Thermoelectric coolers operating on the Seebeck effect are now commercially available. For low temperatures, refrigerated brine or other liquid may be used. For heating, an electrical heater is usually inserted into a metal container in the form of a blade, for small thermostats, or in the form of a coil, for large thermostats. Infrared heating is also used. It has the advantage of low heat capacity and the disadvantage of direct heating of the sensing element.

The heater is operated electrically with a mercury switch or electrical relay turning it on and off as needed. Sometimes two heaters are used, one operating continuously at a level just insufficient to maintain the temperature and the other, smaller, heater going on and off as needed. The relay or switch is controlled by a thermoregulator. This is usually either a bimetallic strip wound in a coil or a mercury regulator. The bimetallic element is a coil composed of two pieces of metal that expand and contract unequally as the temperature is changed. The coil is wound in such a way that the element straightens out on being heated, making or breaking an electrical circuit. Bimetallic elements are usually adequate for temperature control to within 1°C, and some can be used for control within 0.02°C. The mercury thermoregulator is usually more sensitive. It is essentially a thermometer with electrical contacts, one in the bulb and the other in the capillary at the level that corresponds to the desired temperature. When the mercury rises to the cutoff temperature, it reaches the upper contact, activating the relay, which then cuts off the heater. Some regulators have fixed contacts; others have adjustable contacts. One type of mercury thermoregulator has both an adjustable contact and an adjustable mercury level. Mercury is added to or removed from a reservoir until the level is somewhere within the desired range. The contact is then adjusted to a position corresponding to the temperature desired. This type of regulator is commonly used to control temperature to within 0.01°C, but with care can maintain temperature to within 0.001°C. Fixed contact thermometers can control temperature to within 0.05°C or less, depending upon the particular model. Most types of regulators must activate relays or switches which then operate the heaters, since the heater currents are too great to pass through the regulator contacts. The relays may make or break contact mechanically by moving a switch, or they may operate by cutting off the current passing through a tube. There are also hydraulic thermoregulators, which move contacts by the expansion of confined liquids or gases.

Recently solid state devices have been used to provide proportional control of the heater. Instead of the heater operating at the maximum rate or else being completely off, the heater is regulated so that the current through it is gradually decreased as the bath temperature approaches the set temperature. This provides better control with less ripple.

The final element in a thermostat is the stirrer. In the operation of any thermostat, the stirrer, heater, and regulator should be so placed and the stirring speed so adjusted that there are no temperature gradients within the bath due to eddy currents or stagnant portions of the liquid.

References

Benedict, R. P., *Fundamentals of Temperature, Pressure and Flow Measurement*. Wiley, New York, 1969.

Coxon, W. F., *Temperature Measurement and Control*. Heywood, London, 1960.

Weissberger, A., and B. W. Rossiter, *Physical Methods of Chemistry*, Part V, ch. 1. Wiley–Interscience, New York, 1971. (Note: *Physical Methods of Chemistry* is Vol. I of the series *Techniques of Chemistry*, edited by A. Weissberger.)

Chapter 7

Calorimetry

Calorimetry is the measurement of the quantity of heat, expressed either in joules or in calories or, by engineers, in British thermal units (Btu), evolved or absorbed during a process. Until recently, the calorie was defined as the quantity of heat needed to raise the temperature of 1 gram of water 1°C at a specific temperature (e.g., 14.5–15.5°C). Now it is defined in terms of the joule (1.000 calorie = 4.1840 absolute joules). The Btu is still defined as the quantity of heat needed to raise the temperature of 1 pound of water 1°F, from 63 to 64°F.

The calorimeter is the basic instrument in thermochemistry and thermodynamics. In chemical engineering, calorimetric measurements are made to determine heat capacities of materials, fuel values of compounds, and heat changes due to physical changes. In chemistry the emphasis is more on the use of thermochemical measurements to calculate the thermodynamic quantities needed for studies of structure, equilibrium constants, and entropy, enthalpy, and free energy changes of formation, reaction, bonding, and so on. Bioenergetics is especially dependent on accurate thermochemical measurements.

In a thermochemical experiment heat is usually evolved but may be absorbed. There are three main types of calorimeter. In the *adiabatic* calorimeter, heat transfer is prevented and thermal energy stays in the system, changing the temperature. From the temperature change and the heat capacity one calculates the thermal energy produced. In the *conduction* calorimeter, all the heat produced is transferred through a thermal gradient to a heat sink. Calculations are based on the heat flow. In the *isothermally jacketed* calorimeter, some but not all of the heat of reaction is transferred to an outer isothermal heat reservoir. From temperature–time measurements the heat evolved is calculated.

The design of the calorimeter reflects the nature of the process being studied. The slower the evolution of heat, the more stringent are the requirements for controlling heat transfer and for measuring temperature changes. In very slow reactions stirring must be eliminated, since the mechanical heat it generates may be significant. Calorimeters for slow reactions are often designed for specific reactions. Microcalorimeters are preferred, since the smaller the ratio of volume to surface, the more efficient is the internal temperature equilibration and the heat transfer. For fast reactions, such as combustion and neutralization, problems of instrumentation are simpler and general purpose instruments are in use.

The combustion, or bomb, calorimeter is by far the most widely used precision calorimeter. Energies of combustion are determined and heats and energies of formation are calculated. In this instrument the material being studied is ignited in a reacting atmosphere of about 20–25 atm of oxygen, although some special reactions are studied in gases such as chlorine. The reaction chamber, or bomb, is surrounded by a water reservoir equipped with a thermometer and stirrer. The heat generated by the combustion flows into the reservoir. Calculations are made from changes in reservoir temperature. The combination of bomb and reservoir is surrounded by an outer jacket. This jacket may contain water which can be heated or cooled, for adiabatic measurements, or it may be just a thermally insulated water or air jacket, for isothermally jacketed measurements.

Energies and heats of combustion can be routinely obtained to four and even five significant figures. Heats of formation are often smaller than heats of combustion and so, for these, the precision is not quite as good. Heats of reaction other than combustion are usually small and pose a special problem. The obvious method to obtain the heat of reaction would be to combust the reactants and the products, obtain their heats of combustion, and then subtract the heats of combustion of reactants from those of products. However, each combustion results in an experimental error. Two or three combustions would add up to a large absolute experimental error. Since the heats of reaction are usually small differences between two or more large heats of combustion, the experimental precision in the calculated heat of reaction would be rather poor. This is especially true in the case of biochemically significant reactions, such as denaturation, hydrolysis, and enzyme–substrate interactions. In obtaining these from separate combustion measurements on reactants and products, one would lose two significant figures. Therefore, it is essential to run the reactions directly, in appropriately designed calorimeters.

If the reaction is reasonably fast, it can be run in an adiabatic or an isothermally jacketed calorimeter that is provided with some means of mixing the ingredients at the desired time. If the reaction is very slow, it must be run in a specially designed apparatus.

For both adiabatic and isothermally jacketed calorimeters, the basic formula is

(7-1) $\qquad Q = C\,\Delta T$

Q = heat evolved
C = heat capacity of the calorimeter
ΔT = change in temperature

The heat capacity of the calorimeter is calculated from the measured temperature change when a known quantity of heat is transferred in, either electrically or by combusting a standard substance. Care must be taken to avoid heat leakage during the determination of heat capacity. Similarly, for a reaction under investigation, the measured change in temperature must be due to the reaction and not to any heat coming from the outside; at the same time, all the heat of the reaction should go into producing the temperature change rather than leaking out.

Reaction chambers have as low a heat capacity as possible, so a small heat change causes a large temperature change. Metal is used if heat transfer is desired, and glass or styrofoam is used if heat transfer is to be suppressed. The most commonly used glass reaction chamber is the Dewar flask, or vacuum bottle, although styrofoam containers are now being used. The Dewar has a thin glass double wall, with the space between the walls evacuated to reduce heat transfer by convection. The inner surfaces of the walls are usually silvered to minimize radiation. The combination of thin glass walls and vacuum makes the Dewar potentially dangerous, since slight jars or scratches can cause implosions, sending showers of sharp glass over considerable distances. Standard safety procedure is to wrap the Dewar with tape to keep the glass from splintering if an implosion does occur. The high cost of the Dewar and the safety factor make styrofoam reaction chambers preferable.

Temperatures in reaction chambers are measured with a mercury-in-glass thermometer, a thermocouple, a thermistor, or a resistance thermometer.

If the heat capacity of the calorimeter is to be determined electrically, the reaction chamber should contain a low-wattage resistance heater of small heat capacity operating on DC rather than on AC to avoid uncertainties in the meter reading due to phase shifts. Line-operated regulated DC power supplies are available but are usually expensive. An inexpensive, stable, and very convenient power supply can be set up with two or more automobile storage batteries in parallel to avoid the fluctuations in line voltage. For student work, however, regulated line AC may be used. For precise determination of the heat input, a resistance bridge method may be used to measure the current and the resistance with 0.1% error, but for work in which errors of 2% or more may be tolerated voltage and current may be determined with meters.

Adiabatic calorimeters operate with the jacket at the same temperature as the reaction chamber. Since there is no temperature difference, there is no heat transfer, and all the thermal energy of the reaction goes into changing the temperature of the reaction chamber. As the temperature of the reaction chamber changes, the temperature of the jacket is changed correspondingly, by heating or cooling. In actual operation there is usually an unavoidable small temperature difference, but with a little practice this can be kept to a minimum. This type of calorimetric measurement is especially suited, with automatic regulation, to slow reactions during which the heat leakage in a nonadiabatic calorimeter would be as large or larger than the heat of the reaction.

The isothermally jacketed calorimeter operates with the jacket at a fixed temperature, which is usually below that of the chamber. For quick results, the "jacket" can be the surrounding air, as when the reaction is run in a Dewar. Heat is transferred into or out of the calorimeter during the experiment, producing temperature changes in addition to those resulting from chemical or electrical heating. To determine the temperature change that would have taken place had there been no heat transfer to or from the jacket, temperature–time graphs are plotted and an extrapolation procedure used.

Figure 7–1 is a temperature–time graph. The lower portion of the curve shows the result of heat transfer to the cold solution. The upper part of the curve shows the heat loss from the hot solution. The first portion of the graph is extrapolated forward and the final portion backward to a vertical line, t_e,

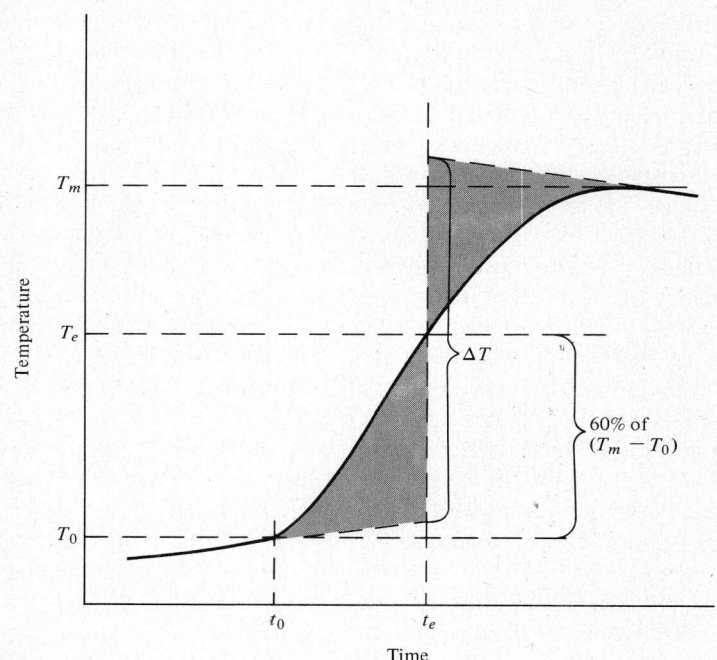

FIGURE 7–1 / Calorimeter temperature vs. time.

located so as to produce equal shaded areas. The vertical distance between the two extrapolated lines at t_e is the temperature change that would have been observed had the temperature changed instantaneously at t_e. The time, t_e, is selected so that during the hypothetical experiment with the instantaneous time change, the amount of heat transferred into the system or out of the system would be the same as in the actual experiment. The area under the temperature–time curve is the total amount of heat added to the system, $Q = k \int (T - T_s)\, d\tau$, where T is the temperature of the system and T_s that of the surroundings. If the shaded areas are equal, the area under the experimental curve is the same as that above the curve and the heat flowing in equals the heat flowing out.

In a series of experiments performed at the National Bureau of Standards, t_e was found to be very close to the time at which the temperature has risen by 60% of the difference between T_0 and T_m. This temperature is shown on the graph as T_e. Unless very precise work is desired, just determine the difference between T_0 and T_m. Then multiply it by 0.60 and add the result to T_0. Next, draw a horizontal line through the curve at this temperature, T_e, and draw the vertical line, t_e, through the intersection of T_e and the experimental curve.

In bomb calorimetry the starting temperature is that at the moment of ignition. In reaction calorimetry, however, there is another troublesome problem in the determination of the starting temperature. Most reactions commence as soon as the reactants are brought together. Usually, by the time the temperature of the reaction mixture can be measured, it has already changed. Generally, either the reactions have to be brought to the same measured temperature inside the reaction chamber and then mixed, or an average temperature must be calculated for the materials at the time of mixing. This average temperature is calculated from the temperature–time data taken in the unmixed reactant solutions.

If the two reactants are kept in the same calorimeter container, they must be separated from each other until thermal equilibrium is reached. One reagent solution may be put into the reaction chamber and the other reagent kept in a container suspended in the solution. After thermal equilibrium is reached, the reactants may be mixed in any one of a variety of ways. If the reagent container is a glass bulb, it may be broken, or a lid can be lifted off or shaken off. One method uses as the container an addition pipet plugged with stopcock grease or covered with a rubber tip (Figure 7-2). The solution can be blown out, or the plug or rubber tip can be poked off from the inside with a thin glass rod.

If the reactants are kept in separate containers, temperature–time data for each solution are taken until just before mixing. One solution is then poured into the other. The average initial temperature can be calculated from the mass and the temperature of each reactant solution at the time of mixing, as obtained from temperature–time graphs. The method requires accurate heat capacity data for the unmixed reagents and calorimeter.

The heat capacity of the calorimetric system and solution may be

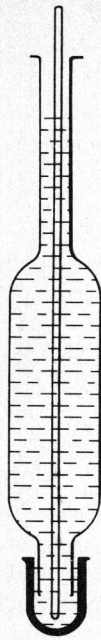

FIGURE 7–2 / Addition pipet.

determined electrically after the reaction mixture has been brought back to the initial temperature. An alternative method is to note the temperature change on mixing known amounts of cold and warm water. The calorimeter has a slightly different heat capacity on every run because of differences in solution volume, extent of immersion of the thermometer and heater, and other factors. However, in most cases, this variation in heat capacity is not significant.

F. E. Condon has developed a simple and inexpensive apparatus for the two-container method with which heats of dilution, solution, and reaction are obtained with a precision of about 1% [Condon et al.]. Condon uses two styrofoam cups, one inside the other, as an inexpensive calorimeter with low mass and low heat capacity. The cover is a sheet of polyethylene about 1 mm thick, with three holes: for the thermometer, stirrer, and addition pipet. The calorimeter contains about 100 ml of one reactant, solution A. The other reactant, solution B, is kept in a test tube immersed in a styrofoam container of water. At the desired time, a 2 ml sample of reactant B is injected below the surface of reactant A. The temperature of solution B is noted to ± 0.01 degree at the sampling time and the usual temperature–time measurements are taken in the calorimeter before and after addition of the sample.

After the measurements are over, the extent of the reaction is determined by a suitable titration. For example, if the reaction is a neutralization, reactant

A is present in excess and its quantity determined before and after the reaction to find the amount neutralized. If the reaction is solution or dilution, the solute concentration is determined by a titration.

In addition to its low cost (except for the thermometer, the entire apparatus can be put together for less than $1), the apparatus has other advantages. The entire calorimeter, including thermometer and stirrer, has a heat capacity of only about 3 cal/degree (determined by mixing known quantities of warm and cold water), or about 3–4% of that of the reaction mixture. The low mass permits the quantities of solution to be determined by weighing the empty and full calorimeter. Finally, the small volume of the calorimeter is not only economical, but permits an ordinary Pasteur pipet to be used to transfer a small sample of reagent. The small volume of sample, in turn, enables the transfer to be completed in seconds.

In *conduction calorimetry*, the heat produced is completely transferred to a heat sink. One special type of conduction calorimeter has a jacket surrounded by an ice–water mixture. The heat of the reaction melts some of the ice. The amount of ice melted is compared with the quantity melted in a blank run using the same calorimeter and volume of solution for the same time interval but without adding the reactants. The difference between the amount of ice melted in the blank and that melted during the reaction is used to calculate the heat evolved in the course of the reaction. The method is indirect, the heat being obtained relative to the heat of fusion of ice. To get different operating temperatures, substances other than water have been used. These include acetic acid, diphenyl ether, and diphenylmethane.

There are now commercially available conduction calorimeters of more advanced design, in which the heat transfer is indicated electrically by the output from a set of thermopiles or thermocouples. The integral of the emf–time curve, over the reaction period, is directly proportional to the heat evolved. The system is calibrated with a reaction of known heat of reaction.

Isothermally Jacketed Reaction Calorimeter: Operating Instructions

DETERMINATION OF ΔT

Pipet the desired volume of one reactant into the calorimeter. Put a 5% excess of the other reactant into the addition pipet. Insert the pipet into the calorimeter.

Stir continuously and allow 15 min for thermal equilibration. Record the temperature and then start the reaction by poking the cover off the tip using a thin glass rod. Raise the pipet as high out of the solution as possible to facilitate draining.

Take temperature–time readings at about 30 sec intervals for at least 5 min after the reaction is over and the temperature levels off and starts to drift at a steady rate because of heat exchange with the surroundings.

DETERMINATION OF HEAT CAPACITY

If necessary, cool the solution down to the original temperature, using a cold finger (a test tube filled with cold water).

Heat electrically at a rate of about 10 W, reading the meters at regular intervals. The total energy is the average power multiplied by the time.

Take temperature–time readings at 30 sec intervals with constant stirring at a rate of about one stroke per second. Start the measurements 5 min before starting the heating. Stop heating when the temperature is about 0.5°C below the maximum temperature observed during the chemical reaction immediately preceding. Continue taking temperature–time readings for at least 5 min after the maximum temperature has been reached and the temperature starts to drop at a steady rate.

Obtain the corrected temperature by extrapolation to the start of testing as shown in Figure 7–1.

NOTES / Select a starting temperature near the lower limit of the thermometer. Tap the thermometer gently before each reading to make sure the mercury does not stick.

It is good practice to disconnect the heater at the end of the heating period.

If necessary, use longer or shorter time intervals.

Condon Two-Container Method

DETERMINATION OF HEAT CAPACITY

Weigh the empty dry calorimeter to ± 0.1 g. Add about 50 ml of water at room temperature. Weigh calorimeter again to obtain the weight of the added water.

Insert the Beckmann thermometer and stirrer and, while stirring, measure the temperature of the water to $\pm 0.01°C$ at 30 sec intervals for about 4–5 min.

Quickly add about 50 ml of water whose temperature (measured to $\pm 0.01°C$ at the time of mixing) is about 4–5°C higher than that of the water in the cup.

Take time and temperature readings with stirring at 30 sec intervals for 5–10 min or until a nearly constant rate of change is apparent. Reweigh to determine weight of warm water that was added.

Calculate the C_p of the calorimeter by equating the heat loss of the warm water to the heat gain of the cold water and calorimeter.

$$(T_2 - T_f)Cm_2 = (T_f - T_1)(Cm_1 + C_p)$$

T_2 = initial temperature of hot water
T_f = final temperature, at equilibrium
T_1 = initial temperature of cold water
C = specific heat of water
    = 1 cal/g-°C
m_2 = weight of hot water
m_1 = weight of cold water
C_p = heat capacity of calorimeter

DETERMINATION OF THE ΔH OF REACTION, SOLUTION, OR DILUTION

Weigh the empty dry calorimeter cup, add about 100 g of reactant A, and weigh again to ±0.1 g.

Place the cover on the calorimeter and take temperature measurements with stirring for 4–5 min at 30 sec intervals.

With a Pasteur pipet, take a sample of about 2 ml of reactant B and inject it beneath the surface of the calorimeter liquid. Record the time of injection. The temperature of B must be known to ±0.1°C.

Continue stirring and taking temperature readings at 30 sec intervals for about 5–10 min, or until the temperature is either constant or else changes at an almost constant rate.

NOTES / Be careful not to get the solution in between the two cups of the calorimeter.

The thermometer should be about 1 cm above the bottom of the cup.

Use a hand lens to read the thermometer and tap it gently before each reading to make sure that the mercury does not stick.

Stir at the rate of about 1 stroke per second.

The reactant to be added is kept in a test tube in a styrofoam container of water. The water temperature (and therefore that of the reactant) is measured with a 50°C thermometer to ±0.1°C.

Keep the dry Pasteur pipet next to the reactant being added so that it is at or near the same temperature.

To sample the reactant, squeeze the air completely out of the rubber bulb and fill the pipet as completely as possible with solution. Then withdraw the pipet and squeeze out one or two drops and release the pressure, raising the lower level of the solution above the bottom of the pipet. To inject the

sample, insert the pipet through the hole in the lid into the solution. Immediately squeeze the reactant out of the bulb and in one continuous motion withdraw the pipet to prevent solution being sucked back.

The Parr Constant Volume (Bomb) Calorimeter

The reaction takes place at constant volume in a stainless steel bomb (Figure 7–3). The sample is ignited in an oxygen atmosphere by contact with a hot wire. The heat generated by combustion raises the temperature of a known amount of water in the bucket. The bucket and bomb together constitute the reaction chamber. The measurement may be either adiabatic or nonadiabatic.

The adiabatic procedure is simply to maintain the temperature of the jacket at the same value as that of the bucket so that there is no heat transfer. In the nonadiabatic procedure temperature–time measurements are taken before, during, and after the ignition, and the initial and final temperatures are obtained by a suitable extrapolation of the temperature–time graph.

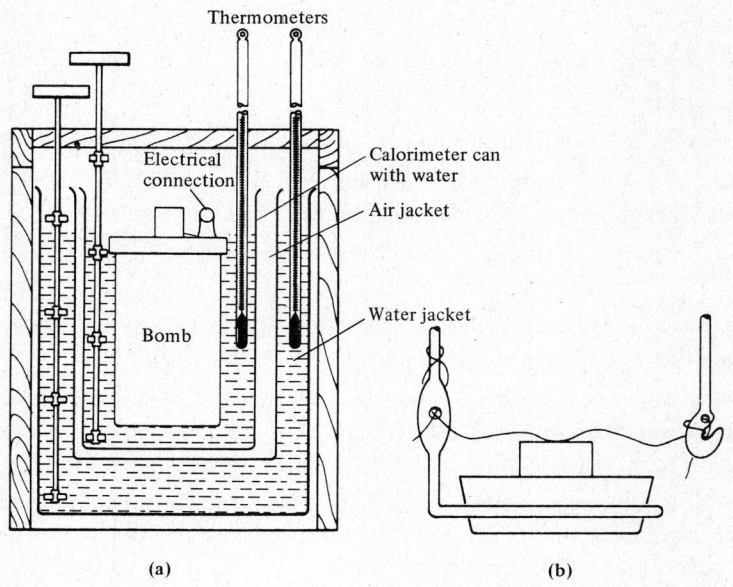

FIGURE 7–3 / Parr bomb calorimeter. (a) Cross section, adiabatic calorimeter: (b) Method of binding fuse wire. (By permission of the Parr Instrument Co.)

OPERATING INSTRUCTIONS

Form and weigh accurately a pellet of about 1 g and place it in a stainless steel cup.

Attach 10 cm of ignition wire to the electrodes. Fasten the ends tightly to the electrodes and let the loop of wire just touch the pellet.

Put 1 ml of distilled water in the cylinder and assemble the bomb, with the sample cup in the supporting ring.

Fill the bomb with oxygen to a pressure of 25 atm. Place it in the water bucket and attach the electrode wire to the slotted contact in the bomb head.

Ignite the charge by pressing the firing button for 5 sec—*and no longer!* Otherwise, the series resistor will burn out and the ignition unit will no longer pass current.

After determining the temperature change due to the reaction open the calorimeter. Replace the bomb in the holder. Open the release or outlet valve to vent the excess gas. Open the bomb and remove and measure the unburned wire.

NOTES / Be sure there are no unburned carbon particles on the capsule or in the bomb before starting. Clean the bomb with steel wool if necessary.

If there are unburned particles of sample after the combustion, repeat with a smaller sample.

Use the following procedure to fill the bomb with oxygen: Close the outlet valve. Attach the union from the oxygen tank to the inlet valve, turning it down *tightly, by hand.* Close the filling valve between the bomb and the gauge (turn clockwise), open the tank valve one-quarter turn, open the filling valve slowly, and allow the pressure in the line to rise slowly to the desired value. Then close the filling valve and the tank valve and open the relief valve under the gauge. The excess pressure in the bomb closes the inlet valve. Unscrew the union.

Remove atmospheric nitrogen before the combustion by putting 10–15 atm of oxygen into the bomb and then releasing it through the outlet valve. Otherwise, formation of nitrogen oxides will contribute to the temperature change.

In closing the bomb, tighten down the bomb cap by hand. Never use a wrench.

If the reaction is to be run adiabatically, keep the jacket and bucket temperatures within 0.03°C of each other by running hot or cold water slowly into the jacket.

If the isothermal jacket method is used, make temperature–time readings at 30 sec intervals for at least 5 min before the ignition, during the period of temperature rise, and for at least 5 min after the temperature has leveled off and started to drift. The corrected temperature change is obtained as shown in Figure 7–1.

Multiply the length of the wire consumed by 2.3 cal/cm. This is part of the heat input.

Be sure that the wire is connected tightly to the binding posts, or the fuse may not ignite.

References

Brown, H. D., *Biochemical Microcalorimetry*. Academic Press, New York, 1969.

Condon, F. E., R. T. Reece, D. G. Shapiro, D. C. Thakkar, and T. B. Goldstein, Influence of hydration on base strength, Pt. V. Hydrazines and oxamines, *J. Chem. Soc. Perkin Trans.* II, 1112–21 (1974).

Cox, J. D., and G. Pilcher, *Thermochemistry of Organic and Organometallic Compounds*, Academic Press, New York, 1970.

Dickinson, H. C., Combustion calorimetry and the heats of combusion of cane sugar, benzoic acid and naphthalene, *Natl. Bur. Std. (U.S.) Bull.* **11**, 189 (1915).

Jesup, R. S., Precise measurements of heat and combustion with a bomb calorimeter, *Natl. Bur. Std. (U.S.) Monograph* **7** (1960).

Oxygen Bomb Calorimetry and Combustion Methods, Technical Manual No. 130. Parr Instrument Company, Moline, Ill., 1960.

Skinner, H. A., *Experimental Thermochemistry*, 2 vols. Wiley-Interscience, New York, 1956, and 1962.

Weissberger, A., and B. W. Rossiter, *Physical Methods of Chemistry*, Part V, ch. 7. Wiley-Interscience, New York, 1971.

Chapter 8

Density Measurement

The density of a liquid can be calculated from the mass required to fill a container of known volume. The container, known as a pycnometer, is calibrated with water. Pycnometric densities are precise to five significant figures with careful technique.

A quick but less accurate determination of liquid density involves the measurement of the buoyant force exerted on an object immersed in the liquid. The instrument used is the Westphal balance. The results are accurate to three or four significant figures. Hydrometers, which are floats calibrated to indicate the density of a liquid by the extent to which they submerge, are rarely used for accurate work.

Solid densities are determined pycnometrically from the volume of reference liquid displaced by the submerged solid.

Pycnometric Determination of Liquid Density

The combination of a large chamber and two capillary ends in the Ostwald–Sprengel pycnometer (Figure 8-1a) makes it possible to measure a large volume of liquid with a high precision (the volume error in adjusting the meniscus to mark is small compared to the total volume). The Weld pycnometer (Figure 8-1b) is less accurate but easier to use. Rubber or glass caps may be fitted over the ends of both types of pycnometer to prevent evaporation of volatile liquids.

PRINCIPLES

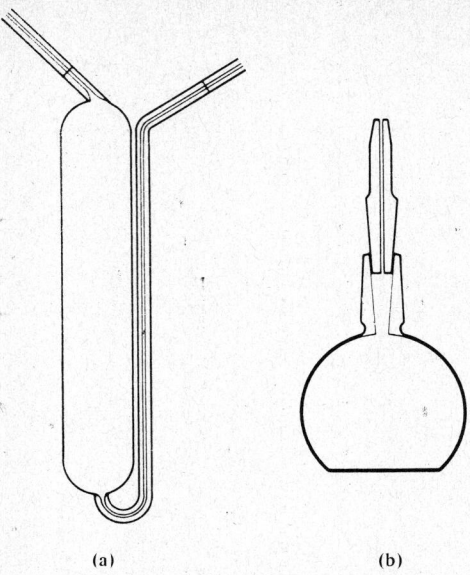

FIGURE 8-1 / Ostwald–Sprengel (a) and Weld (b) pycnometers.

For liquid densities the calculations are trivial:

$$\rho_x = \frac{M_x}{V} = \frac{M_x \rho_r}{M_r}$$

ρ_x = density of liquid being studied
ρ_r = density of reference liquid
m_x = mass of liquid being studied
m_r = mass of reference liquid
V = volume of the pycnometer

OPERATING INSTRUCTIONS FOR OSTWALD–SPRENGEL PYCNOMETER

Fill the pycnometer, using a hypodermic syringe, an aspirator, or a rubber bulb. Fill slowly to prevent a jet of liquid from entering the capillary and trapping air bubbles. Use any convenient mark to level the liquid, but always use the same mark for both liquids.

To remove air bubbles, tilt the pycnometer so that the bubble is at the entrance to the outlet tube. Then aspirate more liquid into the pycnometer. To remove excess liquid, tilt the pycnometer until a drop forms at one end. Then wipe the drop away with filter paper. Add more liquid, if necessary, with a pipet or medicine dropper.

Chapter 8. Density Measurement

EXERCISES

1. Assume that the Ostwald–Sprengel pycnometer has a volume of 3 ml and a capillary radius of 0.30 mm. Calculate the error introduced if enough liquid evaporates during the weighing so that the level of the capillary falls by 0.5 cm. Assume the density of the liquid to be 1 g/ml.
2. Explain why the buoyancy correction will be much less than 1 part per 1000 in pycnometry if a reference liquid of known density is used.
3. Briefly discuss what properties are desirable in a reference liquid. Suggest suitable reference liquids and discuss their advantages and disadvantages.

Pycnometric Determination of Solid Density

To determine the density of a solid with the Weld pycnometer, weigh

(a) The dry empty pycnometer.
(b) The pycnometer filled with a liquid in which the solid is insoluble.
(c) The dry pycnometer containing about 1 g of solid.
(d) The pycnometer containing the sample of solid plus enough of the liquid to fill it completely.

CALCULATIONS

The mass of the solid is (c) − (a). The mass of displaced liquid is [(b) − (a)] − [(d) − (c)]. The volume of the displaced liquid is its mass divided by its density.

EXERCISES

1. Would the experimental results be better if a denser reference liquid were used? Explain.
2. Would powdering the solid increase the accuracy? Explain.

Determination of Liquid Density Using the Westphal Balance

The liquid exerts a buoyant force upward on the immersed plummet (Figure 8-2). The gravitational force on the riders, placed in various positions along the beam, opposes the buoyant force. When the beam is horizontal the force downward (the sum of the rider moments) equals the force upward (the mass of the displaced liquid). The instrument is calibrated so that the sum of the rider

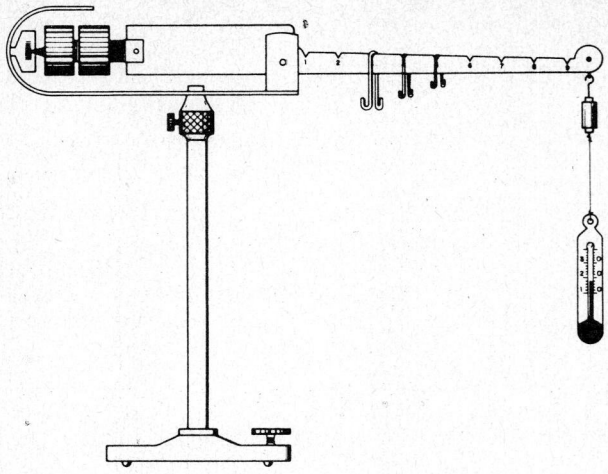

FIGURE 8–2 / Westphal balance.

moments is the density of the liquid. The rider moment is the product of rider mass and the distance from the fulcrum to the rider, as marked on the beam. The rider masses are taken as 1.000, 0.100, and 0.0100.

OPERATING INSTRUCTIONS

Balance the beam in the air by adjusting the screw weights. Then immerse the plummet in the unknown liquid and level the beam again, this time by adjusting the riders. The plummet must be completely immersed and must not touch the wall of the container.

If the plummet has been chipped, check the calibration of the balance, using water as a reference liquid. If the calibration is off, calculate the densities of unknown liquids, relative to water, using the formula

$$\rho_x = \rho_w \frac{M_x}{M_w}$$

ρ_x = density of unknown liquid
ρ_w = density of water
M_x = equilibrium rider moment in unknown liquid
M_w = equilibrium rider moment in water

Reference

Weissberger, A., and B. W. Rossiter, *Physical Methods of Chemistry*, Part IV, ch. 2. Wiley–Interscience, New York, 1972.

Chapter 9

Pressure Measurements, Vacuum Systems, Handling Gases

Pressure is defined as force per unit area. Several pressure units are used, including atmospheres (1 atm is defined as 101,325 N/m²), dynes per square centimeter, pounds per square inch, and bars (1 bar = 10^6 dynes/cm² or 10^5 N/m²). The SI unit is the Pascal (1 Pa is defined as 1 N/m²). In physical chemistry pressure is usually expressed in atmospheres, in centimeters or millimeters of mercury (760 mm Hg = 1 atm), or in torr (1 torr is defined as $\frac{1}{760}$ atm). In high-vacuum work the most frequently used standard pressure unit is the torr.

Pressures are either measured directly or obtained indirectly from the measurement of some pressure-sensitive physical property.

U-Tube Manometers

U-tube manometers are widely used for pressures around atmospheric pressure. They have the advantages of low cost, simplicity of construction, and ease of calibration. The pressure, or pressure difference, if desired, is read directly, and little or no additional computation is necessary. In each limb the total pressure downward is the sum of the gas pressure plus that exerted by the column of liquid, as shown in Figure 9-1.

$$p_x = p_r + h_r - h_x = p_r + \Delta h$$

p_x = pressure of the gas
p_r = pressure in the reference limb, height units
h_r = liquid height in the reference limb
h_x = liquid height in the limb connected to the gas limb

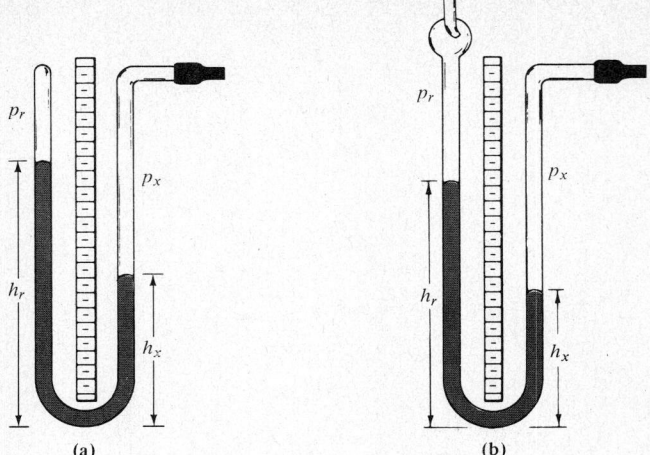

FIGURE 9–1 / Closed-end (a) and open-end (b) manometers.

In closed-end manometers p_r is the vapor pressure of the manometer liquid, which for mercury at room temperature is effectively zero. In open-end manometers p_r is the atmospheric pressure. Since mercury is usually the manometer liquid, p_r is expressed in terms of the "height" of mercury. If a liquid other than mercury is used, the height of the liquid is multiplied by the ratio $\rho_{\text{liquid}}/\rho_{\text{Hg}}$ to obtain the pressure in centimeters of mercury. Besides mercury, nonvolatile liquids such as dibutyl phthalate are used.

Disadvantages include fragility, insensitivity to small pressure changes, and the necessity for careful purification of the manometer liquid. In the closed-end type of manometer the residual air must be carefully removed. In the open-end type of manometer gas bubbles are occasionally trapped under the surface during pressure changes. With mercury, air slowly oxidizes the surface, and a layer of fine oxide powder eventually forms on the glass and obscures the readings. Parallax and thermal expansions and contractions are other important sources of error. At higher temperatures the liquid is less dense, and so a higher column of liquid will be supported by a given pressure. At 20°C, for example, 760 mm of mercury is only 0.996 atm. Equations and factors to correct for temperature effects may be used. These are found in various handbooks. The density of mercury from $-10°C$ to 50°C is given in Appendix 5.

The thermal expansion of mercury gives rise to experimental difficulties in using U-tube manometers to determine the pressures of thermostatted systems. If the manometer is inside the thermostat, corrections must be calculated and applied for each temperature used. On the other hand, if the manometer is outside the thermostat, the gas or vapor in the line between the manometer and the thermostat is not at the thermostat temperature. The problem may be avoided, however, by the use of a mercury isoteniscope, an internal

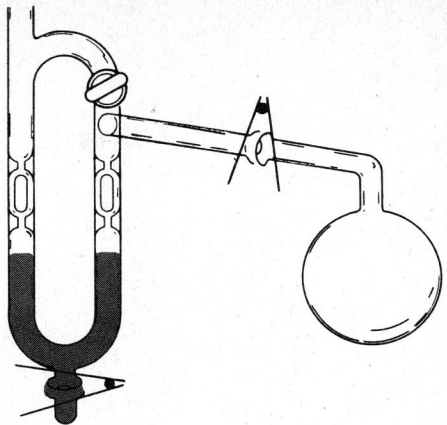

FIGURE 9–2 / Mercury isoteniscope.

U-tube mercury manometer (Figure 9–2) connected to an outside manometer, as shown in Figure 9–3. When the mercury is at the same height in both limbs of the internal U tube, the pressure in the flask is equal to that in the ballast bottle, which may be read with the external manometer. The ground glass or Teflon valves prevent the mercury from being forced out of the U tube by

FIGURE 9–3 / Internal U-tube apparatus.

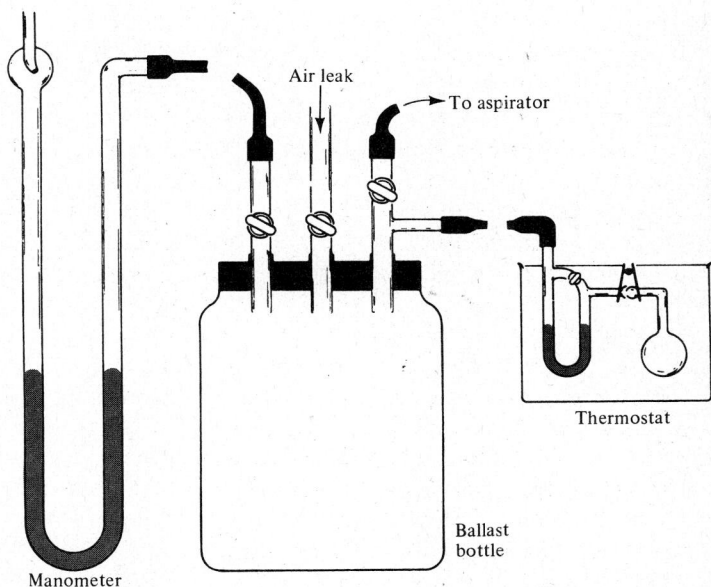

sudden pressure changes. The apparatus can be used to measure homogeneous gaseous equilibria and heterogeneous equilibria such as vapor pressures, sublimation pressures, or decomposition pressures. The fragility of the expensive valves is a drawback of the system.

OPERATING INSTRUCTIONS FOR INTERNAL U-TUBE APPARATUS

Put the material being studied into the bulb and connect the system as shown in the diagram; evacuate air and dissolved gases. To prevent loss of volatile material, chill the flask with a freezing mixture before evacuating.

Close the stopcock and allow at least 5 min for the flask and contents to reach the thermostat temperature before taking measurements. Then adjust the pressure in the ballast bottle until the mercury is at the same height in both limbs of the U tube.

Note and record the reading on the external manometer.

NOTES / Change pressures slowly, to prevent sudden impacts.

Immerse the entire assembly in the thermostat up to the level of the outlet stopcock.

If the valves stick, remove the mercury and clean the U tube with a suitable solvent.

At temperatures above 120°C, correct for the vapor pressure of mercury.

The Barometer

The barometer is basically a closed-end manometer with the gas space above the mercury evacuated except for mercury vapor. At room temperature the pressure in the closed end of the barometer tube is about 10^{-3} torr, which for all practical purposes is negligible. The atmospheric pressure is the height of the mercury column, corrected for changes in the density of mercury and in the length of the metal scale with temperature. Barometers are usually calibrated for 0°C, and the equation for calculating the correct pressure, reduced to 0°C, from the observed pressure is

(9-1) $$p = p_{obs} \frac{1 + \beta(t - t_0)}{1 + \alpha t}$$

p = corrected pressure reduced to 0°C
p_{obs} = observed pressure
t = room temperature
t_0 = temperature at which the scale was calibrated

$\alpha = 1.83 \times 10^{-4}$, coefficient of cubical expansion of mercury
$\beta = $ coefficient of linear expansion of the scale; for a brass scale it is 1.85×10^{-5}.

Appendix 4 is a table of temperature corrections for barometric readings.

OPERATING INSTRUCTIONS

Adjust the level of the mercury in the reservoir until the zero of the scale just touches the mercury surface. Then adjust the indicator until it is exactly at the level of the top of the meniscus of the mercury.

Read the height of the mercury column, using the vernier, and determine the temperature of the barometer.

Electrical Pressure Transducers

A *transducer* may be defined as a device for taking energy in one form and using it to supply energy in another form. For example, the photocell converts light into electric current, the phonograph pickup converts torque into electric current, the loudspeaker converts current into sound.

Usually, the term *transducer* is applied to electrical pressure transducers, which produce an electrical signal. There are two principal types: the strain gauge and the potentiometer. The strain gauge transducer is essentially a taut fine wire, one end of which is fastened to a diaphragm. As the pressure is increased, the diaphragm is pushed outward, stretching the wire, decreasing its cross-sectional area, and increasing its length. The change in dimensions of the wire increases its electrical resistance. The pressure measurement is therefore changed into a resistance measurement, which can be made with high precision and reproducibility (see pages 150–52).

The potentiometer type of transducer consists of a wire of uniform electrical resistance, to which three leads are connected, one to each end and one to a movable contact that slides along the wire (the setup is that of a Wheatstone bridge, Figure 14-9). The sliding contact is connected to a diaphragm that responds to pressure changes. As the pressure changes the contact slides along the resistance wire. To operate the transducer, an emf is connected across both ends of the resistance wire, and the potential across the slider to either end of the wire is measured. This potential is calibrated against pressure, and should be linear over the pressure range at which the transducer operates. One great advantage of this type of transducer is the ability to measure rapidly changing pressures. The output is fed into a strip chart recorder, and a rapid pressure change may be quickly and permanently recorded.

PRINCIPLES

There are now available pressure transducers, such as the baretron, whose capacitance, rather than resistance, changes with pressure. These operate in bridge circuits.

McLeod Gauge

The McLeod gauge (Figure 9–4) gives direct pressure readings, independent of the nature of the gas, provided that the gas does not condense in the gauge. It is the standard gauge for work at moderately low pressures. Indirect gauges,

FIGURE 9–4 / McLeod gauge.

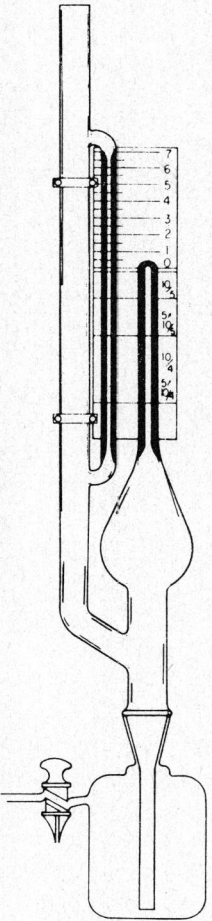

which measure some pressure-dependent property, are calibrated against a McLeod gauge for each different gas used. Some McLeod gauges operate at pressures as high as 2 torr, and others at pressure as low as 10^{-6} torr, although no one gauge covers the entire range.

In the McLeod gauge, a known fixed volume of gas at the system pressure is trapped and compressed to a much smaller volume in a capillary. Boyle's law is used to calculate system pressure from the final pressure and volume of the trapped gas and from the fixed initial volume. There are two methods of operating the McLeod gauge: the *quadratic method*, at lower pressures, and the *linear method*, for pressures at the high end of the range. In the quadratic method mercury is added from the reservoir until the mercury in the reference tube is at the level of the top of the gauge capillary. The final pressure of the compressed sample is the difference in height between the mercury levels in the reference capillary and in the sample capillary. The pressure is given by

$$(9\text{--}2) \qquad p = \frac{hv_f}{v_i} = \frac{ah^2}{v_i} = kh^2$$

p = unknown pressure, torr
h = difference in mercury levels
   = final pressure of sample
v_f = final volume
v_i = initial volume
a = cross sectional area of capillary
$k = a/v_i$

For lower system pressures, the upper section of the gauge scale is calibrated either in centimeters or in terms of the initial pressures corresponding to various values of h.

In the linear method, the mercury in the sample capillary is brought to the level of one of the lines engraved on the scale. Each of these lines corresponds to a fixed ratio of final volume to initial volume. The system pressure is then simply the difference in the level of the mercury in the two capillaries multiplied by the scale factor engraved at the calibration line (i.e., 10^{-1}, 10^{-2}, etc.). The quadratic scale is more accurate, but the linear scale is easier to use.

OPERATING INSTRUCTIONS

Adjust the scale so that the zero line (marked 0) on the scale is at the upper end of the closed capillary. Then, with the system at the desired operating pressure, gently and carefully open the reservoir valve and air inlet and allow the mercury to rise slowly. If the linear method is being used, close the reservoir valve, stopping the mercury, when the level in the *closed* capillary is at one of the engraved lines on the scale (e.g., 10^{-3}). If the quadratic method is being used, stop the mercury when the level in the *open* capillary is at the top end of the closed capillary (i.e., at the zero line).

NOTES / **The mercury level must be adjusted gently to avoid sudden surges that might splash mercury into the capillary or the manifold.**

The McLeod gauge cannot be used for vapors that condense at the final pressure in the capillary.

The McLeod gauge indicates a pressure that is usually too low, because the steady flow of mercury atoms from the gauge to the cold trap transfers momentum to the molecules of the gas above the mercury in the reference capillary. This effect is usually minor.

EXERCISE

1. Explain how mercury can be used in the McLeod gauge to measure pressures of 10^{-4} torr or less, even though mercury itself has a vapor pressure of 10^{-3} torr at room temperature.

Thermal Conductivity Gauges

In the pressure range 10^{-2}–10^{-4} torr several varieties of gauge are used, based upon the thermal conductivity of gas. A filament or a wire is heated and the gas is allowed to impinge upon it, conducting away some of the heat. The rate of cooling is approximately proportional to the gas pressure, but the proportionality constant depends upon the particular gas. In the *Pirani gauge* the filament is part of a resistance bridge circuit, with the filament of another Pirani gauge, containing gas at a known pressure, in the other limb. The circuit becomes unbalanced when the gas in the system is not at the same pressure as the gas in the reference bulb. The change in resistance, or the imbalance of the circuit, may be determined more precisely than the resistance itself and is less affected by changes in room temperature. In some modifications the current needed to rebalance the circuit is measured. Thermistors are sometimes substituted for filaments to increase sensitivity, since they have a greater temperature coefficient of resistance.

The *thermocouple gauge* operates upon the same principle as the Pirani gauge. One junction of a thermocouple is in contact with the filament, and the emf is measured, rather than the resistance.

Ionization Gauges

Ionization gauges usually ionize the gas and measure the ion current, which is approximately proportional to the pressure, the proportionality constant again depending upon the particular gas, as in thermal conductivity gauges. The ionization may be produced by alpha particles, as in the *alphatron gauge*, or by electrons. *Cold-cathode ionization gauges* operate in the range 10^{-3}–10^{-7} torr, while *hot-cathode* or *thermionic gauges* operate between 10^{-3} and 10^{-8} or 10^{-10} torr, with some modifications going down even lower.

In the cold-cathode gauges (*Phillips gauge, Penning gauge*) the electrons are produced by a potential of several kilovolts. As the electrons are accelerated toward the anode, they collide with the ionized gas molecules. The positive ions produced are then discharged at the cathode, resulting in an *ion current* that is directly proportional to the gas pressure.

The great advantage of the cold-cathode gauge over the thermionic gauges is that it will not burn out if air should leak into the system. This property also makes it very useful in safety cutoff switches. A sudden increase in pressure will increase the ion current and activate a relay, operating safety devices. Hot-cathode gauges would burn out if subjected to the same treatment.

The thermionic, or hot-cathode, gauge is extremely sensitive and will detect and measure gases at very low pressures. The gauge consists of a central hot filament surrounded by a positive grid, which in turn is surrounded by a negative plate. The electrons are collected by a grid and the ions are collected by the plate. The ion current is proportional to the pressure and nature of the gas. The pressure, for a particular gas, is obtained from the ratio of ion current to electron current.

Vacuum Systems

The three essential parts of a vacuum system are the vacuum pumps, the gauges and meters (already discussed), and the sample chambers and inlet systems.

For pressures greater than 10^{-7} torr the systems are constructed of pyrex glass, with joints and stopcocks lubricated with special high-vacuum greases. For pressures smaller than 10^{-7} torr, all-metal systems are used, with flanges replacing ground glass joints and valves replacing stopcocks. Metal is necessary so that the system can be baked out at temperatures up to 450°C in order to desorb gases adsorbed onto the walls. Pressures as low as 10^{-13} torr have been reached with all-metal systems. Ceramic materials have also been used, especially for seals around metal gauges and valves, although glass too may be sealed to metal. Glass–Kovar seals and glass–copper seals are commercially available, and for pressures greater than 10^{-7} torr, epoxy resins (trade name Araldite) are adequate.

The usual vacuum system consists of a manifold, or large tube, to which the other parts are connected. These may include pumps, traps, gauges, inlet and outlet systems, reaction chambers, and measuring and analyzing equipment. Wherever possible the various parts should be connected with gastight permanent connections, but it is often necessary to have demountable and sometimes even flexible connections.

In metal systems, permanent joints are made by welding, brazing, or hard-soldering. Welding should be done in an inert atmosphere. Care should be taken to exclude impurities that might produce pinhole leaks and to avoid materials that might volatilize appreciably at high temperatures. Antimony, which is often present in soft solders, is a particularly troublesome impurity. It evaporates and is difficult to remove.

For demountable joints in metal systems, flanges are usually used. The two pieces connected to each other in a flange may be of a variety of shapes. They may be two flat surfaces, or two flat surfaces with a groove, for insertion of a gasket; they may be tapered or machined for one to fit inside the other; and so on. Usually, some sort of gasket is used. For systems that will not be baked out at high temperatures, rubber or Viton gaskets and O rings may be used. For systems that will be baked out at high temperatures, gold seals should be used. These are thin wire rings that are compressed between the two pieces of the flange. For an ultra high vacuum, joints are made of a carefully machined, hard metal edge that is forced into a flat bed of softer metal to form a gastight connection. Such seals, however, must not be subjected to large variations in temperature, or differential expansion may produce leaks.

For flexible connections, the usual material is heavy-walled rubber tubing. This is often somewhat porous, which makes it unsuitable for high-vacuum work. Stainless steel bellows connected with flanges are necessary for an ultra high vacuum.

In all-glass systems, permanent joints are glass sealed to glass. Both glasses in the seal must have the same composition, for differences in thermal expansion will cause the joint to crack or at least to leak. If it is necessary to seal pyrex to soft glass, for example, a graded seal must be used. Graded seals that are commercially available consist of several pieces of glass of intermediate composition sealed one to the other, with properties varying by small steps over the entire range.

To manipulate gases, stopcocks may be used in all-glass systems, but metal valves must be used in all-metal vacuum systems. Glass stopcocks are lubricated by greases. Even special vacuum greases that have a satisfactorily low vapor pressure will often become contaminated with dissolved gases and organic vapors. Greased stopcocks cannot be baked out, even at the relatively low bake-out temperatures for all-glass apparatus. Metal valves of the Hoke or Alpert type are preferred. Great care is necessary in lubricating stopcocks or ground glass joints. The surfaces must fit tightly and be clean and free of dirt, old grease, and grit. Even a speck of grit will produce a leak. One method of lubricating is to spread two thin streaks of grease on opposite sides of the surface and then to rotate the two glass surfaces back and forth until the grease spreads uniformly. A better way is to apply a thin uniform film to the surface before inserting the joints or stopcocks. The film is usually applied to the outer three fourths of the joint, and the applied vacuum then distributes it over the rest of the surface. The surface should look uniform, without streaks or blobs, and must in all cases be tested for leaks.

Leaks may occur anywhere in a system, but they are most frequently found at seals in glass tubing, at welded or brazed metal joints, and at flanges and gaskets. To test for a leak, each portion of the system must be isolated with the appropriate connections and valves. Once the general area is found, for metal systems the procedure is to paint the area with a plastic lacquer. When the pressure stops rising the leak has been covered. For all-glass systems a Tesla coil may be used. The Tesla coil is a high-voltage induction coil with the primary operating on 110 V AC and with one end of the secondary attached to a metal electrode. If the tip of this electrode is brought near a ground, a spark will jump the gap, ionizing the air molecules along the way. The higher the concentration of air molecules, the brighter the discharge. As the Tesla coil is brought near the glass (but without touching it) there will be a diffuse glow in the system, and at the site of the leak, or pinhole, the discharge through the concentrated air will form a bright spot. To find the pinhole, the Tesla coil is moved slowly along the surface of the evacuated system until a bright spot is seen. Care must be taken not to touch the coil to the glass and not to use an excessive voltage, since the high-energy spark may puncture the glass. The coil should be kept moving along the surface of the glass. If no leak is found with the Tesla coil, the leak is probably in a stopcock.

At ordinary pressures and even at moderately reduced pressures, where the mean free path of the molecules is small compared to the dimensions of the apparatus, gases move from a higher pressure to a lower pressure by bulk flow, like ordinary liquids, and may be treated as if they were structureless compressible fluids. (For most gases at around room temperature the mean free path at 1 torr is about 10^{-3} cm and varies inversely with pressure. At 10^{-3} torr the mean free path would be 1 cm.) To transfer all the gas completely, without leaving

FIGURE 9–5 / Toepler pump.

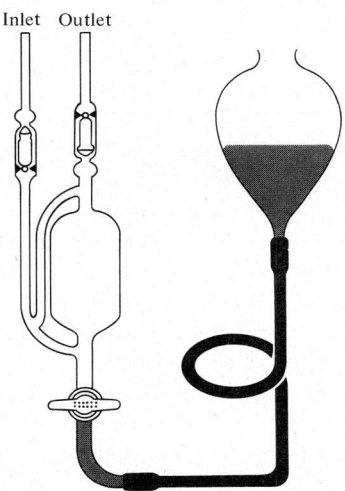

any in the original chamber, a Toepler pump is used (Figure 9–5). Alternately raising and lowering the leveling bulb removes the gas from the chamber to which the pump inlet is connected and transfers it to the chamber connected to the pump outlet. When the leveling bulb is lowered, the mercury level falls in the apparatus and gas expands through the inlet into the pump chamber. When the leveling bulb is raised, the gas in the chamber is trapped by the rising mercury and is pushed out through the exhaust valve.

Gases at low pressures move by molecular flow rather than by bulk flow. Each molecule migrates independently of all other molecules. Therefore, at low pressure, transfer may fractionate a gas because the lighter, faster molecules diffuse more rapidly. Such fractionation takes place through a *molecular leak*, which is simply a very small perforation, of the order of 0.001–0.01 mm in diameter, in a metal foil. When a mixture of components is sampled through a molecular leak, it is necessary to multiply the ratio of gases passing through the leak by the square root of the molecular weight ratio to obtain the composition of the original gas sample:

(9–3) $$\frac{p_a}{p_b} = \left(\frac{M_a}{M_b}\right)^{1/2} \frac{N_a}{N_b}$$

p = pressure before sampling
M = molecular weight
N = number of molecules diffusing through the leak
a, b = different molecular species

In analyses of gases by mass spectrometry (pages 290–91) this fractionation must be taken into account.

Vacuum Pumps

The pumps used in a system are selected on the basis of the ultimate pressure desired, the pressure against which the pump is to operate, and the speed at which the gas is to be removed. Most high-vacuum diffusion pumps will not function unless the pressure in the system being evacuated is below a certain limit, characteristic of the particular pump. For high-vacuum work, therefore, the initial pressure of the system must be reduced by a preliminary "forepump" before using the diffusion pump to reach the final pressure. The most common forepumps are the rotary oil or rotary cylinder pumps.

In a rotary pump (Figure 9–6) a sample of gas is trapped between the cylinder and the wall. The rotating cylinder pushes the gas out through the exhaust and then continues around to trap another sample of gas, which in turn is pushed out. The vanes prevent the gas from leaking back into the inlet. The pump is lubricated with oil to prevent leakage of gas between the wall and the rotating cylinder. This oil presents problems because condensable organic

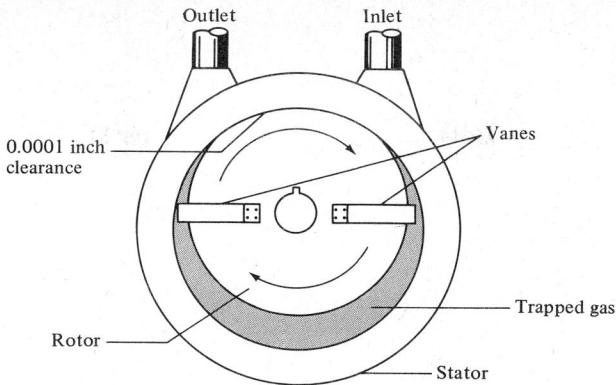

FIGURE 9–6 / Welch Duo-Seal rotary pump (by permission of the Sargent-Welch Instrument Co.).

vapors dissolve in pump oil and interfere with the pump action. The pump should therefore always be protected by a vapor trap and the oil should be changed periodically. The oil must be checked frequently and more should be added whenever necessary. Commercial laboratory rotary pumps will evacuate down to 10^{-4} torr when used with cold traps.

Diffusion pumps (Figure 9–7) are used to obtain the high vacuum after the forepump has removed most of the gas originally present and lowered the pressure down to 10^{-2} torr or less. Some pumps can operate at 10^{-1} torr, but most do not work above 10^{-2} torr. The pumping action comes from a jet of vapor moving in the direction of the exhaust leading to the forepump. Molecules of gas diffusing in from the manifold collide with the fast-moving heavy molecules of mercury or oil in the vapor jet and are accelerated downward toward the exhaust. Usually, there are several stages in the diffusion pump, so that the gas molecule is subjected to repeated downward acceleration before it finally reaches the forepump. For the pump to function efficiently, the pressure must be low enough that the mean free path of the gas molecules is longer than the distance between the inlet and outlet of the pump. Otherwise, the momentum picked up from the collision with the oil or mercury will be dissipated among several molecular collisions, decreasing the rate of evacuation.

Diffusion pumps release small amounts of mercury or oil vapor at the inlet side. To keep the pressure down and to keep this vapor out of the vacuum system, a cold trap or a baffle is placed between the pump and the system.

Mercury pumps are very dangerous, and highly toxic mercury vapors can be released into the room if the system leaks. Oil diffusion pumps are safer, but the hot oil has a tendency to crack and decompose if there is contact with the air. Special bypass systems are needed if air has to be introduced into the system. Pumps designed for oil cannot be used with mercury, and vice versa.

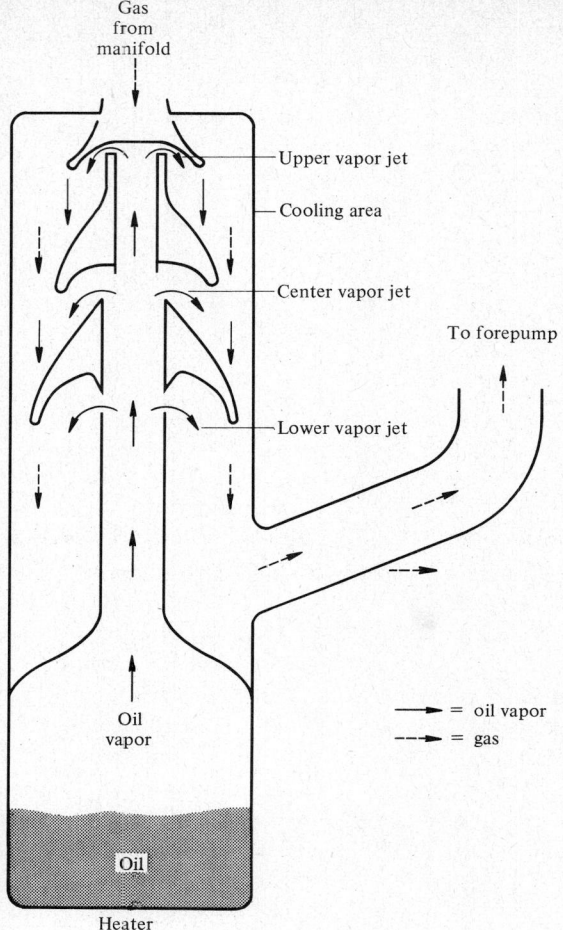

FIGURE 9–7 / Three-stage glass–mercury diffusion pump (schematic).

For ultrahigh vacuum, *getter* and *getter-ion* "*pumps*" are used. The getter pump is a chemically active metal that is evaporated into the system after the pressure has been reduced down to the limit of the diffusion pump. The evaporated metal condenses on the walls of the system, where it either adsorbs the residual gas or reacts with it chemically or even buries the gas molecules under the mass of condensed metal (the gas is said to be "gettered"). Titanium is the most widely used getter. The getter-ion pump is a combination of a getter pump and a vacuum tube. An electron beam ionizes the gas molecules to positive ions that are collected at a cathode. At the same time a film of titanium is deposited on the cathode, burying the discharged gaseous ions.

For routine vacuum apparatus with glass systems, rotary pumps, and oil diffusion pumps, design is not critical. Within minutes, the forepump will usually reduce the pressure to the point where the mean free path of the gas molecules is large compared to the dimensions of the system. The remaining gas will move by molecular flow instead of bulk flow. Molecular flow is the much slower process. Therefore, to have good pumping rates, the system should be constructed of wide-diameter tubing and the connections between different parts should be as short as practicable. It is not usually necessary to estimate the pump loads and evacuation rates. For ultrahigh vacuum systems or for large systems with high pumping loads, where system design is critical, the reader is referred to Dushman and Lafferty [1962] and to Roberts and Vanderslice [1964] for further details.

Cold Traps

Cold traps are necessary to protect pumps and to keep diffusion pump vapor from backing up into the system and increasing system pressure. The cold traps

FIGURE 9–8 / Cold traps.

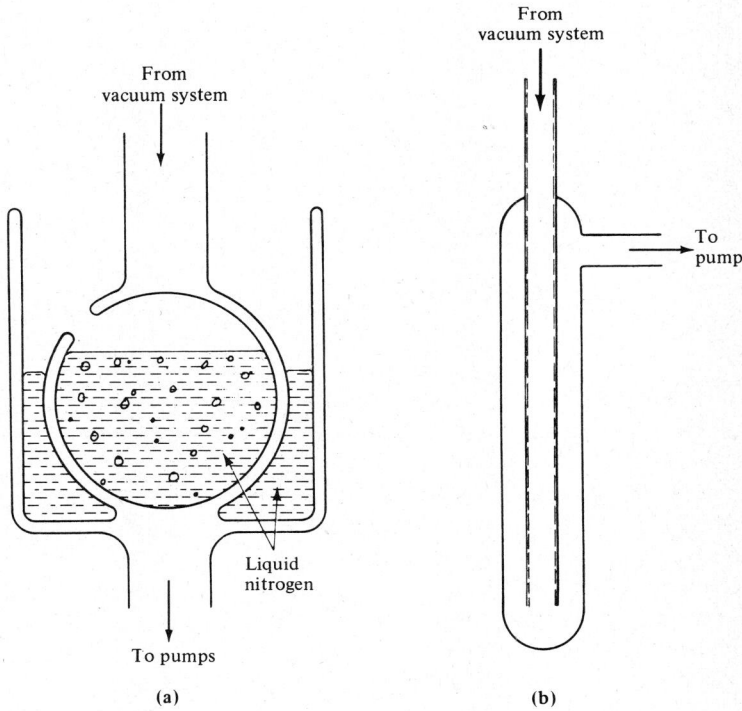

shown in Figure 9–8 are widely used in glass systems between the system and the diffusion pump. The design shown in Figure 9–8a has the advantage that the vapor diffusing back from the pump toward the system condenses on the trap and returns directly to the pump. With the design shown in Figure 9–8b the glass must be cut in order to return the mercury or oil to the diffusion pump. However, the trap in Figure 9–8b can be placed in a Dewar flask, thus conserving the refrigerant, while the refrigerant in the trap in Figure 9–8a must be replenished frequently. Several traps can be cascaded either for better trapping or to take advantage of the good features of each. The refrigerants most frequently used are liquid nitrogen, which is commercially available and relatively inexpensive, and mixtures of solid carbon dioxide and trichlorethylene.

To protect rotary pumps from corrosive or otherwise undesirable vapors, traps of the type shown in Figure 9–8b are used. These may either be chilled or filled with an appropriate absorbent. Since they operate at the high-pressure side of the diffusion pump, such protective traps can be made with ground glass joints, which make them easy to clean without having to cut and reseal glass.

Handling Gases

Many gases are now commercially available in tanks and cylinders at pressures up to several thousand psi (pounds per square inch). These cylinders must be

FIGURE 9–9 / Needle valve (by permission of the Mattheson Chemical Co.).

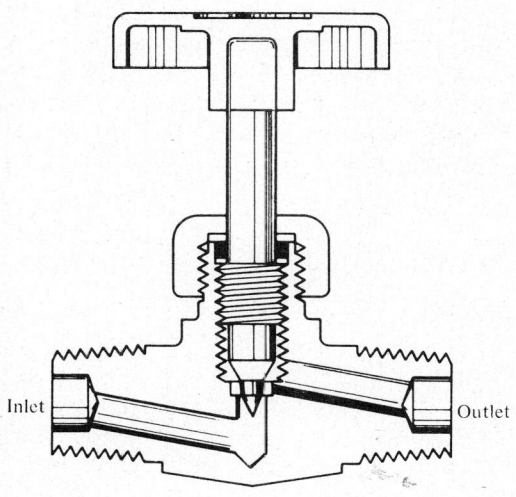

Chapter 9. Pressure Measurements, Vacuum Systems, Handling Gases

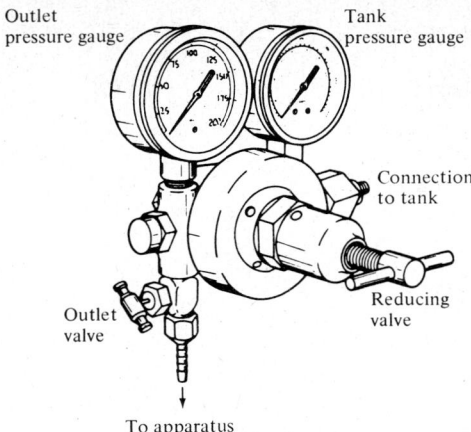

FIGURE 9-10 / Reducing valve (by permission of the Mattheson Chemical Co.).

handled with care to prevent explosions. Tanks are color-coded, with a different color for each gas. However, there is as yet no universally accepted code. Rely on the tank label rather than the color.

Cylinders must always be chained or otherwise fastened to a desk, wall, or other support to prevent them from being knocked over. Whenever a cylinder is moved, the protective cap should be screwed down tight to prevent anything from striking the valve. *Cylinders must not be placed near steam lines, radiators, or other heat sources.*

At moderate pressures thick-walled rubber or plastic tubing connected by metal nipples, by glass or metal adapters, or by T-tubes or Y-tubes is usually adequate for gas lines. For high pressures, such as the 25 atm used in combustion calorimetry, all-metal lines must be used, with metal unions and valves.

The gas pressure in small lecture bottles is usually low enough for the gas flow to be regulated with a needle valve (Figure 9-9). Turning the valve moves the needle into or out of the gas stream, decreasing or increasing the space available for gas flow. In large cylinders, however, the pressure is too great to connect the gas directly to the apparatus, and a reducing valve must be used to control the pressure at which the gas is delivered (Figure 9-10).

OPERATING INSTRUCTIONS FOR REDUCING VALVES

Connect the valve to the fitting on the tank and secure it with a wrench. Avoid overtightening.

Close the outlet valve and turn the handle of the reducing valve counterclockwise until it turns easily and feels loose. This closes the reducing valve.

Slowly open the valve on the tank until the tank gauge needle indicates the tank pressure. Then open the tank valve an additional half-turn.

Turn the reducing valve handle clockwise until the gas is at the desired outlet pressure as shown by the outlet pressure gauge.

Open the outlet needle valve slowly to deliver the gas to the apparatus.

To shut off the cylinder, first close the tank valve and then allow all the gas to flow out of the reducing valve. If necessary, open the reducing valve wider to deliver more gas. When the reducing valve is empty, close it and then close the outlet valve.

NOTES / Any given regulator valve will not fit all tanks. A hydrogen valve, for example, does not fit on an oxygen tank, and vice versa, to prevent the accidental production of dangerous gas mixtures. If you have difficulty in attaching a valve to a tank, it may be that you are using the wrong valve.

Do not open the tank with the reducing valve open, or the sudden impact may break the diaphragm of the reducing valve.

Do not let the gas remain in the reducing valve over long periods of time.

Do not drain all the gas from a cylinder. Leave about 25 psi showing on the gauge.

The rate of gas flow may be determined with either a bubble counter or a flowmeter. A bubble counter is simply a capillary dipping into a soapy liquid.

FIGURE 9–11 / Orifice meter.

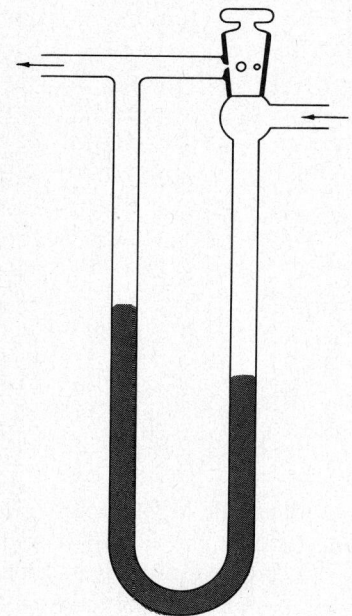

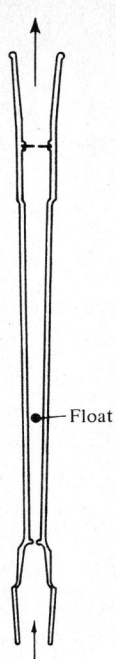

FIGURE 9–12 / Rotameter.

The gas flows through the capillary, producing bubbles of known volume at a known pressure and temperature which may be counted. Gas flowmeters are of several types. The simplest is the orifice meter shown in Figure 9–11. The pressure difference in the two arms of the U-tube is a function of the flow rate. Another simple device is the rotameter, a tapered tube with a float, as shown in Figure 9–12. The gas flows through the tube between the float and the wall. Increased flow raises the float, thus increasing the area through which the gas escapes. The height reached at the steady state depends on the flow rate and the particular gas. Rotameters must be calibrated for each gas used.

Gases are purified by both chemical and physical means, depending on the particular gas and the contaminant.

Water vapor is usually removed by passing the gas through a tube filled with a drying agent such as $Mg(ClO_4)_2$, P_2O_5, KOH, $CaSO_4$, alumina gel, or silica gel. Water vapor may also be removed by freezing it with a liquid nitrogen or dry ice trap or by adsorbing it on activated carbon at low temperatures.

Oxygen is usually removed by passing the gas over hot degreased copper turnings at 300°C. Oxygen may be removed from hydrogen by passing the gas mixture over a palladium catalyst, forming water, or by passing it through a palladium thimble, through which only the hydrogen can diffuse. Carbon dioxide is usually removed by passing the gas through NaOH on asbestos (ascarite) or over KOH.

References

Benedict, R. P., *Fundamentals of Temperature, Pressure and Flow Measurement*. Wiley, New York, 1969.

Dodd, R. E., and P. L. Robinson, *Experimental Inorganic Chemistry*. Elsevier, Amsterdam, 1954.

Dushman, S., and J. M. Lafferty, *Scientific Foundation of Vacuum Techniques*, 2nd ed. Wiley, New York, 1962.

Matheson Gas Data Book. Fifth edition. The Matheson Company, Inc., East Rutherford, N.J., 1971.

Minnar, E. (ed.), *Transducer Compendium*. Plenum Press, New York, 1963.

Neubert, H. K., *Instrument Transducers, An Introduction to Their Performance and Design*. Clarendon Press, Oxford, 1963.

Roberts, R. W., and T. A. Vanderslice, *Ultra-High Vacuum and Its Applications*. Prentice-Hall, Englewood Cliffs, N.J., 1964.

Rondeau, R. E., Vacuum techniques in radiation chemistry, *Sci. Tech. Aerospace Rept.* **1**, 95 (1963).

Weissberger, A., and B. W. Rossiter, *Physical Methods of Chemistry*, Part V, ch. 2. Wiley–Interscience, New York, 1971.

Chapter 10

Vapor Pressures and Boiling Points of Liquids and Solutions

The equilibrium vapor pressure of a liquid is the pressure exerted by the vapor in equilibrium with the liquid. The boiling point is the temperature at which the vapor pressure equals the pressure of the atmosphere above the liquid.

Vapor pressure varies exponentially with the reciprocal of temperature. The Clausius–Clapeyron equation expresses the relation between vapor pressure, temperature, and the heat of vaporization. The *differential forms of the Clausius–Clapeyron equation* are

(10–1a) $$\frac{dp}{dT} = \frac{p \, \Delta H_v}{RT^2}$$

$$\frac{d(\ln p)}{dT} = \frac{\Delta H_v}{RT^2}$$

and $$\frac{d(\ln p)}{d(1/T)} = -\frac{\Delta H_v}{R}$$

p = vapor pressure
T = Kelvin temperature
ΔH_v = molar heat of vaporization
R = gas constant

The first two differential forms give linear graphs over a temperature range of only a few degrees at most. The third differential form and the integrated forms (10–1b) give linear graphs over a much wider range, provided that ΔH_v remains constant. The *integrated forms of the Clausius–Clapeyron equation* are

(10-1b) $$\ln \frac{p_2}{p_1} = -\frac{\Delta H_v}{R}\left(\frac{1}{T_2} - \frac{1}{T_1}\right) = \frac{\Delta H_v(T_2 - T_1)}{RT_1 T_2}$$

$$\ln p = c - \frac{\Delta H_v}{RT}$$

and $$p = a \exp\left(-\frac{\Delta H_v}{RT}\right)$$

$c, a =$ constants

In measuring vapor pressure care must be taken to remove from the liquid volatile impurities which increase the observed pressure. All impurities lower the vapor pressure of the liquid, but this is a relatively minor effect, as shown by Raoult's law in the form

$$\frac{p_0 - p}{p_0} = X_2$$

$p_0 =$ vapor pressure of the pure liquid
$X_2 =$ mole fraction of the solute or impurity

For example, 0.1 mole of an nonionized solute in 1 liter of water will lower the vapor pressure of the water by only 0.2%. A small amount of a nonvolatile impurity will therefore have a negligible effect. A volatile impurity, however, will exert its own vapor pressure in addition to that of the solvent, increasing the observed vapor pressure over the solution.

The Static or Direct Method

The simplest method of measuring the vapor pressure is to connect a manometer to the liquid–vapor system. This is the static or direct method. Foreign gases and volatile impurities are removed by boiling the liquid. Temperature gradients in the gas phase are eliminated by the use of an internal manometer.

A simple device for measuring the vapor pressure of liquids is the *isoteniscope* (Figure 10–1). Essentially, the isoteniscope is an internal manometer as described on page 81, employing the experimental liquid as its own manometric fluid, thus eliminating both valves and mercury vapor corrections. The experimental setup is the same as that shown in Figure 9–3. When the manometric liquid is at the same level in both limbs of the internal U-tube, the pressure observed with an external manometer is the vapor pressure.

Chapter 10. Vapor Pressures and Boiling Points of Liquids and Solutions

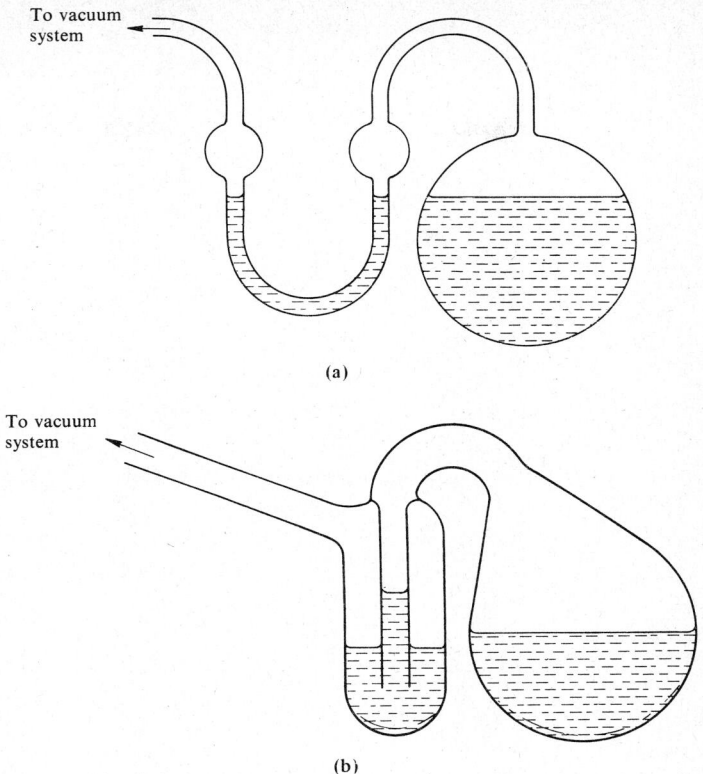

FIGURE 10–1 / Typical isoteniscopes.

OPERATING INSTRUCTIONS FOR THE ISOTENISCOPE

Fill the reservoir of the cleaned isoteniscope with liquid to about the three-fourths level and connect the apparatus as shown in Figure 9–3.

Lower the pressure in the ballast bottle until the liquid in the isoteniscope boils gently, sweeping out air and dissolved gases. Boil for 5 min.

Tilt the isoteniscope and let some liquid flow into the manometer portion. Then adjust the ballast bottle pressure until the levels are the same in both limbs of the internal manometer. Note and record the reading on the external manometer.

NOTES / To fill the isoteniscope, warm it and let the air pull the liquid into the bulb as it cools.

If air gets back into the boiler between readings, it is necessary to boil the liquid again.

Allow at least 5 min for the liquid to reach the thermostat temperature. Rapid boiling cools the liquid.

Work from high temperatures to low temperatures. (Why?)

EXERCISE

1. How could the isoteniscope be modified to measure the vapor pressure of solutions of nonvolatile solutes?

Ramsay–Young Method

In the Ramsay–Young method of determining vapor pressure (Figure 10–2) the pressure in the boiler is set at an arbitrary value and the liquid is heated until it boils. At the boiling point the vapor pressure equals the preset pressure in the boiler. The pressure and temperature are then recorded.

Unfortunately, the phenomenon called boiling does not occur at the equilibrium boiling temperature, but above it. Boiling is the formation of vapor bubbles in the interior of the liquid. Since the pressure in the interior of the

FIGURE 10–2 / Ramsay–Young apparatus.

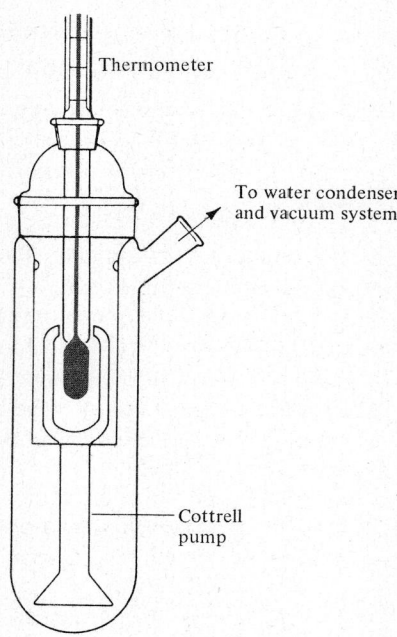

liquid is greater than atmospheric pressure (because of the hydrostatic head) and since surface tension compresses bubbles, the pressure inside a vapor bubble is always greater than atmospheric pressure. Surface tension compresses the small bubbles initially formed more than it does large bubbles. Consequently, the observed temperature of the boiling liquid is actually above the temperature defined as the boiling temperature. The entire phenomenon of excess pressure and increased boiling temperature is called *superheating*.

Superheating can be minimized, but not completely eliminated, by suspending the thermometer in the vapor phase, eliminating hydrostatic head, and taking the temperature of the spray issuing from a Cottrell pump. The Cottrell pump is essentially an inverted funnel with several exits for bubbles. Bubbles of vapor form inside the cone of the funnel and rise into the stem onto the thermometer bulb. During the passage through the stem, the liquid and vapor reach thermal equilibrium, so that the film of liquid hanging down from the bulb is very close to the true equilibrium temperature.

OPERATING INSTRUCTIONS FOR THE RAMSAY–YOUNG APPARATUS

Evacuate the system to the desired pressure. Boil the liquid gently and record the pressure and temperature when they either remain steady or else fluctuate about some mean value.

NOTES/ Do not reflux so vigorously that vapor passes through the condenser. Use boiling chips, and allow time for thermal equilibration.

Do not grease the glass joints in the boiler-condenser apparatus unless you have made sure that the grease is insoluble in the boiling liquid. (How?) If necessary, lubricate the joints with the test liquid.

With solutions, the Ramsay–Young method furnishes only approximate values of vapor pressures and boiling temperatures because of an insuperable experimental difficulty: During the passage of the mixed liquid and vapor through the Cottrell pump, liquid is continuously evaporating and recondensing. Fractional distillation takes place, and as a result the composition of the film of liquid covering the thermometer bulb is not the same as that of the original solution. Therefore, the experimenter has a choice of either making an accurate measurement of the boiling point of a solution of unknown concentration using the Cottrell method or placing the thermometer in the liquid and making an inaccurate determination of the boiling point of the original solution. The usual procedure is to use the Cottrell method and to hope that the concentration of solution on the thermometer bulb is not very different from that in the bulk.

Solutions of nonvolatile solutes have lower vapor pressures and higher boiling points than the pure solvents. The equation for ideal nonionized solutes is, in terms of the mole fraction of the solvent,

(10-2) $$T - T_0 = \Delta T_b = \frac{-RTT_0 \ln X_1}{\Delta H_v}$$

T = boiling temperature of solution, K
T_0 = boiling temperature of pure solvent, K
ΔT_b = observed boiling point elevation
X_1 = mole fraction of the solvent
ΔH_v = molar heat of vaporization of solvent

For very dilute solutions this reduces to

(10-3) $$\Delta T_b = \frac{RT_0^2 m_2}{1000 L_v} = K_b m_2$$

L_v = heat of vaporization, per gram of solvent
K_b = molal boiling point elevation constant
m_2 = molality of the solute

Transpiration Method

Another indirect method of measuring vapor pressures is the transpiration method, which, while slower and more laborious than either of the other methods, is more accurate and is, in fact, the method of choice for solutions. Essentially, the vapor in equilibrium with the test liquid is removed for sampling by a stream of inert carrier gas. A measured amount of inert gas is bubbled slowly through the liquid, the vapor in each gas bubble reaching equilibrium with the surrounding liquid. Then, from the increase in the volume of the gas, or by weighing or chemically analyzing the contents of an absorption tube through which the gas is passed, the quantity of vapor carried off in the gas is measured. In principle, the method is simple, and by appropriate choice of absorbing agents the partial pressure of each component of a solution may be measured directly.

References

Dodd, R. E., and P. L. Robinson, *Experimental Inorganic Chemistry*. Elsevier, Amsterdam, 1954.

Swietoslawski, W., *Ebulliometric Measurements*. Reinhold, New York, 1945.

Swietoslawski, W., *Azeotropy and Polyazeotropy*. Macmillan, New York, 1963.

Weissberger, A., and B. W. Rossiter, *Physical Methods of Chemistry*, Part V, ch. 4. Wiley-Interscience, New York, 1971.

Chapter 11

Thermal Methods of Analysis

Freezing Points

The freezing point of a liquid is the temperature at which the solid and the liquid are in equilibrium. Freezing points of pure liquids are used to identify materials and as criteria of purity. Freezing points of solutions are used in molecular weight determinations and in thermal analysis of alloys and other solids.

Freezing points are usually, but not always, depressed by the presence of solutes. If the solution is ideal and if no solid solution is formed, the *freezing point depression* is

(11-1) $$T_0 - T = \Delta T_f = \frac{-RT T_0 \ln X_1}{\Delta H_f}$$

T_0 = freezing temperature of the pure solvent, K
T = freezing temperature of the solution, K
ΔT_f = observed freezing point depression
X_1 = mole fraction of the solvent
ΔH_f = molar heat of fusion of the solvent

For very dilute solutions the relationship reduces to

(11-2) $$\Delta T_f = \frac{RT_0^2 M_1 m_2}{1000 \, \Delta H_f} = \frac{RT_0^2 m_2}{1000 L_f} = K_f m_2$$

M_1 = molecular weight of the solvent
L_f = heat of fusion, per gram of solvent
K_f = molal freezing point depression constant
m_2 = molality of the solute

In dilute solutions, only association or dissociation of solute, formation of solid solutions, or chemical reactions cause significant deviations from equation (11-2).

For greater ease in calculating molecular weights of solutes from freezing point depressions the equation is sometimes rewritten as

(11-3) $$M_2 = 1000 K_f \frac{g_2/g_1}{\Delta T_f}$$

M_2 = molecular weight of the solute
g_2 = mass of dissolved solute
g_1 = mass of solvent

For organic solutions, benzene is the usual solvent. It has a molal freezing point depression constant of 5.12°C. If there is only a small amount of material, camphor, with a K_f of 40.0, is used as the solvent, and the quantities and apparatus are made much smaller. Actually, the determination of freezing points in camphor solution is considered a semimicro method.

Equation (11-2) can also be used to calculate the heat of fusion of the solvent if the molecular weight of the solute is known. Results are accurate to three significant figures, at best.

The melting point of a pure solid can be determined to within 0.1–0.2°C by the *capillary method*. A few crystals are put into a capillary, which is heated slowly in a bath until the last crystal disappears. This method does not work for mixtures or solutions, since these melt over a temperature range rather than at a given temperature.

Beckmann Method

Two alternative approaches to measuring the freezing point, applicable to solutions, are the *equilibrium* and the *freezing point* methods. In the equilibrium method, the solution is allowed to reach equilibrium with solid solvent and the temperature is recorded. A sample of liquid is then withdrawn for analysis. In the freezing point, or *Beckmann*, method the pure liquid or solution is cooled at a slow steady rate, and temperature–time readings are taken until a solid has deposited. A graph is then plotted and the freezing point obtained from the graph. The precision of these two methods is the same, and the choice between them is based on the experimenter's convenience. For pure liquids (Figure 11-1a) the horizontal portion of the cooling curve is the freezing temperature. The dip below the freezing point is due to supercooling, which almost invariably occurs, especially with materials with complicated crystal structures. The cooling curve for the solution (Figure 11-1b) shows no initial horizontal portion because the solution becomes steadily more concentrated as the

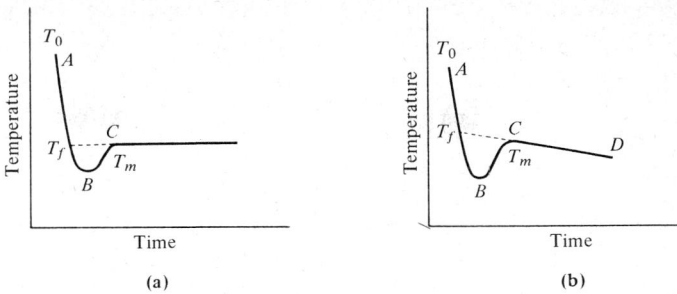

FIGURE 11-1 / Cooling curves for pure liquids (a) and solutions (b).

solvent freezes out and the freezing point drops continuously. Line CD is extrapolated back to intersect line AB. The intersection, T_f, is the temperature at which the original solution would have frozen had there been no supercooling.

OPERATING INSTRUCTIONS FOR THE BECKMANN CRYOSCOPIC APPARATUS

Assemble the apparatus as shown in Figure 11-2. Use a cooling bath about 5°C colder than the expected freezing point. As an alternative method, use an insulating jacket around the sample chamber and keep the bath considerably below the expected freezing point.

Take temperature–time readings at about 30 sec intervals. Continue for at least 5 min after freezing starts.

NOTES / Organic solvents must be dry. If a drying agent is needed, select one that does not dissolve in the solvent (many ionic materials, such as $CaCl_2$, have appreciable solubility in organic solvents, such as alcohol and acetone). Filter the solvent, if necessary, to remove drying agent.

Stir the liquid at a rate of about one stroke per second to maintain thermal equilibrium without mechanical heating.

Stir the cooling bath at frequent intervals. Do not use a very cold freezing mixture without an insulating jacket. Too rapid a rate of cooling makes it difficult to maintain thermal equilibrium.

It is not necessary to use regular time intervals as long as sufficient data are obtained to graph the cooling curve.

Do not allow laboratory air to enter the jacket during the cooling. (Why not?)

Read the thermometer with a lens or magnifying glass. Tap the thermometer stem gently before each reading to prevent the mercury from sticking.

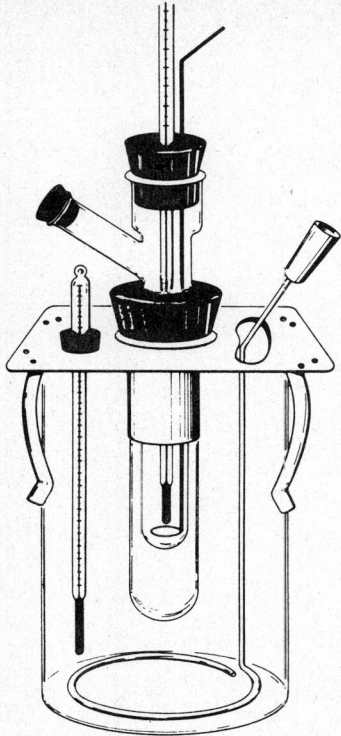

FIGURE 11-2 / Beckmann cryoscopic apparatus.

In addition to the determination of molecular weight, cryoscopy is widely used in the determination of activity coefficients and in the study of solution equilibria.

Thermal Analysis

Thermal analysis (TA), the application of cooling or heating curves to phase studies, is especially valuable for work with mixtures that solidify at high temperatures, such as alloys. Figure 11-3a is the cooling curve of either a one-component liquid or a two-component system at a eutectic or peritectic composition or at the composition of a compound. Figure 11-3b shows the cooling curve for a simple eutectic mixture except at the eutectic composition, or for a peritectic mixture at a composition between those of the pure component and of the compound with the incongruent melting point. Figure 11-3c is the cooling curve of a peritectic system whose composition is intermediate

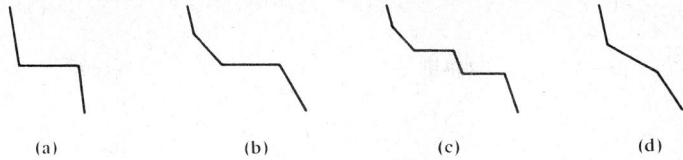

FIGURE 11-3 / Idealized cooling curves.

between the eutectic composition and that of the unstable compound. Figure 11-3d is the cooling curve of a liquid solution that freezes to give a solid solution. It differs from the others in that there is no flat portion.

Experimentally, accurate thermal analysis data are often difficult to measure. Unless the system is in complete thermal equilibrium, the temperatures will scatter randomly about the curve, obscuring the break points. This problem is especially acute at higher temperatures and with large samples.

Differential Thermal Analysis

In *differential thermal analysis* (DTA) the break points are located very precisely. Changes in the time–temperature curve are observed instead of the time–temperature curve itself—hence the name differential thermal analysis. The charges are greatest at the break points, and so transition temperatures are located with great accuracy.

In DTA the sample and an inert reference material are placed in a controlled-temperature environment, such as an air bath, an oil bath, or a metal block. The junctions of a sensitive thermocouple are placed directly in the sample and the reference materials and the potential difference between them, which depends directly on the temperature difference, is amplified and recorded. The output signal is therefore proportional to the temperature difference between the two materials. Since at every transition point heat is either absorbed or evolved, each transition or break point produces an output signal that shows up on the graph, endothermal processes producing a trace in one direction and exothermal processes in the other.

In commercial instruments the temperature of the environment is increased at a controlled rate, using some kind of a temperature programmer. (Theoretically, the environment could be cooled, also, but heating is easier to control than cooling.) The amplified thermocouple signal is fed into a recorder. If, as in most cases, the temperature is varied, the readout is on an X–Y recorder, with the temperature on the X axis. However, in cases where the temperature is kept constant, as perhaps in kinetic runs, a time base recorder is

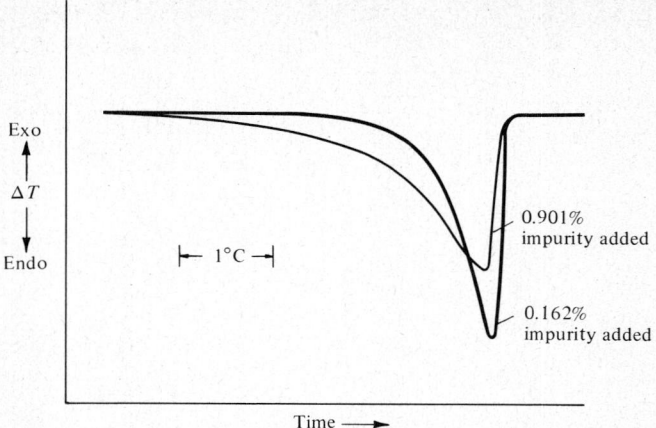

FIGURE 11–4 / Effect of impurity on ΔT vs. time curve. The sample is benzoic acid, heated at 1°C/min. The interval marked on the graph shows the time corresponding to a 1°C change in temperature.

used, with time being the X axis. Figure 11–4 shows a typical set of thermograms, for the melting of benzoic acid. Note the broadening due to very small amounts of impurity. The method is obviously capable of being used to demonstrate impurity levels. The curve in Figure 11–5, taken on a sample of cis-polybutadiene at low temperatures, shows an exotherm at $-72°C$, due to a

FIGURE 11–5 / Low-temperature DTA transitions. The sample is cis-polybutadiene, heated at 20°C/min.

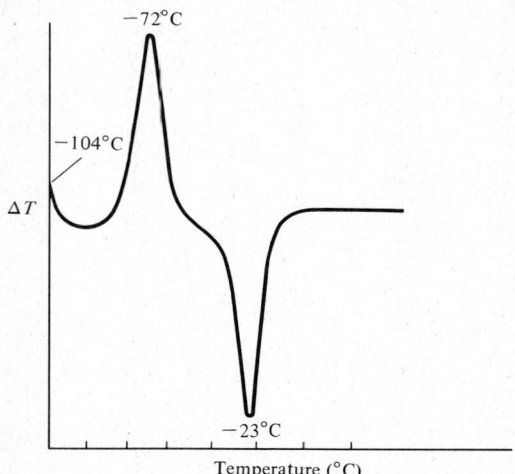

crystallization of amorphous material, followed by an endotherm at −23°C corresponding to melting. Commercial instruments can produce DTA scans in a few minutes covering temperature ranges of up to 1800°C, locating temperatures to within 0.002°C in samples whose mass ranges between 0.1 and 100 mg. As a result, there are a very wide variety of industrial uses for this method.

Differential Scanning Calorimetry

In physical chemistry one is often interested in the energies of transition as well as in the transition temperatures. For such energy determination a technique called *differential scanning calorimetry* (DSC) has been worked out. In DSC the signal recorded is proportional to the energy that flows into or out of the sample at any given time or temperature. To monitor the energy flow, the thermocouples are not placed in the sample and the reference materials. Instead, the sample and the reference material are put into identical pans which are then placed on a support of thermally conducting material of carefully controlled dimensions and shape. The two thermocouple junctions are welded to the support just below the pans, and they measure the temperature of the support at the two points just below the sample and reference. At a transition

FIGURE 11-6 / DSC scan for heating of liquid crystal. The sample weighed 13.85 mg, and was heated at 20°C/min in a nitrogen atmosphere. The melting took place at 134°C. Note that from 170°C on, the scale of the ordinate has been increased tenfold, to emphasize the small changes.

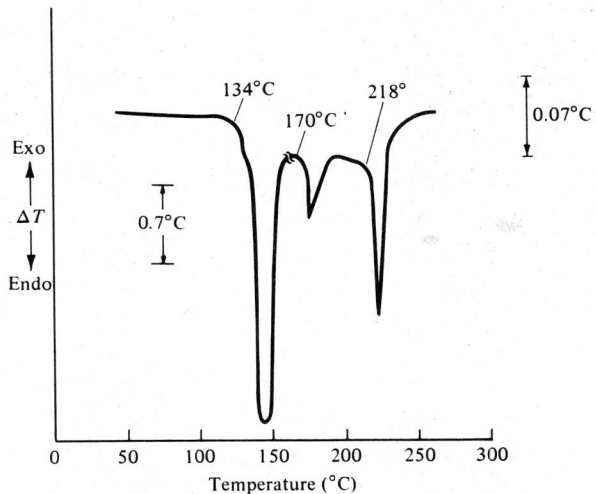

point heat flows into or out of the sample, producing a temperature difference in the two thermocouple junctions. The temperature difference in the thermocouple junctions is directly proportional to the heat flow. The proportionality constant is determined by running a scan with a material having a known heat capacity or a known heat of transition. An auxiliary thermocouple indicates the temperature of the reference. The total heat evolved (or absorbed) is obtained from the area under (or over) the curve. Figure 11-6 shows a DSC scan of a liquid crystal. At 134°C the solid melts, forming the smectic liquid crystal phase. At 170°C there is a change to the cholesteric phase and at 218°C the liquid becomes isotropic. (The scale of the figure is increased tenfold at 170°C to show the low energy liquid–crystal transformations.)

Thermogravimetry

Another thermal analysis technique has been worked out for studies of adsorption and desorption of liquid and gaseous materials and for studies of various pyrolytic combustion reactions. This is *thermogravimetry* (TG). In TG the sample is heated in a cell while suspended from one arm of a sensitive balance. The weight of the sample is recorded as a function of time or temperature. As material is adsorbed or desorbed, the weight changes.

The many industrial applications of these methods have resulted and are resulting in constant improvements and development of new methods and equipment in this rapidly growing field.

References

Cox, J. O., and G. Pilcher, *Thermochemistry of Organic and Organometallic Compounds*. Academic Press, New York, 1970.

Dodd, R. E., and P. L. Robinson, *Experimental Inorganic Chemistry*. Elsevier, Amsterdam, 1954.

Garn, P. D., *Thermoanalytical Methods of Investigation*. Academic Press, New York, 1965.

Levy, P. F., Thermal analysis, an overview, *Amer. Lab.* Jan. 46–58 (1970).

Mackenzie, R. C., *Differential Thermal Analysis*, Vol. 1, "Fundamental Aspects," Vol. 2, "Applications." Academic Press, New York, 1970 and 1972.

Schwenker, R. F., Jr., and P. D. Garn (eds.), *Thermal Analysis*, Vol. 1, "Instrumentation, Organic Materials and Polymers," Vol. 2, "Inorganic Materials and Physical Chemistry." Academic Press, New York, 1969.

Smothers, W. J., and Y. Chiang, *Handbook of Differential Thermal Analysis*. Chemical Publishing, New York, 1966.

Weissberger, A., and B. W. Rossiter, *Physical Methods of Chemistry*, Part V, chs. 3, 7. Wiley–Interscience, New York, 1971.

Chapter 12

Viscosity

When one layer of fluid flows past another, an internal viscous force or drag is exerted that slows down the faster layer and speeds up the slower. To overcome this drag and maintain a constant velocity gradient between layers, an external force must be applied in the direction of motion. The relationship between the external force and the gradient of velocity between layers is

(12–1) $$f = \eta A \frac{du}{dx}$$

f = applied force
η = viscosity (formerly called coefficient of viscosity)
A = area on which force is exerted
du/dx = velocity gradient, perpendicular to direction of motion

Liquid Viscosity

The viscosity of a liquid depends on its molecular size and shape, the intermolecular attractions, and the liquid structure, if any. Many empirical relations exist between liquid viscosity and composition, but there is no completely satisfactory quantitative relation, either empirical or theoretical. Nevertheless, viscosity measurements, especially of solutions, can furnish valuable information on the molecular weight and on the size and shape of the molecules being studied.

For mixtures of small molecules that form ideal solutions

(12–2) $$\log \phi = X_a \log \phi_a + X_b \log \phi_b$$

ϕ = fluidity = $1/\eta$
X = mole fraction
a, b = components of the solution

With polymer molecules the viscosity of the solution depends on the concentration of the solute, its molecular weight, and the shape of the solute molecule. The relative increase in viscosity due to the presence of the solute is called the *specific viscosity*, defined by

$$(12\text{-}3) \qquad \eta_{sp} = \frac{\eta_c - \eta_0}{\eta_0} = \frac{\eta_c}{\eta_0} - 1$$

η_c = viscosity of the solution
η_0 = viscosity of the pure solvent

The specific viscosity is a function of the concentration of the solute, the size and shape of the solute molecules, and the strength of the solvent–solute interactions.

$$(12\text{-}4) \qquad \frac{\eta_{sp}}{c} = [\eta] + k[\eta]^2 c + \cdots$$

c = concentration of polymer, g/cm^3
$[\eta]$ = intrinsic viscosity, cm^3/g
k = characteristic constant

The greater the attraction between solvent and solute, the looser the structure of the polymer molecule and the farther out into the solution it extends. In good solvents the viscosities are higher. In a good solvent a linear molecule forms a loose, flexible chain. The value of k is usually about 0.35. In a poor solvent a linear molecule is arranged into a coil or ball, and k is less than 0.35.

Theoretical studies of solute–solvent interactions are concerned with behavior at infinite dilution. Since the specific viscosity is zero at infinite dilution, the *intrinsic viscosity*, which has a finite value at infinite dilution, must be used. Based upon equation (12-4), this is defined as

$$(12\text{-}5) \qquad [\eta] \equiv \lim_{c \to 0} \left(\frac{\ln(\eta_c/\eta_0)}{c} \right)$$

Another term for the intrinsic viscosity is the *limiting viscosity number*.

For many systems the intrinsic viscosity is related to the molecular weight of the solute by a simple and useful equation. For solutions of polymers with molecular weights up to about 6000–8000

$$(12\text{-}6) \qquad [\eta] = K_i M + K_0$$

M = molecular weight

The two constants are characteristic of each solute–solvent system. Tables of these constants for large number of systems have been prepared and are available in the literature.

For solutions of polymers of molecular weight above 10,000,

$$(12\text{-}7) \qquad [\eta] = K M^a$$

TABLE 12–1 / Molecular Weight–Intrinsic Viscosity Data

Polymer	Solvent	$K (\times 10^5)$	a	Molecular Weight Range ($\times 10^{-3}$)
Polypropylene	Benzene	27	0.71	60–300
Polystyrene	Benzene	722.7	0.72	2–8
Polyvinyl alcohol	Water	20	0.76	6–20
Polyvinyl acetate	Acetone	21.4	0.68	40–300
Polyvinyl acetate	Chloroform	20.3	0.72	40–300
Polyvinyl acetate	Methanol	101	0.5	40–300
Polyvinyl pyrrollidone	Water	67.6	0.55	10–40
Polymethyl methacrylate	Acetone	7.5	0.70	15
Polyethylene glycol	Water	156	0.5	0.2–8

Source: J. Brandrup and E. H. Immergut, *The Polymer Handbook*. Wiley-Interscience, New York, 1966.

where K and a are constants characteristic of the solvent–solute system, and a depends particularly on the shape of the molecule, which of course depends on the solvent–solute interactions. Theoretically, for rigid, rod-shaped molecules, $a = 2$. For flexible, randomly coiled linear chains, a is 0.5–0.8; for hard, spherical molecules, $a = 0$. That is, the intrinsic viscosity is independent of molecular weight. This last prediction has been confirmed for all globular proteins. Table 12–1 is a partial list of constants for equation (12–7).

Although the experimental determination of intrinsic viscosities is quick and easy, the molecular weight calculated is a complicated *average molecular weight*, somewhere between the *number average* and the *weight average* values, but close enough to the latter to furnish a useful approximation. (It will be remembered that for the most part high polymers are mixtures of molecules with different degrees of polymerization. Each particle has a different mass, and the molecular weight is therefore an average value.) The relatively low cost of the equipment and the ease of measurement combine to make viscosity the most frequently used method of determining polymer molecular weights.

Liquid viscosities increase with increasing pressure. They usually decrease exponentially as the absolute temperature increases, following the empirical relationship

(12–8) $\quad \eta = ae^{b/T}$

a, b = constants

The product bR, where R is the gas constant, may be considered as the activation energy for viscous flow.

Absolute methods of determining liquid viscosity measure either the resistance of liquids to the motion of objects being moved or rotated in the liquid under a known force, or the rate at which the liquid flows under a known applied force. Relative methods compare the effect of the viscosity of the liquid to that of a standard or reference liquid. Relative methods are usually quicker and easier, but must ultimately be referred to some precise absolute determination of the viscosity of the reference liquid.

The *capillary method* of Poiseuille is direct and simple in concept. The time is measured for a volume of liquid under a constant pressure head to flow through a capillary. The viscosity is calculated using *Poiseuille's equation* for laminar flow:

$$(12-9) \qquad \eta = \frac{\pi r^4 P t}{8Vl}$$

r = capillary radius
p = pressure head
t = time
v = volume of liquid passed in time t
l = capillary length

The radius must be measured very carefully, since it appears to the fourth power in this equation. The liquid must wet the walls of the entire tube because the derivation assumes that the liquid in contact with the wall remains stationary. The derivation also assumes that at a constant velocity all of the force exerted on the liquid is used to overcome friction.

Ostwald Method

The *Ostwald method* (Figure 12-1) is a variation of the Poiseuille method. The time for a fixed volume of liquid to fall through a capillary into a reservoir under a variable pressure head is a function of the density and viscosity of the liquid and the dimensions of the viscometer. The time and the density are measured, and the viscosity of the test liquid, relative to that of a reference liquid in the same viscometer, is obtained from the relationship

$$(12-10) \qquad \frac{\eta_1}{\eta_2} = \frac{t_1 \rho_1}{t_2 \rho_2}$$

t = time
ρ = density

Although the operations are simple, there is an inherent error in the method. Because the liquid spends more time falling when the pressure head is lower, the average pressure head for each liquid is not the mean between the initial and the final value, but an average over the time of the experiment. If two liquids have different drainage times, they have different average pressure heads, and equation (12-10) becomes an approximation rather than an exact

Chapter 12. Viscosity

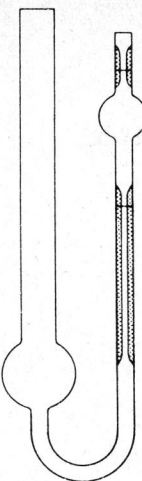

FIGURE 12–1 / Ostwald viscometer.

relationship. This error is of importance only in work of high accuracy, and may be ignored in student measurements. It may be avoided by choosing a reference liquid whose drainage time approximates that of the liquid being investigated.

OPERATING INSTRUCTIONS FOR THE OSTWALD APPARATUS

Pipet either 5 or 10 ml of liquid, depending upon the particular viscometer, into the reservoir and allow it to reach the desired temperature.

Raise the liquid level above the higher mark and release it. Measure the time it takes to fall from the upper mark to the lower mark.

NOTES / Control the temperature to ±0.1°C. The temperature coefficient of viscosity is usually about 1% per degree Celsius.

Make sure that the viscometer is mounted vertically.

Filter the liquid before use to prevent solids from clogging the capillary.

Raise the liquid level by forcing air into the reservoir end rather than by pulling air out of the capillary end. If the level is raised by aspiration at the capillary end, liquid often gets into the rubber tubing, contaminating the system.

In putting the tubing onto the viscometer, hold the end to which the tubing is being attached. Do not hold it by the other end. Even a small torque may snap the fragile viscometer.

PRINCIPLES

In removing the rubber tubing, push it off the tube. Never pull rubber tubing off a tube. Pulling rubber tubing causes it to contract and hold on more tightly. Pushing the tubing causes the bore to expand, and the tube then slips off easily.

Ubbelohde Method

The *Ubbelohde pipet* is a variation of the Ostwald pipet, especially adapted for work with solutions of polymeric materials. These are so viscous that it would take an inconveniently long time to fill and to empty the capillary. The Ubbelohde dilution viscometer in Figure 12-2 allows the liquid to fall through the capillary without a pressure other than its own weight, since the lower end of the capillary is above the level of the solution. The time measured is that for the liquid level to drop from the upper mark on the capillary to the lower mark.

FIGURE 12-2 / Ubbelohde dilution viscometer.

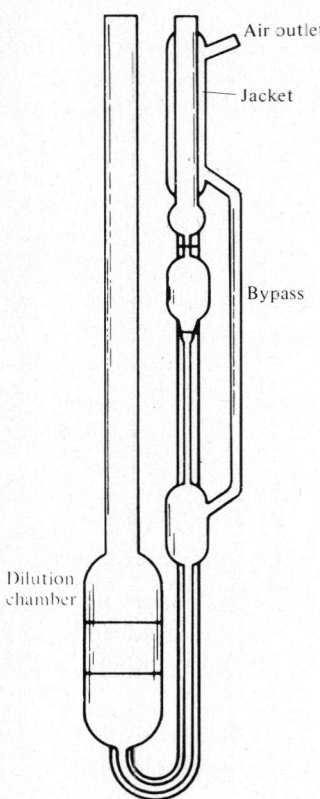

The liquid is forced into the capillary tube by pressure. Additional portions of solvent may be added to the reservoir, providing for in situ dilution, as long as the level of the liquid remains below the bottom of the capillary.

OPERATING INSTRUCTIONS FOR THE UBBELOHDE APPARATUS

Pipet 20 ml of clean, dust-free polymer solution into the large dilution chamber of the viscometer and immerse the viscometer in a thermostat at the desired temperature. When the temperature has become constant, apply gentle pressure at the large end, forcing the liquid up into the capillary side of the viscometer. The liquid will rise much faster in the bypass than in the capillary. When the jacket is almost full, stop the exit vent with a finger. The liquid will then be forced up through the capillary. When the solution is above the upper mark on the capillary tube, release the pressure and then open the exit hole.

Determine the time for the liquid to fall from the upper mark in the capillary tube to the bottom mark. Repeat until four consecutive measurements agree within 0.5 sec (experienced workers can achieve precision to within 0.2 sec, but this takes practice).

Calibrate the pipet with a reference liquid.

To make dilutions, pipet additional solvent into the dilution chamber. Mix the solution by repeatedly forcing it into the outer jacket and then releasing it.

The same precautions should be observed in using the Ubbelohde viscometer as with the Ostwald viscometer, with the additional requirement that all liquids be dust-free to prevent the capillary from becoming clogged.

Other Methods

The *Saybolt viscometer* is one of the most common industrial viscometers. The time is measured for a standard quantity of liquid to flow out of an orifice in a standard cup.

In the *falling ball method*, based on Stokes' law, a metal or glass ball of known density is dropped through the liquid and its time of fall is noted. The method can be used to obtain absolute values of viscosity, but is usually used for relative determinations. The ratio of drop times in two liquids is given by the equation

(12-11) $$\frac{t_1}{t_2} = \frac{\eta_1}{\eta_2} \left(\frac{\rho - \rho_2}{\rho - \rho_1} \right)$$

t_1, t_2 = drop times in the liquids
ρ = density of the ball
ρ_1, ρ_2 = densities of the liquids

In operation the ball does not have to fall freely. It may be rolled down an incline, as in the Hoepler apparatus, which is essentially a cylinder containing the liquid and the ball. The cylinder is inverted and the time is noted for the ball to roll the distance between marks. A series of balls are provided for different viscosity ranges, each ball already having been calibrated against a reference liquid and the factor $\eta_2/t_2(\rho - \rho_2)$ supplied by the manufacturer. For absolute measurements the ball is allowed to fall freely and its steady state velocity is measured. The viscosity is computed from the *Stokes law* relationship

(12–12) $$\eta = \frac{2g(\rho - \rho_2)r^2}{9u}$$

g = acceleration of gravity
ρ_2 = density of liquid
r = radius of the ball
u = terminal velocity

The *rising bubble method* is similar to the falling ball method.

Rotational methods involve the measurement of the torque and the angular velocity of objects being mechanically rotated in a liquid. Although the experimental method is simple, there are serious mathematical difficulties in relating the experimental quantities to each other and to the viscosity. Consequently, these methods are rarely used as absolute methods. Instead, viscosities are obtained relative to those of reference liquids. There are two types of rotational apparatus: one operates at constant torque and measures the speed of rotation, and the other measures the torque required to maintain a constant speed.

Gas Viscosity

When a gas passes through a tube with laminar flow, the distance between the layers flowing past each other is a mean free path. This distance is too great for intermolecular forces to contribute much to viscosity. In gases the observed resistance to flow (which in liquids is largely due to intermolecular attractions and the necessity for displacing molecules) is largely due to the transfer of momentum from the faster layers to the slower ones. Gas viscosities are several orders of magnitude smaller than liquid viscosities and have a different temperature dependence. For ideal gases viscosities should increase with the square root of the Kelvin temperature. The observed increase for real gases is somewhat faster. Liquid viscosities, on the other hand, decrease with increasing temperature.

For a gas whose molecules are hard spheres, according to kinetic molecular theory,

(12–13) $$\eta = \frac{5\pi}{32}\rho\bar{c}l = \frac{(RTM)^{1/2}}{N\pi^{3/2}\sigma^2}$$

ρ = density
$\bar{c}$ = average speed
l = mean free path
R = gas constant
T = Kelvin temperature
M = molecular weight
σ = collision diameter
N = Avogadro's number

(Older texts used the factor 1/3 in the relationship between mean free path and viscosity, but accepted theory now holds that the correct factor is $5\pi/32$.) As indicated by equation (12–13), the viscosity is independent of pressure for an ideal gas. For real gases the forces between colliding molecules cause a variation of viscosity with pressure and affect the temperature–viscosity relationship somewhat. There is no theoretical equation in complete agreement with experimental data.

Gas viscosities are measured by *viscous reactance* (analogous to the rotation methods for liquids) and by *transpiration* methods (analogous to the Poiseuille method).

In the viscous reactance method two objects in the form of cylinders, disks, or spheres are placed close to each other in the gas. One object is rotated or oscillated, moving the gas, which in turn rotates or oscillates the other object. The viscosity is obtained either from the measured force exerted on the second object or from the measured rate of rotation or oscillation of the second object.

In transpiration methods either time–volume measurements are made on gas flowing through a capillary under constant pressure, or time-pressure measurements are made as a vessel is evacuated through a capillary.

To determine absolute values of viscosity from transpiration measurements, the dimensions of the apparatus must be carefully determined. To avoid this, viscosities are determined relative to a standard gas.

Evacuation Rate Method

In the evacuation rate method (Figure 12–3) the gas is evacuated from a large bulb through a capillary and its pressure is measured at various time intervals.

PRINCIPLES

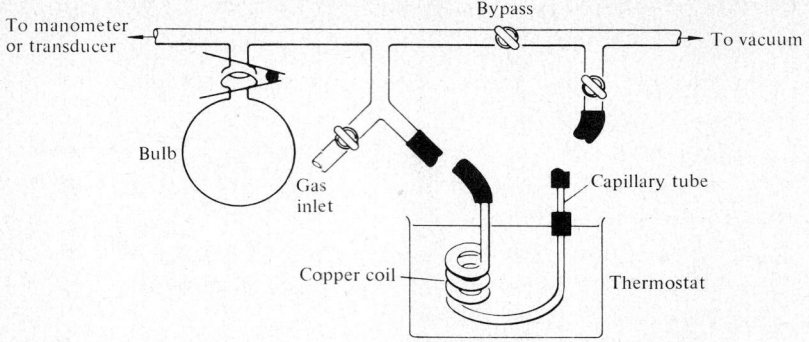

FIGURE 12-3 / Apparatus for measuring gas viscosity by evacuation rates.

The relationship between pressure and time is obtained by combining Poiseuille's equation with the ideal gas law. The volume of gas in the apparatus is assumed to be constant and the pressure at the exit of the capillary is assumed to be negligible:

$$(12\text{-}14) \qquad \frac{1}{p} = \frac{\pi r^4 t}{16 \eta V L} + \frac{1}{p_0} = \frac{kt}{\eta} + \frac{1}{p_0}$$

p = pressure at time t
p_0 = initial pressure at t_0
r = capillary radius
t = time
v = volume of the system
l = length of the capillary
k = constant for the system

To calculate the viscosity, a graph of $1/p$ is plotted against t and the slope is measured and compared to the slope obtained using a reference gas of known viscosity. The apparatus can be used to measure vapor viscosities simply by operating at pressures below the equilibrium vapor pressure of the liquid.

OPERATING INSTRUCTIONS

Connect the apparatus, insert the coils in the thermostat, and evacuate the system.

Close the bypass and fill the system with gas to about atmospheric pressure.

As the manometer pressure drops, record both the pressure and time at 0.5 cm intervals for 20 min, or until the pressure decreases by only 0.1 cm/min.

If a transducer-recorder combination is used, record pressures for 20 min.

NOTES / Evacuate all the air from the gas inlet lines. Flush the system with gas once or twice.

If the reference gas is laboratory air, dry it by passing it through a drying agent.

Use a trap between the pump and the apparatus. If an aspirator is used, do not turn it off until the system has been opened to the air. Otherwise the apparatus will fill up with water.

The volume of the system is not constant, since the level of the mercury in the manometer changes. However, if the bulb volume is 500 ml or more, the volume change is negligible.

Equation (12–14), although derived for gas diffusing into a vacuum, is approximately valid at the exit pressures produced by an aspirator.

To determine the viscosity of a vapor, put 5–10 ml of liquid into the bulb and evacuate the system until the liquid has evaporated. Then proceed with the pressure–time measurements. Note Exercises 2 and 3, however.

If the pressure changes too slowly, use a shorter capillary.

The stopcocks can be replaced by T-tubes, pressure tubing, and screw clamps, if desired, although the system will not be as satisfactory.

EXERCISES

1. Derive equation (12–14) from the Poiseuille equation and the ideal gas law.
2. Is the equation valid at pressures in which the mean free path is longer than the capillary? Explain.
3. At what approximate pressure will the mean free path equal the length of the capillary? Compare this with the average pressure in the capillary.

References

Guggenheim, E. A., *Elements of the Kinetic Theory of Gases*. Pergamon Press, Oxford, 1960.

Moelwyn-Hughes, E. A., *Physical Chemistry*, 2nd ed. Pergamon Press, Oxford, 1964.

Partington, J. R., *An Advanced Treatise on Physical Chemistry*. Longmans Green, New York, 1949.

Weissberger, A., and B. W. Rossiter, *Physical Methods of Chemistry*, Part VI, ch. 2. Wiley–Interscience, New York, 1977.

Chapter 13

Surfaces and Membranes

Molecules at a surface, in contrast to those in the interior, are in an asymmetrical environment. There are no intermolecular attractions from above to counterbalance those from below. Consequently, the intermolecular attractive forces which operate at intermediate distances pull the surface molecules into the interior until the short range repulsive forces balance the attractions and an equilibrium position is reached. To increase the surface area at constant temperature and pressure, reversible work must be done and the free energy of the system increased. This additional free energy is usually called *surface free energy*. The surface tends to contract or to resist expansion, thus keeping the surface free energy at a minimum. This gives rise to the phenomenon known as *surface tension*.

Surface Tension

To increase a surface, work must be done. The reversible work, at constant temperature and pressure, is the *surface free energy*:

(13–1) $\quad dW = dG_s = \gamma \, dA$

dW = isothermal reversible work done *on* the system
dG_s = increase in surface free energy
γ = surface tension
dA = change in surface area

The factor γ, the surface tension, has the dimensions ergs per square centimeter or Newtons per meter, and is therefore a force per unit length.

Surface tension decreases with increasing temperature, as the kinetic energy of the molecules overcomes the forces causing contraction. At 6°C below, the critical temperature there is no longer a well-defined surface, and the surface tension becomes zero. This is expressed by the *Ramsay–Shields–Eötvös equation*:

(13–2) $\quad \gamma(Mv)^{2/3} = k(T_c - T - 6)$

M = molecular weight
v = specific volume, the volume of 1 g
k = empirical constant, about 2.1 for nonassociated liquids
T_c = critical temperature
T = temperature

Surface tension is usually determined from the work done in expanding a surface (direct methods) or the force exerted by the liquid in resisting expansion of the surface (static methods).

Du Noüy Method

The *du Noüy tensiometer* (Figure 13–1a) measures the force needed to expand a surface. A horizontal planar ring immersed in the liquid is pulled through the surface by a force supplied by a torsion balance (Figure 13–1b). The surface tension is the applied force divided by the length of contact between the wire and the surface. The length of contact depends upon the thickness and the diameter of the ring. To a close approximation, the ring, just before it breaks through, holds up a hollow cylinder of liquid. The perimeter of contact is $2 \times 2\pi r$, where r is the radius of the ring, since contact is made at both the inner

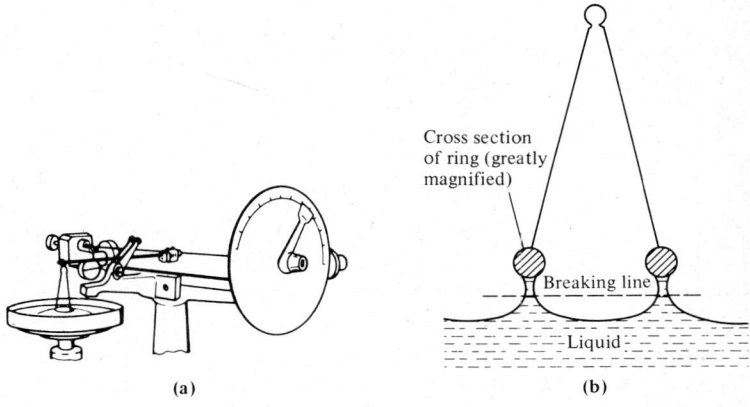

FIGURE 13–1 / (a) Simplified du Noüy tensiometer (courtesy of The Central Scientific Co.). (b) Ring holding up surface.

and outer edges of the ring. For a thick wire, however, the circumferences of the inner and outer rings are not equal. The perimeter is therefore $2\pi(r + r')$, where r and r' are the inner and outer radii, respectively. If the ring is distorted or bent out of a plane, the applied force is no longer equal to the surface tension, and the measurements are not valid.

The tensiometer is rapid and accurate for pure liquids. For solutions the results are accurate enough for student use, but not for research. The commercial instrument is equipped with a platinum–iridium ring of standard dimensions for which the dial has been calibrated to read directly in dynes per centimeter. The ring may be quickly and easily cleaned by flaming, avoiding the laborious purification required by some of the other methods.

OPERATING INSTRUCTIONS

Measurement of the Surface Tension. Immerse the ring below the surface of the liquid, level the beam, and set the dial at zero. Then slowly increase the torsion on the wire supporting the ring. At the same time lower the platform to keep the beam horizontal. Record the dial position corresponding to just enough torsion to pull the ring through the surface.

Calibration of the Dial. Level the beam in the air and note the reading. Then place a small piece of paper with a 100 mg weight on the ring and rebalance the beam. Record the reading, weigh the paper, and measure the radius of the ring.

Calculations. The surface tension is the final dial reading, unless the dial needs recalibration, in which case the surface tension is the dial reading multiplied by a correction factor. This factor is mg/Rl, where m is the mass of the paper and weight, g is the acceleration of gravity, l is the perimeter of contact, and R is the difference in readings with and without the paper and weight.

NOTES / The pointer must remain horizontal during the measurement, or the surface tension will be opposed by only a component of the torsion.

The ring must be cleaned, either by flaming or by dipping it into sulfuric acid–potassium dichromate cleaning solution.

Make sure that the ring is circular and planar. Do not bend or distort the ring.

Check the dial calibration using distilled water.

The buoyant force on the ring is so slight that it can be ignored. Nevertheless, it is good practice to zero the beam in the liquid rather than in air.

Wettable Blade Method

The wettable blade method (Figure 13-2) is a static, or equilibrium, method that is used for measuring surface tensions of liquids and solutions and for direct measurements of the surface pressure π. The wettable blade, a sandblasted platinum plate a few centimeters in perimeter, is touched to the surface of the liquid. As the liquid wets and spreads out over the surface of the blade, the surface tension, acting at the edges of contact between the liquid and the blade, pulls the blade down into the liquid. An external force is then applied from an analytical balance or a torsion balance, restoring the blade to the original position. At equilibrium,

(13-3) $$\gamma = \frac{f}{s}$$

f = external force
s = contact perimeter, twice the sum of the width and thickness of the blade

There are no corrections necessary such as those in the du Noüy method, and since the measurement is made at equilibrium, it may be used for solutions. (In the du Noüy method, the expansion of the surface disturbs the equilibrium distribution, and the surface breaks before equilibrium can be reestablished.)

FIGURE 13-2 / Simplified Rosano tensiometer (courtesy of Biolar Corporation).

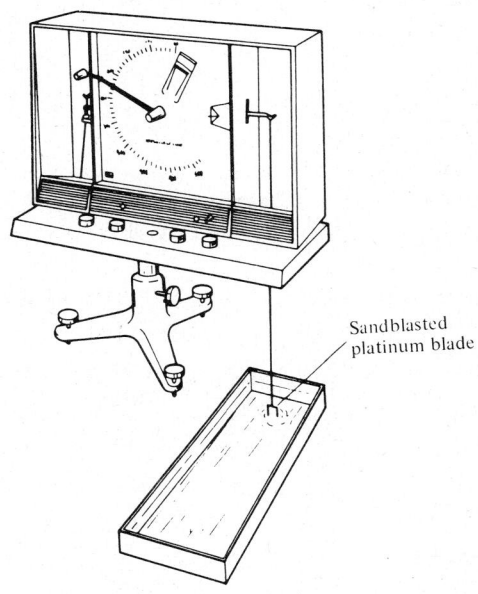

OPERATING INSTRUCTIONS

Surface Tension Measurements. Adjust the torsion until the beam holding the plate is horizontal.

Lower the balance or raise the sample until the plate just touches the liquid surface.

Adjust the balance until the beam is again horizontal and the plate is just touching the surface. Record the required restoring force.

Surface Pressure Measurements. Adjust the torsion until the beam is horizontal, with the plate *just touching* the surface of the pure liquid.

Spread the film over the surface and adjust the torsion again, restoring the plate to its original position.

CALCULATIONS

Surface tension: $\gamma = f/s$, where f is the restoring force.

Surface pressure: $\pi = f'/s$, where f' is the decrease in the restoring force after the film has been spread.

NOTE / Clean the blade by flaming it. If necessary, dip it into concentrated nitric acid and then flame it, repeating the process until the flame is colorless.

Capillary Rise Method

The capillary rise method (Figure 13-3) depends on the tendency of a fresh surface to contract in order to reach equilibrium. If a liquid wets the glass of a tube, a thin film spreads out on the glass surface, increasing the liquid surface area. Surface tension, acting to decrease the surface area, pulls the liquid upward over the glass surface. The liquid rises until the force of surface tension pulling upward is matched by the force of gravity pulling downward. At equilibrium,

(13-4a) $$\gamma = \frac{rh\rho g}{2 \cos \theta}$$

r = capillary radius
h = height of rise
ρ = liquid density
g = acceleration due to gravity
θ = wetting angle

Chapter 13. Surfaces and Membranes

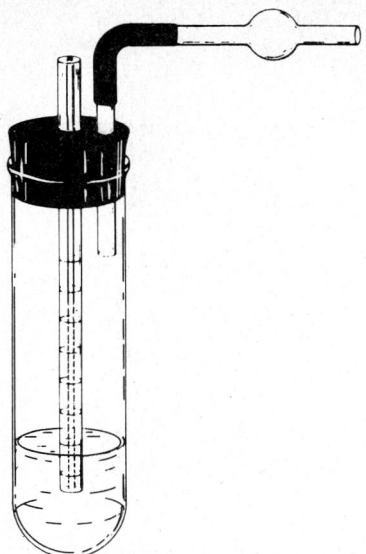

FIGURE 13-3 / Capillary rise apparatus.

If θ is small, $\cos \theta$ is close to unity, and for approximate measurements

(13-4b) $\qquad \gamma = \tfrac{1}{2} r h \rho g$

This method has the advantages of simplicity of apparatus and ease of temperature control. It may be used for both pure liquids and solutions. However, impurities will concentrate at the liquid surface, and unless there is careful purification, results are likely to be low. The capillary may be calibrated with a reference liquid, but this in turn must be prepurified. It is better to calibrate the capillary by filling the tube with mercury and calculating the radius from the measured length and mass of the mercury column. In the capillary rise apparatus the outer container must be wide enough so that the outer liquid surface is not significantly curved. The sensitivity of this method is limited by the difficulty of reading the difference in liquid levels inside and outside the capillary. It is rarely used in research because of lack of sensitivity and accuracy.

Bubble Pressure Method

The bubble pressure method is based on the tendency of the stretched surface to contract. To form a gas bubble in a liquid, the pressure inside the bubble must be greater than the pressure outside the bubble. The relationship between the radius of the bubble and the pressure difference across the interface is

$$(13\text{-}5) \qquad p_i - p_o = \frac{2\gamma}{r}$$

p_i = inside pressure
p_o = outside pressure
r = radius of curvature of the surface

The excess pressure on the inside of the bubble is needed to balance the force of surface tension. From the maximum pressure developed when a bubble is formed by blowing air into a liquid from a capillary of known radius, the surface tension may be obtained.

Adsorption

Adsorption is defined as the increase in concentration of material at a surface or an interface over that in the bulk phase (a decrease in concentration at the surface is termed *negative adsorption*). Materials that lower the surface free energy are adsorbed, while materials that raise the surface free energy are negatively adsorbed. The adsorbent surface may be solid or liquid. The adsorbate may be gas, liquid, or solid.

ADSORPTION ON LIQUID SURFACES

Adsorption on liquid surfaces is classified in terms of the solubility of the adsorbed material. Insoluble layers or films are formed when insoluble materials are wet by the liquid and spread out over the surface, decreasing the surface free energy. (If the insoluble material is not wet, it simply remains as a droplet or a particle of solid.) Soluble surface layers or films are formed when the dissolved material has a surface concentration greater than that in the bulk (i.e., when it is adsorbed). The decrease in surface tension of a liquid when it is coated with a film, either soluble or insoluble, is called the *surface pressure* or *spreading pressure*:

$$(13\text{-}6) \qquad \pi = \gamma_0 - \gamma_s$$

π = surface pressure
γ_0 = surface tension of the liquid
γ_s = surface tension of the liquid covered by the film

Measurements of surface pressure are useful in determining molecular weights of insoluble materials, molecular length, and molecular cross sectional area, using a two dimensional analog of the ideal gas law:

$$(13\text{-}7) \qquad \pi = \frac{kT}{\sigma} = \frac{nRT}{A} = \frac{RT}{A_m}$$

k = Boltzmann's constant
T = Kelvin temperature
σ = area per molecule
n = number of moles adsorbed
R = gas constant
A = observed geometrical area
A_m = area per mole

The surface pressure may be obtained from surface tension measurements using equation (13–6), or it may be measured directly using the wettable blade method or the barrier method. In the barrier method, a surface film presses against either a mica barrier or a silk thread. To maintain the position of the barrier, an equal and opposing force is applied by some form of torsion balance. The surface pressure of the film is then read directly from the dial of the torsion balance. The apparatus is commercially available.

For insoluble films the area per mole is easily obtained from the number of moles of adsorbed material and the geometrical area of the film. If the molecular weight is unknown, it can be easily calculated from the surface pressure and the geometrical area when a film is formed from a measured weight of material, using equation (13–7).

If the material is soluble, the area per mole must be obtained indirectly from the *Gibbs adsorption isotherm*:

$$(13\text{-}8) \qquad \Gamma = -\frac{c}{RT}\frac{d\gamma}{dc} = -\frac{1}{RT}\frac{d\gamma}{d\ln c}$$

Γ = difference between the surface concentration of adsorbate and the concentration at a plane in the interior of the phase, moles/cm^2
c = interior concentration
R = gas constant, J/mole K

Although the derivation of this equation places no restriction on the nature of the bulk phase (i.e., solid or liquid), in practice, the Gibbs isotherm is applied only to liquid surfaces, since the surface tension of solids cannot be measured directly. For exact work, activities are used rather than concentrations. The negative sign indicates that positive adsorption is accompanied by a decrease in surface tension.

In using the Gibbs equation to determine the molar surface area of soluble films, surface tensions are measured at a series of concentrations and a graph of γ vs. the natural logarithm of concentration is plotted. From the slope of the graph Γ is obtained at each concentration being studied. Since Γ has the dimensions moles per square centimeter, $1/\Gamma$ is the area per mole, A_m.

Graphs of π vs. A_m sometimes show discontinuities at increased π, analogous to the breaks in p vs. v graphs of gases. These indicate the formation of a liquid or solid condensed phase. When the molecules in the film are compressed to the smallest area they can occupy without the film crumpling and collapsing, the surface pressure rises sharply without decrease in A_m. From the area per mole at this point the cross sectional area of the molecules may be obtained. The calculated area will depend on the orientation of the molecules. Many long-chain organic molecules are oriented perpendicularly to an aqueous surface, and therefore the cross sectional area determined from the π vs. A_m graphs is the cross sectional area of the hydrocarbon chain.

From the cross sectional areas and the experimentally determined densities the length of the molecules can be determined and used to evaluate bond angles and bond distances.

ADSORPTION ON SOLID SURFACES

There are two distinct limiting types of adsorption on solids and, of course, many cases of intermediate behavior. *Chemisorption* is the formation of a chemical bond between the surface and the adsorbate. Typically, only one layer is formed, at most, since the forces operate at short ranges, and in some cases only a portion of the surface contains active bonding sites. The heats of chemisorption are of the order of 20 kcal/mole or more and decrease with increased surface coverage. *Physical adsorption*, on the other hand, is much less specific and results from attractions, such as van der Waals attractions or dipole–dipole attractions, rather than bond formation. Heats of physical adsorption are of the order of 10 kcal/mole or less. More than one layer may be adsorbed, and sometimes many layers are.

The *Langmuir adsorption isotherm* applies to adsorption of single layers (monolayers) on a homogeneous solid surface:

(13–9a) $$a = \frac{k_1 c}{1 + k_2 c}$$

or, for gases,

(13–9b) $$a = \frac{k_1 p}{1 + k_2 p}$$

a = amount adsorbed per unit surface area
c = concentration of adsorbate in the external phase (the fluid phase)
k_1, k_2 = constants
p = pressure

The Langmuir isotherm is derived on the basis of an equilibrium between adsorbed and unadsorbed molecules. The surface is assumed to be uniform and homogeneous and the adsorbed molecules are not supposed either to attract or to repel each other. Therefore, the heat of adsorption does not change as the surface is covered. These are rather drastic assumptions because in most cases surfaces are heterogeneous and adsorbed molecules do attract or repel each other. Nevertheless, there are many cases where the fact that the surface is heterogeneous and the heat of adsorption decreases with surface coverage is compensated for by attractions between molecules. In other words, as the surface holds the molecule less strongly, other adsorbed molecules hold it more strongly. In such cases, as well as in the ideal case, the Langmuir isotherm holds.

Usually, in testing the adsorption isotherm in terms of the Langmuir equation, the reciprocal of a is plotted against the reciprocal of p or of c. If the data give a straight line, the Langmuir equation holds and

(13-9c) $$\frac{1}{a} = \frac{1}{k_1 c} + \frac{k_2}{k_1}$$

The *Freundlich adsorption isotherm* is an empirical equation of the type

(13-10) $$a = kc^n \quad \text{or} \quad a = kp^n$$

n = empirical constant whose value is between 0 and 1

For a limited range of concentrations it is frequently impossible to decide which of these two equations fits experimental data better. The Langmuir equation is somewhat better for adsorption of gases, while the Freundlich isotherm is somewhat better for the adsorption of a solute from a solution. There are, however, many exceptions, and many examples are known which fit neither of these two equations.

For multilayer adsorption there is no accepted theoretical isotherm. The Brunauer–Emmett–Teller semiempirical isotherm is most frequently used, especially in the determination of the surface area of powders and irregular solids.

(13-11) $$\frac{p}{v(p_0 - p)} = \frac{1}{v_m C} + \frac{C-1}{v_m C}\left(\frac{p}{p_0}\right)$$

p = gas pressure
v = volume of adsorbed nitrogen at STP
p_0 = saturation pressure
v_m = volume of gas at STP needed to form an adsorbed monolayer
C = a constant, characteristic of the adsorbent-adsorbate pair

Nitrogen at room temperature is allowed to diffuse onto a sample of powder, kept at the boiling point of liquid nitrogen. At equilibrium the pressure and volume of the gas are measured. From the results of a series of such measurements, a graph of $p/v(p_0 - p)$ vs. p/p_0 is plotted. From the intercept and slope of the graph, C and v_m are obtained, and the number of moles of nitrogen required to form a monolayer is calculated. The surface area is then calculated taking the cross sectional area of the nitrogen molecule to be 12 Å^2.

Membranes

Membranes are films separating two fluid phases which may be either liquid or gaseous. Most membranes are selectively permeable, although they need not be. Examples range from visking tubing used for dialysis to the membrane bounding living cells. Whether they are artificial membranes composed of one substance or complex natural membranes built of over 100 different molecules, most membranes are highly organized, with considerable short range order. The molecules comprising the surface of some membranes can move about freely, in this respect resembling two-dimensional liquid or solid films. On other membranes, the molecules have little lateral motion. Some membranes, especially biological membranes, are mosaics of both solidlike and liquidlike regions. Three types of membranes will be described here, briefly.

ION EXCHANGE MEMBRANES

Ion exchange membranes are either sheets of ion exchange resin or particles of such resin embedded in a film. The ion exchange resins are water-insoluble polymers with either acidic or basic functional groups. The acidic resins exchange cations with an aqueous solution, picking up metallic ions and releasing other, loosely bound, metallic ions or hydrogen ions. Basic resins pick up anions, releasing other loosely bound anions or hydroxide ions. Typical ion exchange resins are polystyrene sulfonic acid for cations and aminopolystyrene for anions. When tap water is forced through two consecutive columns, one containing a cation exchanger that releases hydronium ion and the other an anion exchanger that releases hydroxide ion, the only ions left in solution are hydroxide and hydronium, which then form water. The product of these consecutive exchanges is deionized or demineralized water, which is identical with distilled water except for nonionic impurities, bacteria, and often traces of amines from the resins.

Ion exchange membranes, if made of an anion exchange resin, are permeable to anions and if made of a cation exchange resin are permeable to cations. They are essentially impermeable to water and to ions of the opposite charge. One use for such membranes is in desalting water by electrolysis. The

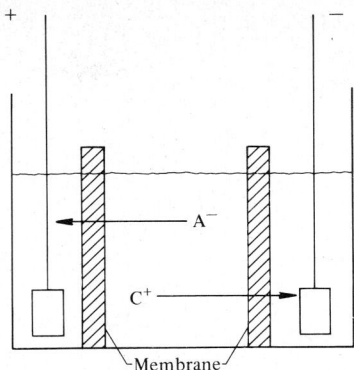

FIGURE 13-4 / Apparatus for electro-desalting, using ion exchange membranes.

principle of desalting under the influence of an electric field is shown in Figure 13-4. Anions from the central compartment travel through the anion exchange membrane, moving toward the positive electrode. Cations travel in the opposite direction through the cation exchange membrane. Since neither water nor the counter-ion passes through a membrane, there can be no back diffusion and so salt is removed from the central compartment. The water cannot be completely deionized, however, because there must be some electrolyte left in the central compartment to carry the current. Nevertheless, enough salt can be removed to make the water fit for drinking and for irrigation.

For each equivalent of salt removed from the central compartment, there must be one equivalent of oxidation at the anode and of reduction at the

FIGURE 13-5 / Electrodesalting with several ion exchange membranes in series.

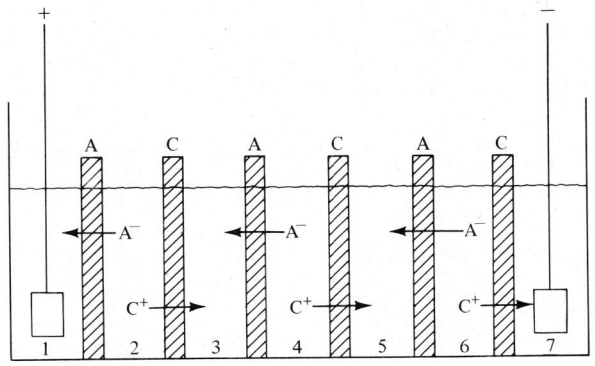

cathode. (In the simple system shown in Figure 13-5, these electrode reactions would probably produce oxygen and hydrogen.) Even with special "reversible" electrodes, this electrolysis wastes energy. The energy lost by electrolysis, per equivalent of salt removed, can be reduced by having several compartments in series, as in Figure 13-5. In this configuration, for every faraday of charge passed through the system, one equivalent of salt is removed from compartments 2, 4 and 6 and added to compartments 3 and 5; one equivalent of anion is transferred to compartment 1, where oxidation takes place; and one equivalent of cation is transferred to compartment 7, where reduction takes place.

BILAYER LIPID MEMBRANES

Bilayer lipid membranes, also known as *black lipid membranes*, are composed of two back-to-back layers of lipids which have hydrophobic or nonpolar hydrocarbon chains on the inside, acting as the center of the membrane, and hydrophilic polar groups facing outside on the surfaces of the membrane. The membranes are formed by depositing a drop of solvent containing the lipid on a small hole in a Teflon sheet. The hole is about 1-2 mm in diameter and the sheet about a millimeter thick. The sheet is then immersed in an aqueous solution. As the solvent leaves the drop, the lipid forms a film which at first shows interference fringes. Then, piecemeal, the fringes disappear and the membrane becomes dark as the hydrophobic portions of the lipid zip together into the bilayer, which is too thin for interference fringes.

These membranes have been studied intensively as simple models of naturally occurring membranes. Some properties are summarized in Table 13-1. The high electrical resistance is of particular interest because it shows that the hydrophobic interior is quite resistant to the passage of ions. The addition of certain antibiotics and polyethers can lower the resistance by up to four or more orders of magnitude. (Surfactants will also do this, but they usually make the membrane unstable.) This drastic drop in resistance is considered to be due to pore formation in some cases. In other cases, some of the added molecules, acting as carriers for ions, form complexes with sodium or potassium ions. The complexes are believed to be hydrophobic, with the ions held in the interior. Pores (channels) are responsible for transport in biological membranes. Their opening and closing control many important biological processes. Measurements of the rates of such processes show them to be high enough to explain both the lowering of resistance in artificial membranes and the observed transport of ions through natural biological membranes. Complexes diffuse through the bilayer lipid membrane only when it is liquid. They cannot transport ions when the membrane is cooled below a "phase" transition temperature and becomes a "solid." Recent results have raised questions with regard to the fluidity of the membrane, its physical state, the arrangement of the hydrocarbon chains, and the oscillations of the membrane as a whole.

TABLE 13-1 / Properties of Some Bilayer Lipid Membranes

Property	Typical Value
Resistance	10^8–10^9 ohm-cm^2 (pure membrane[a])
Capacitance	0.3–1 × 10^4 μF/m^2
Thickness	5–7 nm
Young's modulus	2 × 10^7 N/m^2 (pure membrane[b])

[a] Suitable carrier or pore-forming materials lower the resistance by several orders of magnitude.
[b] Measured values are about 2 × 10^5 because of lenses that have not thinned down.

PLASMA MEMBRANES

Plasma membranes, the membranes bounding cells, resemble bilayer lipid membranes in having surfaces that are loosely composed of polar lipids with their hydrophilic groups facing out. Protein molecules, the other major component of these membranes, are much larger than lipid molecules, with molecular weights of 10^4–10^5. These huge molecules may be pictured as protein icebergs in a sea of lipid. There are also protein molecules and chains of protein molecules extending from one face to the other. Bilayer lipid membranes are usually prepared from one lipid, for example, glycerol esterified with two stearic acid molecules and one phosphoric acid molecule, or at most a few lipids. Plasma membranes are extraordinarily complex and may contain over 100 different lipids derived from fatty acids, in addition to cholesterol, proteins, and possibly complex carbohydrates. It is believed that this complexity results in regions with different properties that enable the membrane to transport selectively many substances into and out of the cell at controlled rates, while excluding other materials.

References

Adamson, A. W., *Physical Chemistry of Surfaces*. Wiley-Interscience, New York, 1960.

Burdon, R. S., *Surface Tension and the Spreading of Liquids*. Cambridge University Press, New York, 1949.

Davies, J. T., and E. Rideal, *Interfacial Phenomena*. Academic Press, New York, 1961.

Flood, E. A. (ed.), *Solid–Gas Interfaces*, Vols. 1 and 2. Marcel Dekker, New York, 1967.

Good, R. S., R. R. Stromberg, and R. L. Patrick, *Techniques of Surface and Colloid Chemistry and Physics*. Marcel Dekker, New York, 1972.

Hwang, S., and K. Kammermeyer, *Membranes in Separations* (Vol. VII of A. Weissberger (ed.), *Techniques of Chemistry*). Wiley-Interscience, New York, 1975.

Rose, J., *Dynamic Physical Chemistry*. Wiley, New York, 1961.

Starzak, M. E., Ion Fluxes through membranes, *J. Chem. Educ.* **54**, 200 (1977).

Tien, H. T., Bilayer Lipid Membranes. Dekker, New York, 1974.

Weissberger, A., and B. W. Rossiter, *Physical Methods of Chemistry*, Part V, ch. 9. Wiley–Interscience, New York, 1971.

Young, D. M., and A. D. Crowell, *Physical Adsorption of Gases*. Butterworth, London, 1962.

Chapter 14

Electrical Measurements and Circuits

Some knowledge of electrical theory and measurements is essential in all branches of science. In many areas of physical chemistry, such as electrochemistry, the thermodynamics of electrolytes, and electrode kinetics, electrical theory and measurements are an integral part of the subject. Electrical or electronic equipment is required for all studies that cannot be made with stopwatches and glassware. We assume that the student has studied, at some time, elementary electrical theory and has some idea of simple DC circuits and such concepts as current, potential or voltage, resistance, capacitance, and inductance. Table 14–1 lists the symbols and practical units of the most commonly encountered electrical quantities. See also Appendix 7.

Although potential is often referred to as *electromotive force* (emf), it is *not* a force as force is defined in physics. The force on a charge is the product of the charge and the electric field, where the field is defined as the negative gradient of potential. (In the one-dimensional case the field is therefore

TABLE 14–1 / Symbols and Units for Electrical Quantities

Term	Symbol	Unit (abbreviation)
Potential	E, V, e	Volt (V)
Current	I, i	Ampere (A)
Resistance	R	Ohm (Ω)
Reactance	X, X_C, X_L	Ohm (Ω)
Impedance	Z	Ohm (Ω)
Capacitance	C	Farad (F)
Inductance	L	Henry (H)
Charge	Q, q	Coulomb (C)
Power	P	Watt (W)

$-dE/dx$, where x is the distance in the one dimension.) The absolute potential cannot be known, but the difference in potential between two points has meaning. This difference is referred to some reference that is defined as having zero potential.

For convenience, reciprocal electrical quantities are frequently used, especially reciprocal resistance (conductance), which is measured in *mhos*, or reciprocal ohms. The number of mhos is the reciprocal of the number of ohms. The mho is now being replaced by a new unit, the *Siemen*.

DC Circuits

DC circuits can be quite complicated, and it is often necessary to find the different currents, potentials, and resistances in the various portions of the circuit. This can be done using three basic relationships: Ohm's law and the two conservation laws of Kirchhoff.

Ohm's law is easily expressed as

(14–1) $$I = \frac{E}{R}$$

I = current
E = potential
R = resistance

The two Kirchhoff relations are

1. The sum of currents entering any junction point in a circuit is equal to the sum of the currents leaving the junction point:

$$\sum I_{in} = \sum I_{out}$$

2. The sum of the voltage drops around any closed path in a circuit is equal to the sum of the applied potentials:

$$\sum E_{drop} = \sum E_{applied}$$

In applying Kirchhoff's first law, one must arbitrarily assume a direction for the current in each portion. The direction assumed is the one that causes the current to flow in closed paths. When all the current values have been calculated, a negative sign will then indicate that the flow is opposite to the assumed direction.

In applying Kirchhoff's second law, the values of the applied potentials are those of the power sources, but the voltage drops across each portion of the

Chapter 14. Electrical Measurements and Circuits

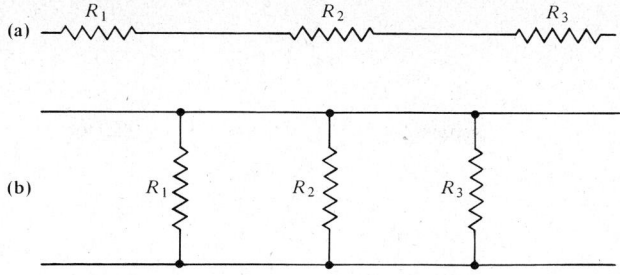

FIGURE 14–1 / Resistances in series (a) and in parallel (b).

circuit must be calculated from the resistances. The total resistance of a number of resistances in *series* (Figure 14–1a) is

(14–2) $$R_{\text{series}} = R_1 + R_2 + R_3 + \cdots$$

When a number of resistances are connected in *parallel*, however, as in Figure 14–1b, the resistances are not additive, however. Instead, the *reciprocals* of the resistances are additive, adding up to the reciprocal of an *equivalent resistance*.

(14–3) $$\frac{1}{R_{\text{parallel}}} = \frac{1}{R_1} + \frac{1}{R_2} + \frac{1}{R_3} + \cdots$$

Equations (14–2) and (14–3) are both consequences of Ohm's law and Kirchhoff's laws.

A typical DC circuit problem is shown in Figure 14–2. From Kirchhoff's first law, at either point A or point B,

(i) $\quad I_1 + I_3 = I_2$

From Kirchhoff's second law, in the left-hand loop

(ii) $\quad I_1 R_1 + I_1 R_2 + I_2 R_3 = E$

FIGURE 14–2 / Circuit with parallel and series resistances.

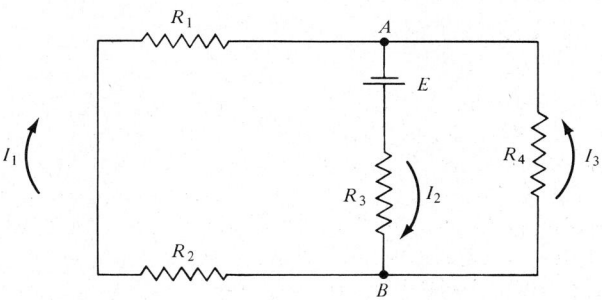

and in the right-hand loop

(iii) $\qquad I_2 R_3 + I_3 R_4 = E$

There are, therefore, three equations that can be solved for the three unknowns I_1, I_2, and I_3.

Ohm's law is not so much a law of physics as it is a definition of resistance. For many circuit elements it is found that the current is directly proportional to the applied voltage, as demanded by Ohm's law, regardless of the polarity of the applied potential. These elements are termed *ohmic* resistors. This is not true for other circuit elements, especially vacuum tubes and transistors. Indeed, rectifiers and all electronic devices depend on *nonohmic* circuit elements.

The power in a DC circuit is given by

(14-4) $\qquad P = EI$

$\qquad P =$ power, watts

Applying Ohm's law gives two other expressions for power:

(14-4a) $\qquad P = I^2 R$

(14-4b) $\qquad P = \dfrac{E^2}{R}$

In DC circuits, these three expressions are equivalent.

Measurement of Direct Current and Voltage

Most common devices for detecting or measuring direct current are modified galvanometers. (Digital meters, which operate on different principles, are discussed in Chapter 15). In the *d'Arsonval galvanometer*, a coil is suspended in a magnetic field by a fine wire which provides a restoring torque when twisted. When current flows through the coil, the coil is deflected. A small mirror is attached to the coil and a beam of light is reflected from the mirror onto a scale. A system of mirrors may be used to lengthen the optical lever and increase the sensitivity.

The deflection is not proportional to the current, unless the magnet is designed to produce a uniformly radial field, because, as the coil moves, it changes its orientation with respect to the field. Therefore, this type of instrument is used primarily to detect, rather than to measure, current. Usually, it is used in bridge circuits as a null-point indicator. Portable box-type d'Arsonval galvanometers may be used to detect currents as low as 0.01 microampere (μA).

To measure current, several types of ammeters are used. In moving vane ammeters, the magnetic field generated by the current attracts a soft-iron armature attached to a pivoted vane or needle, causing the indicator tip to move.

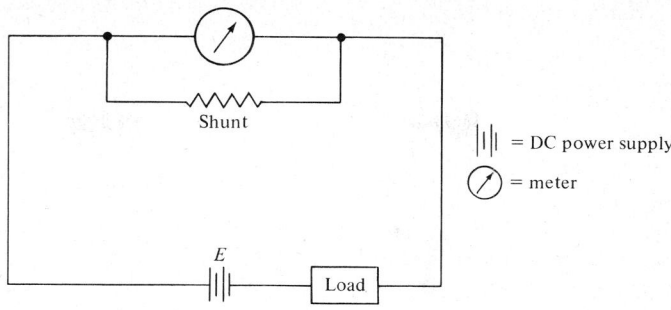

FIGURE 14-3 / Simple ammeter circuit.

The attraction of the field is opposed by a spring which provides a restoring torque. The observed deflection, however, is not directly proportional to the current, because at greater field strengths the vane is pulled closer to the coil, where the effect of the field is greater. This design is rugged and inexpensive, but it is rather insensitive and not as accurate as some other types of movement. More sensitive and accurate ammeters employ a modified d'Arsonval movement, with the coil pivoted between jeweled bearings instead of being suspended by a wire. A restoring torque is provided by a spring, and the mirror is replaced by a pointer. The magnet is modified by soft-iron pieces so that the coil moves in a uniformly radial field. As a result of the modification of the magnetic field, the deflection is directly proportional to the current. D'Arsonval movements operate only on DC.

Most commercial movements operate at or below the milliampere range. To measure larger currents, a resistance, known as a *shunt*, is connected across the milliammeter movement, as shown in Figure 14-3. The current flowing through the circuit is the sum of the current in the meter and that in the shunt. Since the current in each loop of the circuit is inversely proportional to its resistance, the relationship between the circuit current and the meter current is

(14-5) $$I = I_m\left(1 + \frac{R_m}{R_s}\right)$$

I = circuit current
I_m = meter current
R_m = meter resistance
R_s = shunt resistance

For a meter with an internal resistance of 99 ohms, a parallel resistor of 1 ohm would mean that only 1% of the current flows through the meter, or that conversely, the actual current is 100 times that observed on the meter. With the simple circuit in Figure 14-3, a shunt with a very low-resistance is needed

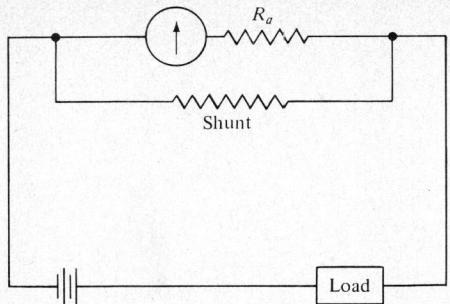

FIGURE 14-4 / Ammeter with Ayrton shunt circuit.

for large currents. The difficulties of making an accurate low-resistance shunt can be avoided by placing a second resistance in series with the meter (Figure 14-4). This circuit is known as the *Ayrton shunt circuit*. The relationship between the circuit current and the meter current, in the Ayrton shunt is

$$(14\text{-}6) \qquad I = I_m\left(1 + \frac{R_m + R_a}{R_s}\right)$$

Most inexpensive commercial meters have an absolute error of about 2% of the full-scale reading; for example, a 0–1 A meter will have a 0.02 A error. The percentage of error in making a reading is therefore greater in the lower part of the scale. For more accurate readings it is best to use a meter that will be deflected almost full scale by the current. Thus, multirange meters are not only more convenient but also more accurate. Precision meters can be obtained with errors of only 0.25% of the full-scale readings. At such high accuracies, however, the errors introduced by temperature changes become important. The meter should be used at constant temperature and the current should be computed with the aid of temperature factors supplied by the manufacturer. To minimize parallax error, accurate meters have a mirror behind the pointer, and the reading is taken when the pointer coincides with its image.

Never connect an ammeter directly across a source of potential without a load (resistance) in series with the meter. The resistance of the meter itself is so small that without an additional resistance, the current passing through the meter will be large enough to do damage to the meter movement. In addition, the potential source may be damaged by the virtual short circuit.

Direct currents may be determined indirectly with an error of less than 0.1%. A precision resistor is placed in series with the circuit and then the potential drop across the resistor is measured with a potentiometer or digital voltmeter. The accuracy of the measurement is limited by the calibration of the resistor and the sensitivity and precision of the potentiometer or digital voltmeter. If the voltage drop across the load is less than several volts, the poten-

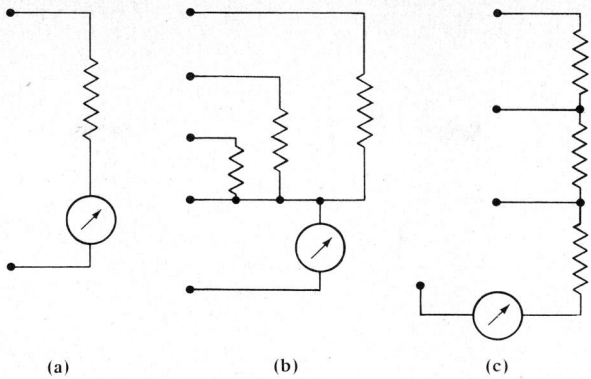

FIGURE 14–5 / Voltmeters with multipliers.

tiometer can be used; if not, a digital voltmeter must be used. (Electronic voltmeters with analog meters as readout can also be used, but their precision is limited by the sensitivity of the meter, which is usually not as good as 1%, even in expensive instruments.)

In measuring voltage, the choice of instruments depends on the accuracy desired, the time that can be spent, and the money available. Rapid, moderately accurate measurements of DC potential can be made by placing a *voltmeter* in parallel with the potential difference. A voltmeter is simply a galvanometer in series with an external resistance called the *multiplier* (Figure 14–5a). The relationship between the voltage across the meter movement and the voltage being measured (i.e., the total voltage across movement and multiplier) is

(14–7) $$E = E_m\left(1 + \frac{R_e}{R_m}\right)$$

E = measured voltage
E_m = meter voltage
R_e = multiplier resistance
R_m = meter resistance

Since current is proportional to voltage, the meter current is directly proportional to the potential difference. Any type of meter movement that can be used for an ammeter can be used for a voltmeter. Multirange voltmeters have several multipliers, each connected to a separate terminal (Figure 14–5b,c). The lower limit of measurement is determined by the minimum current to which the meter responds. Ordinary galvanometer–voltmeters can measure currents down to the millivolt range. For voltages less than this, it is necessary to use electrometers or electronic voltmeters.

The precision of commercial DC voltmeters is the same as that of DC ammeters, 0.25–2% of full scale, depending on the particular meter. Since a voltmeter has a high resistance multiplier in series with the galvanometer movement, it can be placed directly across a source of potential without an additional load.

One difficulty associated with any type of ammeter or voltmeter is the fact that placing it in a circuit disturbs that circuit. The measured current or voltage is the current or voltage with the meter in the circuit, which may be different from the current or voltage that would exist in the circuit without the meter. An ammeter reduces the current by placing an additional resistance, the equivalent resistance of the meter and shunt, in the circuit. For a meter placed in series with the total load, the relationship between the measured current, I', and the current before the meter was placed in the circuit is

$$(14\text{-}8) \qquad I = I'\left(1 + \frac{R_{ms}}{R_{lt}}\right)$$

I = current without the meter
I' = measured current
R_{ms} = equivalent resistance of meter and shunt
R_l = equivalent resistance of total load, including resistance in the potential source

Because all potential sources have internal resistance, a voltmeter reduces the external potential of a potential source, or the potential drop across a load by increasing the flow of current. If there is no load across the potential source, the relationship between the measured potential E' and the true potential E is

$$(14\text{-}9) \qquad E = E'\left(1 + \frac{R_s}{R_m}\right)$$

R_s = internal resistance of source
R_m = total resistance of meter, including multiplier

Therefore, unless the meter resistance is very high, compared to the internal resistance of the potential source, the potential measured by a voltmeter is appreciably less than the true potential.

These systematic errors in the measurement of either current or potential can be circumvented by placing the meter in the circuit and then adjusting the current, or the voltage, to the desired value. Both adjustments cannot be done at the same time. If meters are used to measure potential and current simultaneously, one of the two circuits in Figure 14-6 must be employed. If the configuration in Figure 14-6a is used, the measured current is the current through the load, but the measured voltage is the potential drop across the load plus the potential drop across the ammeter. If the configuration in Figure 14-6b is used, the measured voltage is the potential drop across the load, but the measured current is the current through the load plus the current through the voltmeter. The appropriate configuration for a particular application is obtained by an analysis of the circuit.

Chapter 14. Electrical Measurements and Circuits

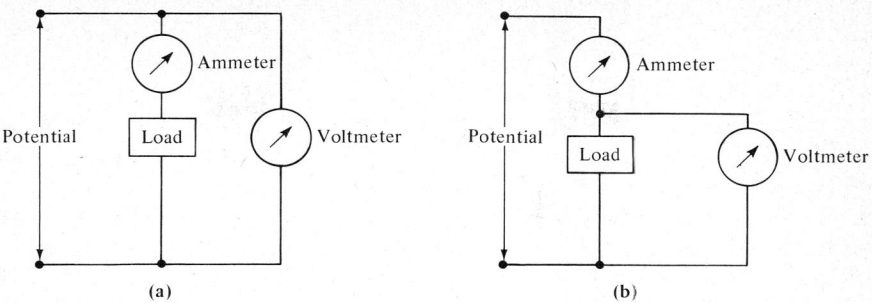

FIGURE 14-6 / Circuits to measure current and voltage.

The preferred instrument for accurate measurement of DC potentials is the electronic voltmeter, which draws currents in the microampere range at most, with some models going down into the femtoampere (10^{-15} A) range. This is equivalent to having the meter resistance, R_m in equation (14–9), extremely large, so that R_s/R_m may be neglected. For measurements in circuits with very high input resistance, the *electrometer* is used. This is a specially designed electronic voltmeter going as far as 10^{16} ohms, the region where one has to worry about the resistance of the insulation. Teflon insulation is good to about 10^{14} ohms, but in the 10^{16} range even air has too high a conductance and so special vacuum devices are needed. Note that at 10^{16} ohms, 1 V produces 10^{-16} A, or 1600 electrons per second. For such low currents, electron counting devices become useful. Note that at 10^{16} ohms, 1 V produces 10^{-16} A, or 1600 electrons per second. For such low currents, electron counting devices become useful.

FIGURE 14–7 / Schematic diagram of the essential parts of a null point potentiometer. E_x, unknown emf; E, working battery; G, galvanometer. The standard cell is not shown.

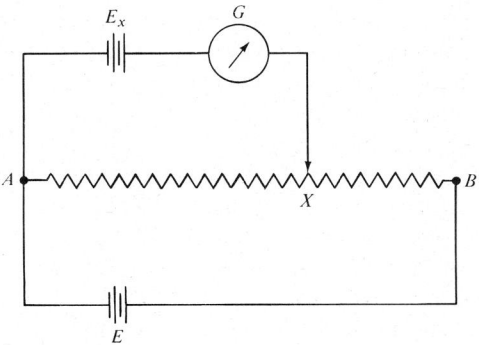

147

The traditional instrument for accurate potential measurement was the *null point potentiometer*, which is still used in many laboratories, especially in schools. This instrument draws only enough current to operate its galvanometer, about 10^{-8} A, as a current sensing device. Such a small current avoids the error introduced by the finite input resistance of the voltmeter. A simple potentiometer circuit is shown in Figure 14–7. The slidewire AB is the heart of the instrument. It is a carefully made uniform wire with a constant resistance per unit length. The sliding contact can be moved to make connection at any point on the slidewire. The unknown emf, E_x, is connected in parallel to the working battery across AX, a portion of the slidewire. If the potential across AX produced by the battery is not equal to E_x, current will flow through the galvanometer in one direction or the other. If, however, no current flows through the galvanometer, the unknown emf has the value of the potential fall across the slidewire between points A and X. Since the slidewire is uniform, the potential drop across AX is proportional to the distance. The unknown emf is

(10–10) $$E_x = \frac{E(A-X)}{A-B}$$

E = potential across AB
$A-X$ = distance between points A and X
$A-B$ = total length of slidewire

To avoid having to measure the potential E, the rheostat is adjusted until the ratio $E/(A-B)$ is 1 V per unit of length. The emf is then numerically equal to the length of the wire between A and X. To make this adjustment (i.e., to standardize the potentiometer) a standard reference cell of known emf is connected across a length of the slidewire proportional to the cell voltage (e.g., a standard cell of 1.0185 V would be connected across 1.0185 units of wire). The rheostat is then adjusted until the battery supplies a potential that balances against the emf supplied by the standard cell, so that no current flows through the galvanometer.

Although commercial potentiometers are equipped with additional refinements, the principle remains the same. The slidewire may be wound on a drum instead of being stretched out. Instead of one uniform slidewire, there may be two in series, one a high-resistance wire equipped with a set of shunts, to permit the potentiometer to be used over several voltage ranges. There may be several rheostats in series, for coarse and fine adjustments. The galvanometer may be protected by a series of shunts and tap keys, with several switches for the battery and the cells. To prevent drain on the cells, contact between the standard cell or the cell being measured and the slidewire is made by tapping a tap key so that only a momentary surge of current flows through the galvanometer. In operation, the sliding contacts are moved back and forth until there is no deflection of the galvanometer needle when the tap key is depressed.

The standard emf is usually a Weston cell (page 193) with an emf of about 1.0185 V at 25°C. Each cell has its own emf, differing from that of other Weston cells only in the fourth decimal place.

Many good potentiometers have an additional adjustment, a temperature compensator to allow for variations of the standard emf with temperature. In theory the potentiometer draws no current, but in practice it is necessary to draw enough current to move the galvanometer needle. Near the balance point, the potential difference across the slidewire is so close to that of the unknown emf that there is not sufficient current to cause a noticeable deflection of the galvanometer needle. Therefore, in working back and forth to determine the balance point, there will be a range of dial positions over which the needle will not move noticeably when the tap key is depressed. The smallest observable needle deflection is of the order of 0.2 mm, which for good galvanometers requires a current of about 10^{-9} A. For circuits with series resistances less than 10^4 ohms, this requires an emf of less than 10^{-5} V, or 0.01 mV. The balance point is then uncertain by only ± 0.01 mV, which is usually negligible. With less sensitive galvanometers, such as those in portable potentiometers, the errors in determining the balance point are greater. With series resistances of a megohm or more, the potentiometer must be replaced by an electronic voltmeter.

The accuracy of the potentiometer decreases more or less in proportion to the source resistances and the potentiometer cannot be used for high resistances, or, as stated earlier, at voltages above several volts. In the microvolt range, potentiometers can be used, but extreme care must be taken to guard against pickup of stray voltages, thermal emf's, contact emf's, or inaccuracies due to aging of resistors and standard cells. For most research uses, potentiometers have been replaced by electronic digital voltmeters. Strip chart and $X-Y$ recorders frequently have self-balancing potentiometers to move the pen.

OPERATING INSTRUCTIONS FOR A NULL POINT POTENTIOMETER

Connect all leads to the appropriate terminals and close the switch in the battery circuit.

Calibrate the slidewire dial by putting the standard cell in the circuit and by adjusting the rheostats until no galvanometer deflection is observed when the tap key is depressed.

Balance the unknown emf against the working battery by adjusting the dial until there is no galvanometer deflection when the tap key is depressed.

Restandardize the instrument to make sure that the emf of the battery has not changed during the measurement.

NOTES / If the potentiometer uses an external standard cell, set the dial at the standard-cell potential before adjusting the rheostats to reach a null point.

PRINCIPLES

To prevent current drain on either the standard cell or the unknown emf, do not hold down the tap keys. Just tap them.

If no null point is found during the standardization, change the working battery.

If no null point is found after standardizing the battery, the emf's are not opposed. Reverse the leads to the unknown emf.

If no deflection is observed on the battery over a wide range of slidewire changes, check to make sure that the circuit is not open.

Keep the battery switch open except when measurements are being made.

Balance the potentiometer, starting both from high values and from low values on the slidewire dial, to find the upper and lower limits of the unknown emf.

Since the eye can detect small motions more easily than small differences in position, a better balance can be obtained by looking for a deflection when the key is tapped than by holding down the key and trying to bring the needle to zero on the dial.

If the potentiometer has two working rheostats, adjust the coarse rheostat first.

Some potentiometers have two tap keys, one with low sensitivity for an approximate balance, the other with high sensitivity for a fine adjustment.

Measurement of Resistance

Approximate measurements of resistance can be made with an *ohmmeter*. A simple ohmmeter (Figure 14–8) consists of an emf in series with an ammeter and several resistors, for different ranges. With points A and B short-circuited, the variable resistor R is adjusted so that the meter deflects full scale at the zero

FIGURE 14–8 / Simple ohmmeter.

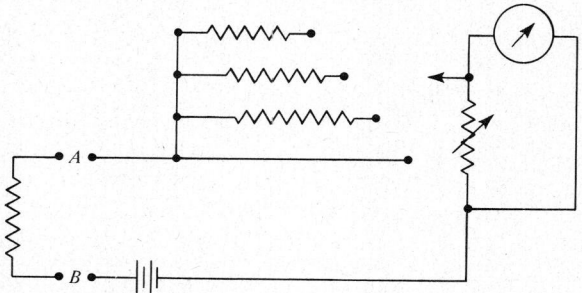

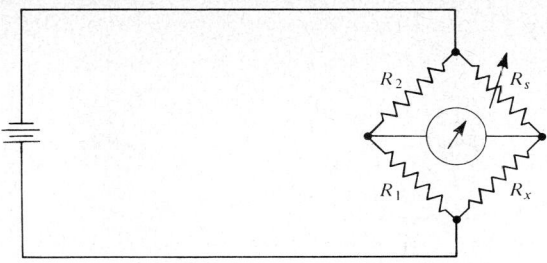

FIGURE 14-9 / Wheatstone bridge.

ohm reading. The unknown resistor, R_x, is then inserted between A and B, decreasing the current and changing the meter reading. The meter scale is calibrated to read directly in ohms. With this circuit, in addition to the errors ordinarily inherent in meter reading, observed resistance depends on the emf and internal resistance of the battery. A change in these changes the meter reading.

For accurate measurements, the unknown resistor is compared with a variable known resistor, using a *bridge circuit*. For DC, the simple Wheatstone bridge shown in Figure 14-9 is adequate. The variable resistor is adjusted until a null point is reached (i.e., no current flows through the meter). At the null point,

(14-11) $$R_x = R_s \frac{R_1}{R_2}$$

R_x = unknown resistance
R_s = standard resistance
R_1, R_2 = fixed resistors in bridge arms

The *slide-wire bridge* is a common modification of the Wheatstone bridge. A potentiometer slide wire is used for resistors R_1 and R_2. The sliding contact is moved to change the ratio R_1/R_2 until the bridge is balanced.

Moderate priced resistors, accurate to 0.1%, are available commercially. More expensive resistors, accurate to 0.01% or better, are also available. These precision resistors, and bridges made with such precision resistors, must be thermostatted because fluctuations in ambient temperature and resistive heating can change the resistance by more than 0.01%. For a precise balance the meter must respond to a small potential drop. A good portable galvanometer will give a deflection on the passage of 10^{-8} A. If the voltage source is 10 V and the resistance as large as 10^7 ohms, the meter current will be at most 10^{-6} A. A change of 1% in meter current would then correspond to 10^{-8} A, the limit of meter sensitivity. When resistances reach the megohm range, it is always necessary to check to see if the meter is sensitive enough.

Bridge circuits have other uses besides measuring resistances. AC bridges can measure capacitance. Because they can indicate small imbalances, they are used with sensors, such as thermistors or strain gauges, that have a resistance which varies with the temperature, pressure, strain, or other quantity being measured. In such applications, R_x in Figure 14-9 is replaced by the sensor and R_s by a balancing resistor or a matched sensor maintained at constant reference conditions. Moreover, the bridge is not balanced. Instead, the meter is calibrated to indicate the measured quantity.

The traditional resistance bridges are rapidly being replaced by electronic instruments, whose accuracies are limited principally by cost. Precision digital voltmeters, at the present time, are often less expensive than null point bridges of comparable accuracy.

Measurement of Charge: The Ballistic Galvanometer

For most applications, whether as galvanometers, ammeters, or voltmeters, meter movements are damped to prevent oscillations. The damping is provided by dissipation, in the external circuit, of currents induced in the moving coil. If, however, a movement is undamped, or lightly damped, the coil will have an appreciable velocity when it reaches the equilibrium position. Because of its inertia it will then overshoot, be stopped by the restoring torque, be brought back to the equilibrium position with an appreciable velocity in the other direction, undershoot, be stopped by the torque from the current in the coil, overshoot the equilibrium position again, and so on. The *ballistic galvanometers* employ this effect to measure charge. In these instruments, the moving coil is designed so as to have a large moment of inertia and the damping is kept as small as possible. When a charge, such as that from a discharging capacitor, is passed through the instrument the coil receives an impulse that is proportional to the charge.

$$Q = \int i \, dt$$

Because of the large amount of inertia, the displacement of the coil is negligible during the time taken to receive this impulse. Consequently, no correction is needed for change of torque with deflection. The impulse produces a deflection that is proportional to the charge. The movement then oscillates slowly, because it is undamped. The oscillations can be stopped, when desired, and the meter brought back to zero by placing a resistance across the terminals, after the initial deflection has been noted. Although the deflection corresponding to a given charge can be calculated, it is easier to calibrate a ballistic galvanometer by passing a known charge through it. Except in special cases, and for didactic purposes, the ballistic galovanometer has been superseded by electronic instruments such as the electrometer.

Elementary AC Theory

Unlike DC, where the current and the voltage do not vary with time, except for transients when a circuit is opened or closed, the current and voltage in an AC circuit vary periodically with time. The simplest AC signal is a sine wave (Figure 14–10). All the more complex periodic signals can be resolved into an unvarying DC component plus a sum of sine waves with various amplitudes and phases and with frequencies that are integral multiples of the fundamental frequency. As Figure 14–10 shows, the voltage and current in an AC circuit need not be in phase. If the voltage is given by

(14–12) $e = E_0 \sin(2\pi ft)$

e = instantaneous voltage
E_0 = peak voltage
f = frequency

the corresponding current is given by

(14–13) $i = I_0 \sin(2\pi ft + \phi)$

i = instantaneous current
I_0 = peak current
ϕ = phase angle

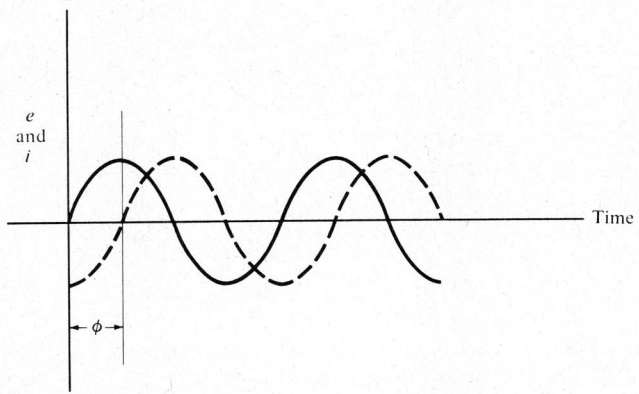

Figure 14–10 / AC signal, showing current and voltage as ordinate and time as abscissa.

From the diagram, it is quite clear that ϕ is the angle by which the current *leads* or *lags* the voltage. Lower case i and e are used for instantaneous current and voltage; capital I and E for root mean square (rms), or DC, current and voltage. *Root mean square*, defined as the square root of the sum of the squares, is used because where positive and negative values are equally probable, such as in AC signals, the average value is zero (i.e., the average AC line voltage is zero but the rms = 110 V).

The phase angle can vary from $+\pi/2$ to $-\pi/2$. Two other expressions for a sinusoidal current or voltage are common. The first involves the trivial substitution of ω, the angular velocity, for $2\pi f$ to give the expressions

$$i = I_0 \sin \omega t \quad \text{and} \quad e = E_0 \sin \omega t$$

The second represents the instantaneous current (or voltage) as a complex number. For the current

(14-14) $\quad \mathbf{i} = I_0 e^{-j\omega t} \equiv I_0(\cos \omega t + j \sin \omega t)$

where $\mathbf{i}$ is the complex† instantaneous current (the symbol i is used for the real instantaneous current), and j is the square root of -1. There is an analogous expression for voltage. The equations have only mathematical, not physical meaning. The actual instantaneous current or voltage is given by either the real or the imaginary part of the expression, depending on the convention used by the author or the material in which it appears.

Because AC voltage and current vary sinusoidally, the average current is zero and is replaced by the rms value. Since the current and voltage are continuously varying, the sum of the squares is replaced by the integral of the instantaneous value with respect to time, over one complete cycle.

(14-15) $\quad I = \left[f \int_0^{1/f} i^2 \, dt \right]^{1/2} = \left[f \int_0^{1/f} I_0^2 \sin^2(\omega t) \, dt \right]^{1/2}$

From this integral

(14-16) $\quad I = I_0/\sqrt{2}$

The rms value is usually given in specifying AC; that is, 110 V AC is 110 V rms.

The instantaneous AC power is the product of the instantaneous current and the instantaneous voltage. The average power over one cycle is therefore

(14-17) $\quad P = f \int_0^{1/f} ei \, dt$

For a sinusoidal waveform and a phase angle ϕ, this becomes

(14-18) $\quad P = E_0 I_0 f \int_0^{1/f} \sin(\omega t)\sin(\omega t + \phi) \, dt = EI \cos \phi$

† Boldface (**i**, **e**, etc.) is used to represent complex quantities.

The quantity cos ϕ is called the *power factor*. In both AC circuits and DC circuits, the power dissipated in resistance is calculated by $P = I^2R$ (equation 14–4a), where, as usual, I is the rms current regardless of the phase angle. For any device with a given power output and operating at a fixed voltage, for example, an electric motor operating on 110 V AC, the poorer the power factor, the greater the current and consequently the greater the resistive losses and the greater the chance of overheating.

CAPACITANCE, INDUCTANCE, REACTANCE, AND IMPEDANCE

A capacitor (or condenser) consists of two conductors separated by an insulator, for example two parallel plates separated by air. If a DC potential is impressed across the conductors, there will be a transfer of charge from one conductor to the other. The relation between the applied voltage and the charge is

(14–19) $\qquad Q = CE$

$\qquad\qquad C =$ capacitance

Neglecting edge effects, the capacitance of a condenser consisting of two identical parallel plates is

(14–20) $\qquad C = \dfrac{k_0 \varepsilon A}{d}$

$\qquad\qquad k_0 =$ permittivity of free space, 8.85×10^{-12} farad/meter
$\qquad\qquad \varepsilon =$ dielectric constant of the insulator
$\qquad\qquad A =$ area of one plate
$\qquad\qquad d =$ distance between plates

If a varying potential is placed across a capacitor, a current will flow. The relation between current and potential can be derived by differentiating equation (14–19) with respect to time, giving

(14–21) $\qquad i = \dfrac{dQ}{dt} = C\dfrac{de}{dt}$

A most important case occurs when a sinusoidal AC potential

$\qquad e = E_0 \sin(\omega t)$

is used. Differentiating this expression for the potential and substituting gives

(14–22) $\qquad i = \omega C E_0 \cos(\omega t)$

This simple expression reveals several important properties of AC circuits:

1. Unlike a DC circuit, where an insulator completely blocks the flow

of current, there is a flow of current in an AC circuit containing a capacitor.
2. If the impressed voltage is sinusoidal, the current is also sinusoidal.
3. The instantaneous current and voltage are not in phase. Since $\cos(\omega t) = \sin(\omega t + \pi/2)$, the current leads the voltage by $90°$.
4. The current and voltage are proportional:

(14–23) $\qquad I_0 = \omega C E_0$

A comparison of this expression with Ohm's law in the form $I = E/R$ suggests that the quantity

(14–24) $\qquad X_c = \dfrac{1}{\omega C}$

is a measure of the restriction of current flow by a capacitor in an AC circuit, just as R is a measure of the restriction of current flow by a resistor in a DC circuit and that (14–23) can be written as

(14–25) $\qquad I_0 = \dfrac{E_0}{X_c}$

X_c is called the *capacitive reactance*. Although capacitive reactance has the same dimensions as resistance and is measured in ohms, it is not resistance. Because current and voltage are $90°$ out of phase, there is no dissipation of energy across a perfect capacitor (equation 14–18), whereas there is a power loss of I^2R (equation 14–4a) across a resistor. If the current and voltage in equation (14–23) are expressed as complex quantities (equation 14–14) with the proper phase relation and the convention that the real part represents the physical current or voltage, then from the rules of algebra for complex numbers X_c must be an imaginary number:

(14–26) $\qquad \mathbf{X}_c = -\dfrac{j}{\omega C}$

The formula (equation 14–20) for the capacitance of a parallel plate condenser suggests the rules for calculating the equivalent capacitance of a network of capacitors. For capacitors in *parallel*,

(14–27) $\qquad C_{\text{parallel}} = C_1 + C_2 + C_3 + \cdots$

For capacitors in *series*,

(14–28) $\qquad \dfrac{1}{C_{\text{series}}} = \dfrac{1}{C_1} + \dfrac{1}{C_2} + \dfrac{1}{C_3} + \cdots$

From these expressions and the relation between capacitance and capacitive reactance (equations 14–24 and 14–26), it follows that capacitive reactances combine like resistance; in *series*

(14–29) $\qquad X_{C,\text{series}} = X_{C_1} + X_{C_2} + X_{C_3} + \cdots$

Chapter 14. Electrical Measurements and Circuits

and in *parallel*

(14-30) $$\frac{1}{X_{C,\text{parallel}}} = \frac{1}{X_{C_1}} + \frac{1}{X_{C_2}} + \frac{1}{X_{C_3}} + \cdots$$

A changing current produces a changing magnetic field, which induces a potential that opposes the change. The relation between the induced potential and the current is

(14-31) $$e = -L\frac{di}{dt}$$

$L = $ inductance

To increase L, an inductance is generally made in the form of a coil. (It may also have an iron core to increase L further if it is designed for low frequency applications.) However, a conductor need not be coiled to have an inductance. A length of straight wire has a slight inductance which may be ignored for most purposes except at high frequencies. Equation (14-29) for the inductance resembles equation (14-22) for the capacitance with current and voltage interchanged and with a negative sign. If a sinusoidal AC current $i = I_0 \sin(\omega t)$ is passed through an inductance, the induced voltage will be

(14-32) $$e = -2\pi f L I_0 \cos(2\pi f t) = -\omega L I_0 \cos \omega t$$

This expression is comparable to equation (14-22) for the AC current across a capacitor. It shows that if the current is sinusoidal, the induced voltage is also sinusoidal; (2) the current lags the voltage by 90°; and (3) even though there is no resistance, the voltage drop is not zero. For induced (or applied) voltage, the corresponding current is limited and proportional to the voltage, according to

(14-33) $$I_0 = \frac{E_0}{\omega L} \qquad I = \frac{E}{\omega L}$$

The quantity

(14-34) $$X_L = 2\pi f L = \omega L$$

is called the *inductive reactance* and is a measure of the restriction of current flow by an inductance, the relationship being

(14-35) $$I = \frac{E}{X_L}$$

Again, although inductive reactance has the same dimensions as resistance and is measured in ohms, it is not resistance. Because current and voltage are 90° out of phase, there is no dissipation of energy across a perfect inductance. If complex notation is used for current and voltage, then $\mathbf{X}_L$, like $\mathbf{X}_C$, must be an imaginary number.

(14-36) $$\mathbf{X}_L = j\omega L$$

Note the difference in sign between X_L and X_C (equation 14-26) when the complex representation of current and voltage is used.

The formulas for finding the equivalent inductance or the equivalent inductive reactance of a network of inductances are the same as those for combining resistances (equations 14-2 and 14-3). The equivalent reactance of a network of capacitors and inductances can be computed with the same formulas, if the sign convention X_L positive, X_C negative, is retained. For reactances in series,

$$(14\text{-}37) \qquad X = X_{L_1} + X_{L_2} + X_{L_3} + \cdots - X_{C_1} - X_{C_2} - X_{C_3} - \cdots$$

For reactances in parallel,

$$(14\text{-}38) \qquad \frac{1}{X} = \frac{1}{X_{L_1}} + \frac{1}{X_{L_2}} + \frac{1}{X_{L_3}} + \cdots - \frac{1}{X_{C_1}} - \frac{1}{X_{C_2}} - \frac{1}{X_{C_3}} - \cdots$$

If the total reactance is positive, the circuit has inductive reactance and the voltage leads the current by 90°; if the total reactance is negative, the circuit has capacitive reactance and the voltage lags the current by 90°. In a purely reactive circuit there are no other possibilities for the phase angle. (In any real circuit, of course, there will be some resistance.)

Two important cases of a circuit containing one capacitor and one inductance are shown in Figure 14-11. In circuit (a), because the reactive elements are in series, the reactance is

$$X = 2\pi f L - \frac{1}{2\pi f C}$$

There is one frequency, the *resonance frequency*, given by

$$(14\text{-}39) \qquad f = \frac{1}{2\pi\sqrt{LC}}$$

FIGURE 14-11 / Series (a) and parallel (b) LC circuits.

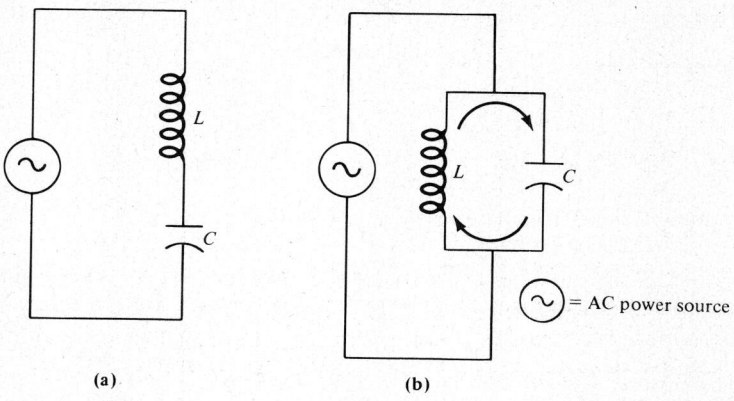

Chapter 14. Electrical Measurements and Circuits

at which the reactance is zero. The only restriction to the flow of current would be any resistance in the circuit. In circuit (b), where the reactive elements are in parallel,

$$\frac{1}{X} = \frac{1}{2\pi f L} - 2\pi f C \quad \text{or} \quad X = \frac{2\pi f L}{1 - 4\pi f^2 LC}$$

In this case, the reactance is infinite at the resonance frequency given by equation (14–39). In an idealized circuit, without resistance, this combination blocks current from an AC source at the resonance frequency. Note, however, that the capacitor and inductance are in series in the loop indicated by the arrows and that, therefore, in an idealized circuit without resistance, at the resonance frequency AC will surge back and forth indefinitely in this loop.

Any real circuit (if we neglect superconductivity) contains resistance, and so the idealized conditions above do not hold. The resistance in real circuits may be combined with the reactance to give the quantity called *impedance*, whose symbol is Z. Impedance replaces resistance in the generalized form of Ohm's law that applies to AC circuits.

(14–40) $$I = \frac{E}{Z}$$

The most general and powerful method for solving the problems of AC circuitry is to consider Z, E, and I to be complex quantities, with Z in the form

(14–41) $$\mathbf{Z} = R + jX$$

In equation (14–41) **Z** is the complex impedance, R the equivalent resistance, and X, which may be either positive or negative, is the equivalent reactance of a circuit or of some portion of a circuit. Because **Z** is complex and not a pure imaginary number, the phase angle (the amount by which the current leads or lags the voltage) is less than 90°. Instead of using the complex representation for **E**, **I**, and **Z** in equation (14–40), the magnitude of **Z** may be used to calculate the magnitude of the current from the voltage or vice versa, and the phase angle may then be calculated separately. Z, the magnitude of the complex quantity **Z**, is

(14–42) $$Z = \sqrt{R^2 + X^2}$$

The phase angle, ϕ, is

(14–43) $$\phi = -\arctan \frac{X}{R}$$

Note that R and X are the equivalent resistance and equivalent reactance of the portion of the circuit under consideration. That is, if resistance and reactance elements with these values were placed in series and substituted for the portion of the circuit being studied, the impedance would be the same.

PRINCIPLES

Complex impedances may be combined like resistances. In series,

(14-44) $$Z_{\text{series}} = Z_1 + Z_2 + Z_3 + \cdots$$
$$= (R_1 + jX_1) + (R_2 + jX_2) + (R_3 + jX_3) + \cdots$$

In parallel,

(14-45) $$\frac{1}{Z_{\text{parallel}}} = \frac{1}{Z_1} + \frac{1}{Z_2} + \frac{1}{Z_3} + \cdots$$
$$= \frac{1}{R_1 + jX_1} + \frac{1}{R_2 + jX_2} + \cdots$$

To illustrate the use of complex numbers, the series and parallel circuits in Figure 14–12 are analyzed below. These are the circuits of Figure 14–11, with resistance added. For the series circuit (Figure 14–12a),

$$Z = R + j(X_L - X_C) = R + j\left(2\pi f L - \frac{1}{2\pi f C}\right)$$

The resonance frequency, $f_{\text{res}} = 1/2\pi\sqrt{LC}$, is the same as that for the series circuit without resistance (equation 14–39). At resonance, the current and voltage are in phase and the current is limited by the resistance R. The same holds true for the loop indicated by the arrows in Figure 14–12b. For the parallel circuit (Figure 14–12b),

$$\frac{1}{Z} = \frac{1}{R + jX_L} - \frac{1}{jX_L} = \frac{1}{R + j2\pi f L} + j2\pi f C$$

and

$$Z = \frac{(R + jX_L)(-jX_C)}{R + jX_L - jX_C}$$

FIGURE 14–12 / Series (a) and parallel (b) RLC circuits.

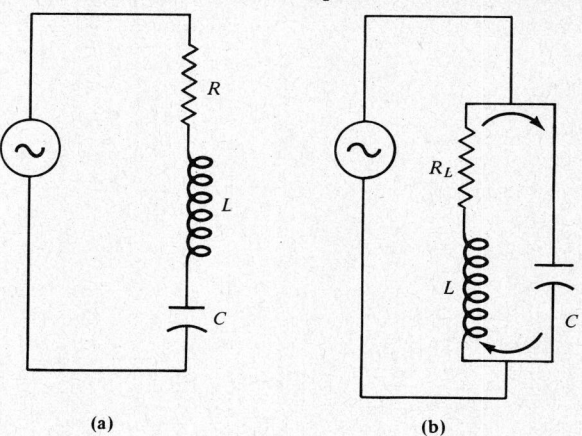

(The reader may, if he or she wishes, convert $\mathbf{Z}$ into real and imaginary parts, substitute $2\pi fL$ and $1/2\pi fC$ for X_L and X_C, and differentiate to find the frequency for which the magnitude of $\mathbf{Z}$ is a maximum.)

Because of the resistive terms Z will not become infinite ($1/Z = 0$) at any frequency. The frequency at which the impedance is a maximum will, in general, not be equal to $1/2\pi\sqrt{LC}$; however, if the resistances are small compared to the reactances, the difference will be small. At low frequencies most of the current will pass through the inductive branch and $\mathbf{Z} \approx R + jX_L$; at high frequencies most of the current will pass through the capacitative branch and $\mathbf{Z} \approx -jX_C$. Resistance associated with a capacitor would parallel the capacitance but can usually be neglected.

In some simple cases it is easier to compute the absolute value of impedance and phase angle. For the series circuit (Figure 14–12a)

(14–46) $\qquad Z = \sqrt{R^2 + (X_L - X_C)^2}$

and

(14–47) $\qquad \phi = -\arctan \dfrac{X_L - X_C}{R}$

For the parallel circuit (Figure 14–12b),

(14–48) $\qquad Z = \sqrt{\dfrac{(RX_C{}^2)^2 + [X_C X_L(X_C - X_L) - R^2 X_C]^2}{[R^2 + (X_C - X_L)^2]^2}}$

and

(14–49) $\qquad \phi = \arctan \dfrac{X_L(X_C - X_L) - R^2}{RX_C}$

Filter Circuits

Many electrical signals consist of AC plus DC or a mixture of AC of various frequencies, and it is often necessary to separate the AC from the DC or to separate the high frequency and low frequency components. One way of doing this is with filter circuits (Figure 14–13) that make use of the dependance of reactance on frequency. Consider the circuit shown in Figure 14–13a with an input across AC and the output across BC. From Ohm's law the voltage across AB is $E_{AB} = IR$. Similarly, $E_{BC} = IX_C = I/\omega C$. Since AB and BC are in series, the current through each must be the same and in phase. The voltage E_{AB} must be in phase with I because R is a pure resistance; the voltage E_{BC} must lag I by 90° because C is a pure capacitance. The relation between E_{AB}, E_{BC} and their vector sum, the input E_{AC}, is shown in the vector diagram (Figure 14–14). At low frequencies, E_{BC} approaches E_{AC} (Figure 14–14a). As the frequency is increased,

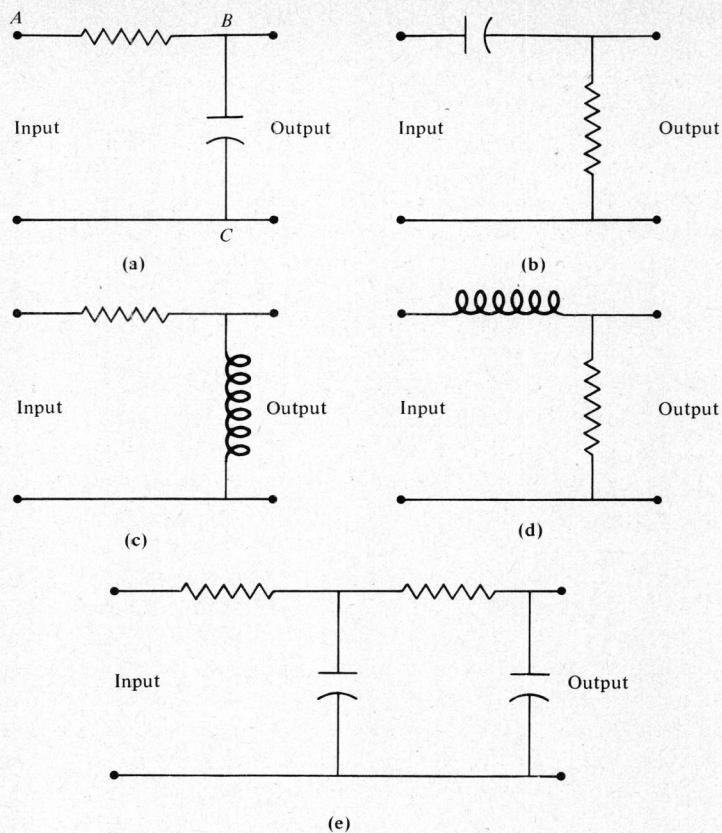

FIGURE 14–13 / Filter circuits.

FIGURE 14–14 / Vector diagrams for low-pass filter shown in Figure 14–13a: (a) low frequency; (b) high frequency.

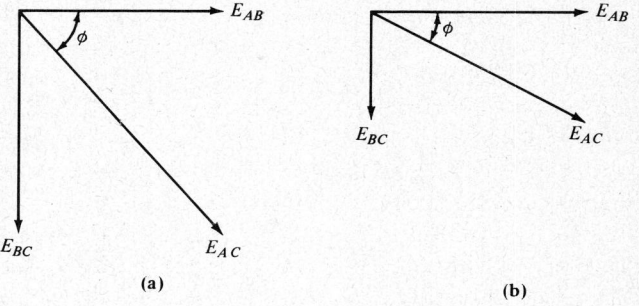

E_{BC} becomes increasingly small compared with E_{AC} (Figure 14–14b). Physically, the capacitor in this *low-pass filter* acts as a shunt for high-frequency signals but not for low-frequency signals. Similar analyses will show that the circuit shown in Figure 14–13d is also a low-pass filter, and that the circuits in Figure 14–13b and c are *high-pass filters*. The sharpness of the cutoff between high and low frequencies can be improved by cascading filters as in Figure 14–13e, but at the expense of increased loss of signal. Active filters, especially those using operational amplifiers, are used where sharp cutoff without loss is required.

Measurement of AC Current, Voltage, Power, and Resistance

Many AC instruments, especially at low frequencies, resemble DC counterparts in that they consist of some form of galvanometer movement with a suitable shunt or multiplier. Much of the theory of DC instrumentation can be applied to AC with little modification. The resistors in shunts, multipliers, and bridges must be carefully wound with the wire doubling back on itself to be noninductive. This problem obviously does not arise with DC instruments. D'Arsonval-type (moving coil) galvanometer movements cannot be used directly because the magnetic poles of the coil reverse as the current reverses to give an average torque of zero. There are d'Arsonval-type instruments with the input through a low-impedance bridge rectifier and with a movement with enough inertia to smooth out the ripple. The precision of these instruments is comparable to that of their DC counterparts. Moving vane movements are widely used for AC measurements because the attraction on the iron armature does not depend on the polarity of the magnetic coil. Like their DC counterparts, they are rugged, inexpensive, rather insensitive, and rather inaccurate. Many AC instruments have an electrodynamometer movement. This is similar to a moving coil DC movement, with the essential difference that the fixed magnet is replaced by a stationary coil. In a voltmeter the two coils are in series with each other and with the multiplier resistance. Because the polarity of the two magnetic coils reverses synchronously, there is a net torque on the moving coil. At zero displacement this torque is proportional to the square of the rms current through the coils. In an ammeter, the two coils are in series with each other and in parallel with the shunt. Again, the net torque at zero displacement is proportional to the square of the rms current through the coils. Electrodynamometer movements are also used in *wattmeters*. Here, the fixed coil is connected like an ammeter into the circuit and therefore has a magnetic field proportional to the instantaneous current. The moving coil is placed in series with a multiplier resistance and connected across the potential like a voltmeter, and therefore has a magnetic field proportional to the instantaneous voltage.

The torque is proportional to the product of *e* and *i*, and consequently the deflection indicates the power. The power factor can be estimated by dividing the power measured by a wattmeter by the product of the current and voltage measured separately.

There are no AC potentiometers. However, electronic instruments are available whose accuracy is limited only by noise.

FIGURE 14–15 / Schering (a) and Wien (b) bridges.

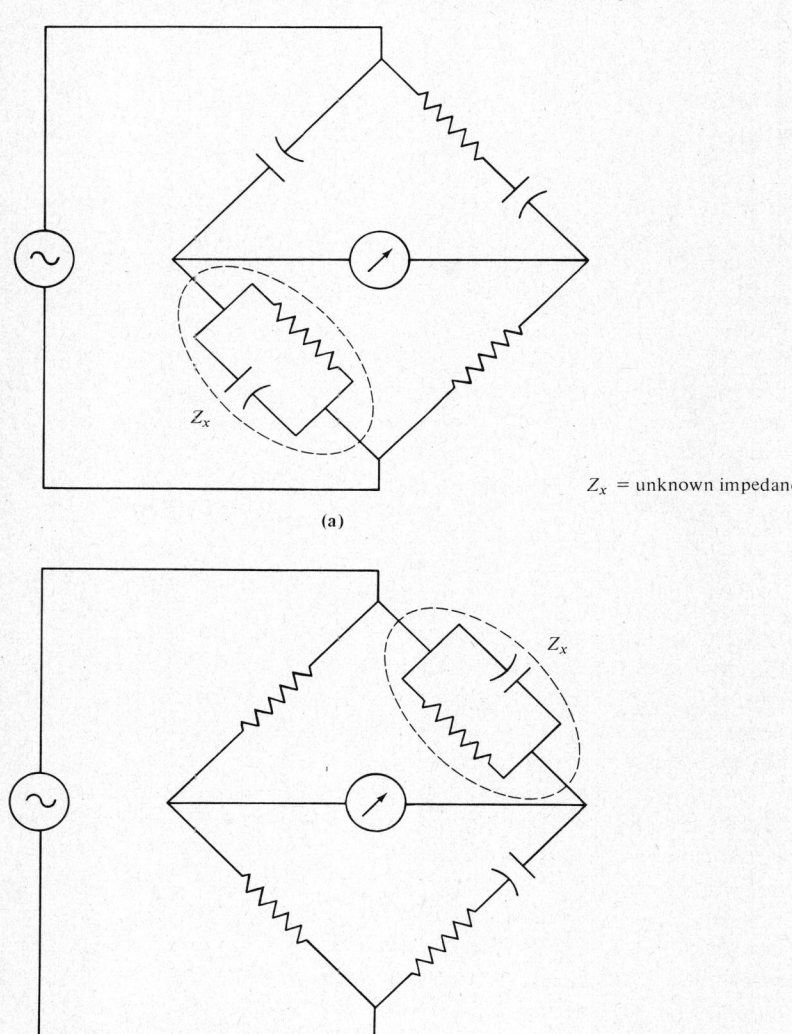

Z_x = unknown impedance

Resistance is commonly measured with DC. In some cases, for example when the ohmic resistance depends on the frequency, AC measurements may be necessary. To measure the resistance of electrolytic solutions AC, commonly 1000 Hz, is used in order to minimize polarization and electrolysis at the electrodes. For an approximate value of the resistance a Wheatstone bridge with noninductively wound resistors and an AC galvanometer can be used. Because of the capacitance associated with the cell, the current through the galvanometer does not go to zero at balance. Instead, there is a region where the current through the galvanometer does not change, the principal effect of adjusting balance being to change the phase angle. Impedances must be matched to achieve a true balance, and therefore adjustable capacitors as well as adjustable resistors are included in the bridge. Two common AC bridge circuits that can be used to measure both resistance and capacitance are shown in Figure 14–15.

Measurement of Inductance, Capacitance, and Dielectric Constant

Inductance and capacitance may be measured with impedance bridges designed for the purpose. As with an AC resistance bridge, the impedance must be matched because both inductances and capacitors inevitably have some associated ohmic resistance. Inductance measurements are of little importance in most branches of physical chemistry. Capacitance measurements are important, partly because they are used to determine the dielectric constant.

In operating a capacitance bridge, the unknown capacitor is connected to the appropriate terminals of the bridge and both the variable resistor and the variable capacitor are adjusted until the detector signal is at a minimum. The capacitance is the reading on the dial of the variable capacitor. For changes in the instrument range, additional fixed capacitors may be switched into or out of the appropriate bridge arms. (This is the same procedure that is used to determine the value of an unknown resistor with a bridge, except that in such a case, resistors are switched in and out.)

An alternative method of determining capacitance is with a tuned or resonant LC circuit, such as the one shown in Figure 14–16. This circuit contains a calibrated standard variable capacitor. The frequency of the oscillator is adjusted until the circuit is in resonance (equation 14–39), as shown by a minimum impedance and a maximum current. The unknown capacitor is then connected in parallel, increasing the capacitance and throwing the circuit out of resonance. The standard capacitor is then adjusted to restore the resonance. The capacitance removed from the variable capacitor is equal to that of the unknown capacitor.

In a more accurate version of the resonance method, called the *heterodyne*

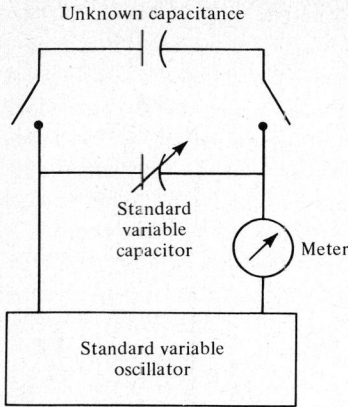

FIGURE 14–16 / Resonant LC circuit for measuring capacitance.

beat method, capacitance is determined by a frequency measurement. The output of a variable oscillator is tuned and coupled with that of a standard oscillator, such as a tuning fork or crystal, which has a fixed frequency. The difference, or beat frequency, between the two is measured with an appropriate meter. The relationship is

(14-50) $\quad f_b = |f_t - f_0|$

f_b = beat frequency
f_t = variable frequency
f_0 = standard fixed frequency

The beat frequency is much lower than either of the other two, usually close to zero, so that even if it is not determined accurately, the frequency of the variable oscillator is known almost as accurately as that of the standard oscillator.

The output of the variable oscillator depends upon capacitance, the capacitance being supplied by a standard variable capacitor and the unknown capacitance. With the unknown out of the circuit, the variable standard capacitor is first adjusted until the desired beat frequency is obtained. Then the unknown capacitance is placed in parallel with the variable standard, which is then readjusted to give the same beat frequency. The decrease in the variable capacitance is equal to the capacitance added from the unknown.

In all these measurements, the observed capacitance includes contributions from the leads and connections and from the edges of the plates. These must be either determined and subtracted from the observed capacitance or eliminated from the measurement. Only if the capacitance is very large may the capacitance of the leads be ignored. In physical chemistry, the dielectric constant is often more important that the capacitance per se. The dielectric constant is the ratio of the capacitance of a capacitor immersed in the medium to that of the capacitor in a vacuum. (The capacitance in air is close enough to that in a vacuum to be substituted for the latter in most dielectric constant measurements not involving gases.) Dielectric constants are important for calculating interionic attractions, ionic mobilities, dipole moments, and the like.

The differential capacitance is the change in capacitance when the area of the plates is varied without changing the leads or varying the edge effects. With overlapping, leaf-type parallel plate capacitors, this may be done by rotating the plates to change the overlap. With the concentric cylinder type, the inner cylinder may be withdrawn to a greater or lesser extent. When the capacitances obtained with two different overlap positions are subtracted from each other, the capacitance of the leads and the effect of the edges will cancel out, since these are constant.

Of the various methods of measuring capacitance described here, the resonance method is the least accurate, but the equipment is the least expensive. The heterodyne beat method is the most accurate. In fact, the accuracy in measuring the beat is probably greater than the reproducibility of the positions of the capacitor plates. In general, however, unless the laboratory is going to specialize in precision dielectric constant measurements, the impedance bridge method is much preferred. It is quick, easy, and accurate to at least 0.1%, depending on the bridge. Moreover, the bridge itself is a valuable general purpose tool.

References

Brophy, I., *Basic Electronics for Scientists*, 2nd ed. McGraw-Hill, New York, 1972.

Herrick, C. N., *Instruments and Measurements for Electronics*. McGraw-Hill, New York, 1972.

Malmstadt, W. V., C. G. Enke, and E. C. Toren, Jr., *Electronics for Scientists*. Benjamin, New York, 1962.

Malvino, A. P., *Electronic Instrumentation Fundamentals*. McGraw-Hill, New York, 1967.

Spitzer, F., and B. Howarth, *Principles of Modern Instrumentation*. Holt, Rinehart and Winston, New York, 1972.

Weissberger, A., and B. W. Rossiter, *Physical Methods of Chemistry*, Part IA and Part IV, chs. 5, 6. Wiley–Interscience, New York, 1971 and 1972.

Chapter 15

Electrical and Electronic Devices

In this chapter several types of electrical or electronic devices of importance to chemists will be discussed. These include power supplies, amplifiers, devices to measure, display, or record transient signals, and digital meters. For current specifications of power supplies, operational amplifiers, and so on, the manufacturers' catalogs provide the best source of up-to-date information.

Power Supplies

The chief DC power sources are batteries and AC-to-DC converters, such as rectifiers and electronic power supplies. Batteries (electrochemical cells) are a convenient source of low voltage DC. Since transistors and integrated circuits require low voltages and need little power, batteries are often used to operate devices using these components. The advantages of batteries include portability, simplicity, and low noise. Moreover, with the advent of the alkaline cells, nickel/cadmium cells, and mercury cells, a much wider range of properties is available than was the case when only zinc/carbon dry cells and lead/acid storage cells were available. On the other hand, batteries are bulky and heavy, have low output per unit cell, and lose large amounts of energy dissipated in voltage dividers or by the series resistors needed to vary the voltage. They also have a limited lifetime, tend to run down toward lower voltages with use, and have become very expensive.

The voltages of the main types of cell are

Cell	Voltage
Zinc/carbon	1.5
Alkaline	1.5
Mercury	1.35–1.40
Nickel/cadmium	1.25
Lead/acid	2.0

Cells may be connected in series to provide higher voltages and either made larger or connected in parallel to provide greater capacity and to increase the maximum allowable current. Capacity is usually measured in ampere-hours, the product of the current at which the battery is operating and the time that the battery can operate, at that current, before the voltage falls below a minimum acceptable value, usually about 80% of the rated value.

All batteries have internal resistance, which may be thought of as a resistor in series with the voltage source. Consequently, as the current increases, the voltage across the terminal drops. The internal resistance often increases drastically as the battery reaches the end of its useful life or if it is abused by excessive currents being drawn.

The lifetime of batteries is greater at low discharge rates, except in alkaline cells, which are therefore preferred for high current applications. (The silver oxide/zinc cell has a very high capacity at high discharge rates, but it is expensive and therefore limited to specialized uses.) Battery lifetime is always finite, even on the shelf. Zinc/carbon dry cells have a relatively short shelf life. Mercury batteries, on the other hand, have a moderately long shelf life and an almost constant voltage during the cell lifetime. At the end of the cell lifetime, however, the voltage drops precipitously. Because of this voltage–cell life relationship, mercury batteries are widely used to provide secondary reference voltages in pH meters and other devices. (Even mercury batteries, although they remain usable for years, should be kept in a cold dry container or room, if they are to be stored for more than a few months.) Alkaline batteries and lead/acid cells also have a relatively constant voltage. Nickel/cadmium and

FIGURE 15–1 / Basic rectifier circuits: (a) half-wave rectifier; (b) bridge rectifier; (c) transformer rectifier.

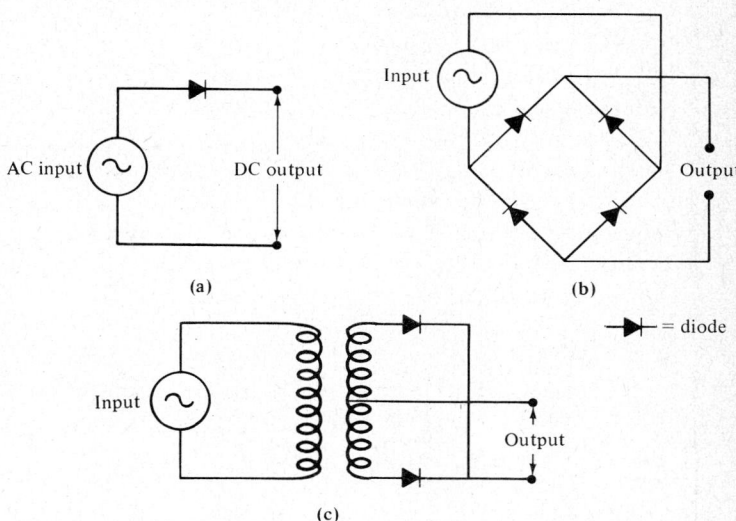

169

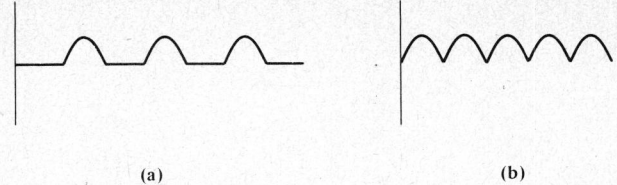

FIGURE 15-2 / Rectifier output: (a) half-wave rectified current; (b) full-wave rectified current.

lead/acid batteries have the important advantage of being rechargeable. Most other common types of battery cannot be recharged, for all practical purposes.

There are a number of circuit components, such as solid state semiconductors, vacuum tubes, gas-filled tubes, and so on, that permit current to pass in only one direction. Such components are used to *rectify* AC to DC. Figure 15-1 shows the three basic rectifier circuits. Figure 15-1a is the *half-wave rectifier*, Figure 15-1b is the *full-wave bridge rectifier* and Figure 15-1c is the *full-wave center-tap transformer rectifier*. Figures 15-2a and 15-2b show the output of these devices, *half-wave* and *full-wave* rectified current, respectively. The variations in output current can also be considered as *ripple*, or varying DC, and must have the variations removed by appropriate filter devices to produce constant DC.

Voltage Regulation

Electronic power supplies can be designed to a very wide variety of specifications. They can be made extremely stable, even with large fluctuations in the input power. Supplies with an adjustable output of several thousand volts, a ripple of less than 0.01% and a drift of less than 0.0001% per day, operating on 60 Hz AC between 100 and 125 V are available. Electronic supplies of various types have made possible much of today's sophisticated research equipment.

With ordinary, nonelectronic power supplies, there is often a problem in maintaining the current or the voltage when the resistance of the load is changed. Figures 15-3, 15-4, and 15-5 illustrate several methods of control.

Figure 15-3 shows two *passive* circuits. In the first, Figure 15-3a, the resistance of the ballast in series is much larger than that of the load. Consequently, moderate changes in load resistance will not have much effect on the current or resistance. In Figure 15-3b, if the resistance of the load is much greater than R_1, the divider resistance in parallel with it, the fraction of the source voltage applied to the load will be primarily determined by the ratio

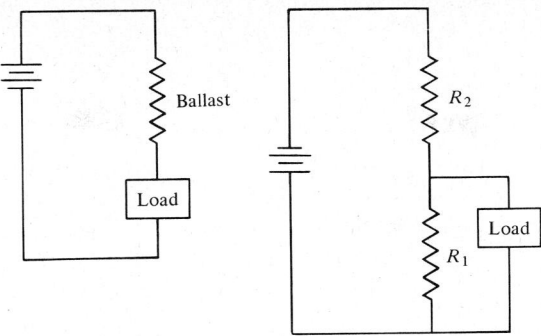

FIGURE 15–3 / Circuits for passive regulation: (a) constant current source; (b) constant voltage provided by a voltage divider.

$R_1/(R_1 + R_2)$ and will not change greatly with moderate changes in load resistance. These passive circuits can also be used with AC sources.

The battery-ballasted power supply (Figure 15–4) was widely used as a source of stable high current, low voltage DC before reliable inexpensive electronic power supplies became available for this purpose. When the load circuit is closed, there is a small steady drain on the battery, most of the current coming from the charger. Because the battery is being discharged very slowly, its voltage and internal resistance characteristics will remain virtually unchanged for hours. When the load circuit is open, the charger slowly recharges the battery.

Figure 15–5 is a block diagram of a DC power supply operating from line power and stabilized with a *Zener diode*. A Zener diode is a solid state device that has two very different resistance characteristics, depending on the voltage applied across it. At voltages exceeding a fixed value, the *breakdown voltage*, the resistance is very low, and at voltages less than the breakdown voltage, the resistance is very high. Therefore, the voltage supplied to any load placed in parallel with a Zener diode will be maintained very precisely at the breakdown voltage. The output of a power supply, such as that shown in Figure 15–5, can be regulated to better than 1%, even though the line voltage changes by 10%.

FIGURE 15–4 / Battery-ballasted power supply.

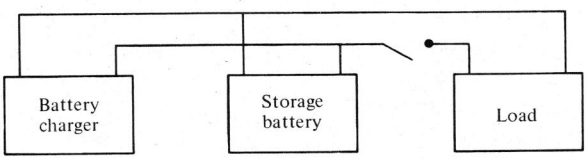

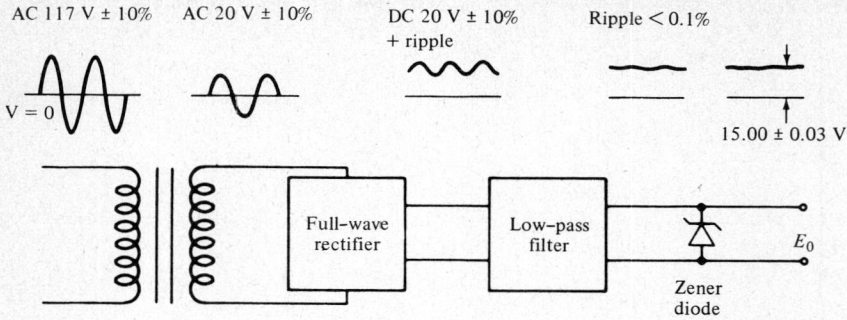

FIGURE 15-5 / Direct current power supply operating from line voltage. This Zener diode has very high resistance for voltages less than 15 V and essentially zero resistance for voltages above 15 V. Hence the voltage drop is always 15 V across the diode (with about a 2% change for a 100% change in input voltage, hence a 0.2% change for a 10% change, giving 15.00 $\pm$ 0.03 V).

To vary the output voltage, the transformer is varied to produce AC of other voltages and the Zener diode is changed. Zener diodes are available with breakdown voltages ranging from a few volts to a few hundred volts, but those that provide high voltages are rather expensive. However, several can be connected in series to provide high voltages. If they are connected in parallel, the voltage will be that of the diode with the lowest breakdown voltage. Zener diodes are very useful secondary sources of reference voltage and are gradually replacing mercury cells for this purpose.

FIGURE 15-6 / Fixed ratio transformer (a) and variable ratio autotransformer (b). The same coil is used for both primary and secondary.

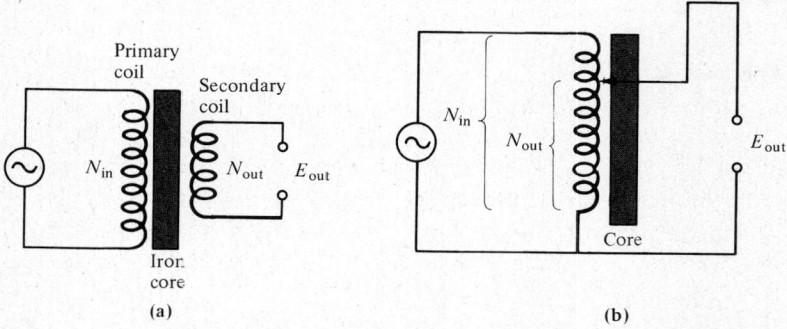

The electric power commonly available in the United States is 110–120 V rms (nominally 115 V), 60 Hz, single phase. If the facility is wired for three-phase current, electricity at 208 or 230 V will also be available. (The standard in Europe is 230 V, 50 Hz.) The frequency is maintained accurately, but the voltage varies from place to place and fluctuates irregularly during the day as the load on the system changes. To minimize line losses, the power factor in the system is maintained close to 1. However, if there are large reactive loads, it may be considerably less than 1, locally. Fixed or variable transformers (Figure 15–6) are used to change the voltage. The current in the primary cell produces a changing magnetic flux which induces a voltage in the secondary coil. The induced voltage is related to the inducing voltage by the ratio of the number of turns in one coil to that in the other.

(15-1) $$E_{out} = E_{in} \frac{N_{out}}{N_{in}}$$

N = number of turns in the coil

Because of self-inductance, the power in the primary depends on the load across the secondary. On open circuit, transformers draw very little power. Most transformers have a fixed number of turns with possibly several terminals to permit several input and output voltages. Continuously variable transformers with ratings up to 10 A (Variac, Varitran, Powerstat) are extremely useful for delivering any desired voltage from 0 to 130 V, using a 115 V house line as source. The efficiency of such power transformers under rated load is 99% or better. Constant voltage (Sola) transformers, which automatically supply an almost constant output voltage despite variations of 15–20% in the input voltage, are used as heavy duty power supplies for devices that require a fixed input voltage. These are much larger, much more expensive, and much less efficient than ordinary power transformers.

DC-to-AC Conversion

There are no simple circuit components, analogous to rectifiers, to convert DC to AC. Such a conversion is done mechanically, by a DC motor driving an alternator, a DC motor driving a rotary switch that periodically reverses the potential, a chopper or interrupter that periodically opens and closes the circuit, as in a doorbell, or by electronic voltage to frequency converters. Usually this is to measure the DC accurately or for transmission of information. DC-to-AC conversion is rarely part of a power supply, although AC-to-DC-converted power supplies are used regularly.

Amplifiers

An *amplifier* is a device to produce a large signal, the *output*, from a small signal, the *input*. An ordinary laboratory thermostat is an example of an amplifier, in that the very small current through the mercury thermoregulator is used to turn the heater current on or off. For many purposes the on–off yes–no amplifier of a thermoregulator is insufficient. Instead, it is necessary to have an output that is a more or less faithful copy of the input.

In the following discussion, electronic amplifiers, with one exception, will be considered as "black boxes" with certain measurable properties. How they function is not of interest here. What concerns the physical chemist is what amplifier characteristics are and what they can do.

The properties of interest are the gain, input impedance, output impedance, frequency response, output (current, voltage, and power), stability and noise, linearity, and distortion.

POWER GAIN

This is the ratio of the output power to the input power. The power gain of solid state devices can be anywhere between unity and 10^{13}, depending on the particular amplifier. Voltage gain and current gain are similarly defined.

INPUT IMPEDANCE

As shown in Figure 15–7, the input impedance is in parallel with the source impedance. A high input impedance is advantageous because it reduces the danger of overloading the source. In general, an input impedance of 100 times the source impedance is sufficient to prevent loading. The input impedance and source impedance are analogous to the load resistance and internal resistance of a battery. At the time of writing, the input impedance in DC circuits (which is purely resistive) can be as high as 10^{13} ohm with solid state circuits and over 10^{16} ohm with special electrometers. The problem of appropriate insulation, as stated in Chapter 14, arises with such high resistances.

In AC circuits, obtaining high input impedance is much more difficult because of the capacitance of the wires, the effect being known as *distributed*

FIGURE 15–7 / Equivalent circuit for input to an amplifier.

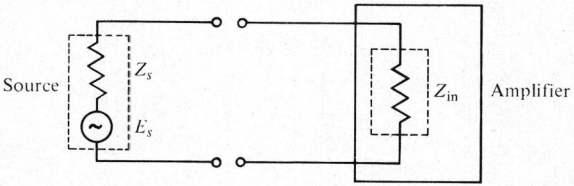

capacitance. Since each pair of leads has some capacitance in parallel with the input resistance.

$$\frac{1}{|Z|} = \left(\frac{1}{X^2} + \frac{1}{R^2}\right)^{1/2}$$

If R is much greater than X, $Z = X$. If the leads have 10 pF capacitance, the capacitative reactance

$$X_C = \frac{1}{2\pi f \times 10^{-11}} = 1.6 \times 10^{10} f^{-1}$$

If, for example, the frequency is 1.6 kHz, then $X_C = 10^7$ ohms and cannot be made larger, no matter how large the input resistance is. This may or may not be sufficient. For some special applications in which it is necessary to operate in the megahertz range, the input impedance drops to 10^3–10^4 ohms.

OUTPUT IMPEDANCE

This should be low enough to prevent being loaded by whatever circuitry follows. Fortunately, it is usually no problem to provide for a low output impedance. However, the load impedance should be *matched* to the output impedance to get maximum useful *power* output. Consider the schematic circuit in Figure 15–8. With DC, impedance and resistance are identical and the load power is

$$P = IE_{\text{load}} = IE_{\text{out}}$$

From elementary circuit theory,

$$P = \frac{E_{\text{out}}^2 R_{\text{load}}}{(R_{\text{out}} + R_{\text{load}})^2}$$

Assume that E_{out}, the output voltage, and R_{out}, the output impedance, are fixed and are characteristic of the amplifier, so that P depends only on R_{load}, the load

FIGURE 15–8 / Amplifier with load.

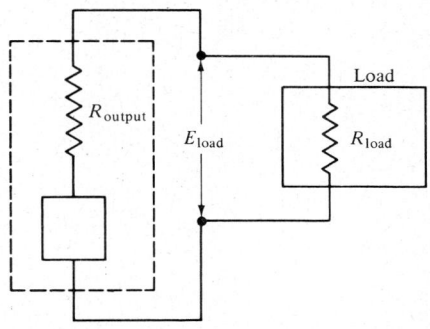

impedance. The maximum of this expression and therefore the maximum useful power from the amplifier occurs when

(15–2) $\qquad R_{out} = R_{load}$

Note that this does not mean that the full output voltage is seen by the load when $R_{load} = R_{out}$. The voltage transfer could be greatly increased by having R_{load} much greater than R_{out} but under these conditions, the useful power supplied by the amplifier would be much less than the maximum.

FREQUENCY RESPONSE

There are limitations to the frequencies that any amplifier can handle. There is always an upper limit and, unless DC can be amplified, a lower limit. The limits, called the *cutoff* frequencies, are arbitrarily taken as that frequency at which the power gain has dropped by a factor of 2; that is, the voltage gain has dropped by $\sqrt{2}$. The bandwidth is the difference between the upper and lower limits.

For general purpose amplifiers and pulse amplifiers, wide bandwidth is desirable. The advantages of wide frequency response include the ability to amplify a wide assortment of signals, including single pulses, with a minimum of distortion. The main disadvantage is that a wider band of noise is included, along with the signal. Sometimes an amplifier circuit is tuned to a single frequency, with all other frequencies attenuated.

MAXIMUM POWER OUTPUT

The power requirements of ordinary electronic circuits may vary from the 50 W or more needed to operate high-fidelity loudspeakers to the 10 mW required to run a meter. In the interests of economy, it is desirable to match the amplifier to its application. It is always important to prevent overloading and drawing excess power, which may cause distortion or even burn out the amplifier.

STABILITY

Except for systems in which the output can be readily and repeatedly compared with a reference, stable characteristics, especially constant gain, are extremely important in most applications. The stability of an amplifier can be improved with negative feedback (see p. 180). Although almost anything can be done for a price, it is more difficult to make drift-free DC amplifiers because a DC amplifier will respond to a slowly drifting DC input. (Any component with a frequency of 1 Hz or less may be arbitrarily defined to be DC.) To avoid the problems of DC amplifiers, the DC may be chopped, that is, converted to AC (400 Hz is commonly used), amplified, and the amplified signal rectified. The amplifier is then designed for the chopper frequency. For example, in a double beam visible, uv or ir spectrophotometer, an AC signal may be generated by some mechanical device by alternately training first the reference beam and then the sample beam on the detector. In a pH meter, the signal is often interrupted by a stable mechanical chopper, such as a commutator turned by a synchronous motor.

LINEARITY AND DISTORTION

The output of the amplifier should be an accurate, amplified replica of the input signal. If, for example, the input was at 1 kHz and 80% of the output was at 1 kHz, 5% at 2 kHz, and 15% at 3 kHz, the output signal would give a distorted view of the input signal. This type of distortion is called *harmonic distortion* since it shows up at harmonics of the input frequencies. Another type of distortion is amplification that varies with the frequency. If, for example, the input had equal components at 1 and at 10 kHz, the output might have a 70:30 ratio. Other distortions include phase shift and phase blockage, or transmission of only part of a signal. Another type of nonlinearity is an output proportional to the square or other power of the input voltage, which sometimes occurs at high power levels. All these problems generally become more serious at high power drains. All must be guarded against, and the output signal must be checked to make sure that it is an accurate replica of the input.

A signal may sometimes be amplified so that the output is not proportional to the input; that is, the gain is not constant but instead has a known functional relation to the input. Logarithmic amplification is a common and important example. It is very necessary in these cases that the amplifier be stable and that the functional transformation be accurate.

NOISE

All signals are accompanied by random fluctuations in voltage, which are called *noise*. If the signal is small, it may be obscured by the noise. In addition to the noise occurring with the signal, there is also noise in the amplifier and other circuit components. Noise, therefore, is the limiting factor in measuring small signals. Normally, it is convenient to discuss noise in terms of the power per hertz. The power per unit bandwidth as a function of the frequency is termed the *power spectrum* of the noise.

There are several sources of noise. *Thermal noise* is always present. This corresponds to one-dimensional blackbody radiation by a resistor. At ordinary and elevated temperatures, the mean square noise voltage per hertz is $4kTR$, where k is the Boltzmann constant, T is the absolute temperature, and R is resistance. *Pickup* is 60 Hz hum from the line AC that is to be found everywhere. Wires, any wires, act as antennas. As a result, almost every circuit picks up spurious 60 Hz signals, unless precautions are taken. Short wires, preferably twisted together, are helpful. The lower the impedance, the less the pickup. In addition, harmonics of the 60 Hz noise at 120 Hz, 180 Hz, and so on, may be picked up. Sometimes, especially near radio transmitters, radio frequencies may be picked up. The best prevention is shielding. *Contact noise* arises at the contacts, either from thermal fluctuations (the Seebeck effect) or from poor soldering. *Drift* usually refers to long-term shifts, taking place over periods of seconds or longer and affects only DC characteristics of the component parts.

ELECTRON MULTIPLIER

The *dynode*, or *electron multiplier*, is a high-vacuum device for amplifying extremely weak currents. It has a number of plates, each of which is kept appreciably more positive than the preceding one. The current, which may be either a stream of electrons from the usual filament or a stream of positive ions from a mass spectrometer, hits the first plate with sufficient energy so that each incoming charged particle knocks out several electrons. These electrons are attracted to the second plate with sufficient energy so that each in turn knocks out several electrons, each of which in turn strikes the next plate. The effect cascades from plate to plate. The current collected from the last plate can be 10^6 times the entering current. Photomultiplier tubes work on a similar principle, except that the electrons for the first stage are produced by photons hitting the first plate.

SELECTION OF AN AMPLIFIER

There is no ideal all-purpose amplifier. Different circuits must be designed for different applications. A circuit that would properly amplify a 1 V signal to 1 kV should not be expected to amplify a 10^{-3} V signal to 1 V. As an example of the selection of an amplifier, consider the requirements for a pH meter that uses a glass electrode.

The voltage between the two electrodes is usually from a millivolt to a volt. The glass electrode typically has a 10^7 ohm impedance. The power produced is therefore as low as 10^{-10} W. This is much too small to run a d'Arsonval meter, so it is fed into an amplifier and the amplified signal is used to run a meter or to move the pen on a recorder. Since the glass electrode has a 10^7 ohm impedance, the necessary input impedance is usually 10^9 ohms or more. Since pH measurement is a slow process, there is no need to have a high frequency response. Therefore, a DC amplifier can be used and the input capacitance can be relatively high. To operate a meter or a recorder, the output impedance should be 10^2 ohms, so a 1 mA output would generate 10^{-4} W. In sum, the input impedance should be at least 10^9 ohms, the output impedance should be 10^2 ohms, the amplifier can have a high input capacitance, it need not respond to rapidly varying signals, and the power gain should be at least 10^6. An alternative approach, as mentioned above, is to chop the input and use an AC amplifier. The input and output impedance requirements are the same.

FEEDBACK

A system with *feedback* is any system with the property that the results of its operation are used to control its operation. A thermostatted water bath is an example of a system with feedback because the bath temperature is measured and used to control the heaters that maintain the temperature. Another example is an infrared spectrophotometer, where the difference in intensity between the measuring beam and the reference beam is used to attenuate the reference beam, making it match the measuring beam. The degree of attenuation is recorded, usually in the form of a graph of absorption of radiation versus wavelength, the spectrum. In another example, during an automatic titration, the change in pH per increment of titrant is used to slow down the addition of titrant as the endpoint is approached. Feedback is necessary in all control systems from a simple thermostat to an opera singer matching vocal sound with that of the orchestra.

Figure 15-9 is a schematic diagram of an amplifier circuit employing feedback. The relations among the amplifier gain, the signal, input and output voltages, and feedback, to a good approximation for large G, are

(15-3) $$G = \left| \frac{E_{out}}{E_{in}} \right|$$

(15-4) $$E_{in} = E_{sig} + BE_{out}$$

(15-5) $$\frac{E_{out}}{E_{sig}} = \frac{G}{1 - BG}$$

(15-6) $$E_{in} = \frac{E_{sig}}{1 - BG}$$

G = amplifier gain
B = fraction of the output that is fed back

FIGURE 15-9 / Amplifier employing feedback.

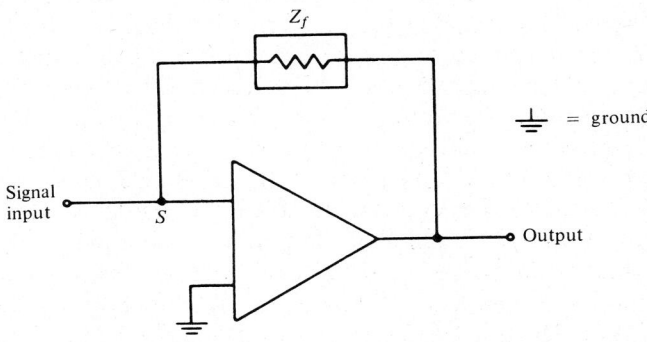

Feedback can be either negative ($B < 0$) or positive ($B > 0$). Negative feedback is more commonly used because it opposes the operation of the system, thereby increasing stability and providing control. From equation (15–5), the effective amplification is decreased by negative feedback. With high-gain amplifiers, where $-BG$ is much greater than 1, the effective amplification,

$$(15\text{–}7) \qquad \frac{E_{\text{out}}}{E_{\text{sig}}} = -\frac{1}{B}$$

becomes independent of amplifier gain. Under these conditions, the effective amplification is insensitive to fluctuations in the gain. Furthermore, it may be easily varied by varying B, the feedback, which is done, in practice, by altering Z_f, the feedback impedance. With negative feedback, the actual input voltage is decreased (equation 15–6). To maintain the same input current, the input impedance must be decreased proportionately; that is, it must be divided by the factor $(1 - BG)$.

Negative feedback increases input impedance and decreases output impedance, while positive feedback decreases input impedance and increases output impedance. In some amplifiers, the product of gain and bandwidth is constant, so that as the bandwidth goes up the gain drops off. Even in cases where the product of gain and bandwidth is not constant, the bandwidth still increases with negative feedback. Conversely, it decreases with positive feedback.

Positive feedback decreases the stability of the system. It is inadvisable for a discrete two-valued system, such as a thermostat, because it introduces instability (i.e., turning the heat *on* when the temperature goes *above* a set point). As shown by equations (15–5) and (15–6), an amplifier circuit with positive feedback will run away when $BG = 1$. When $BG < 1$, it will have a higher effective amplification than the corresponding amplifier without feedback, but at the cost of decreased stability. One use of positive feedback is in an oscillator. The action may be visualized by considering a mechanical counterpart, the swing, where small impulses in phase with the oscillations will gradually build up a large amplitude.

The amplifiers used in the type of circuit shown in Figure 15–9 are generally chosen for high gain and are called *operational amplifiers*. Because of the very high gain and the feedback, practically no current can enter the amplifier. Therefore, the sum of the currents at S must be zero and the feedback ensures that this is so. Any current increase from the input is immediately balanced by current fed back through Z_f. The signal in Figure 15–9 is a *current*, which may be provided from a voltage signal and an input impedance, Z_i. The output voltage, E_{out}, divided by the feedback impedance, Z_f, is equal to the feedback current, I_f, which balances the input current.

$$(15\text{–}8) \qquad \frac{E_{\text{out}}}{Z_f} = -I_{\text{sig}}$$

If the signal current is provided by a voltage source and input impedance,

(15-9) $$I_{in} = \frac{E_{sig}}{Z_{in}} - \frac{E_{out}}{Z_f} \approx 0$$

and

(15-10) $$E_{out} = -E_{sig}\frac{Z_f}{Z_{in}}$$

The gain in voltage, E_{out}/E_{sig}, is therefore determined only by the ratio of feedback impedance to input impedance. Combining equations (15-10) and (15-7),

$$B = \frac{Z_{in}}{Z_f}$$

In other words, this circuit amplifies voltage, with the gain being determined only by the choice of resistors. Without any input impedance, the input current can still be converted to a voltage,

$$E_{out} = -I_f Z_f$$

and it is generally easier to measure and manipulate voltage than current. Furthermore, Z_{in} and Z_f are not limited to resistors. Frequency dependent impedances, with capacitors, are used for active filters, oscillators, analog computers, and so on. The operational amplifier has become a versatile and valuable tool.

Oscilloscopes and Recorders

Transient signals fall into major categories, those slow enough to be recorded and displayed with mechanical devices, such as recorders, and those too fast for equipment with moving mechanical parts in real time. These latter must be displayed with an oscilloscope. Alternatively, they can be converted to digital form, stored in a computer memory, and used for computation or to control a plotter at a later time.

The oscilloscope is a versatile and useful device for observing the variation of a signal with time or with another signal. The shape of the voltage–time trace, the magnitude of the horizontal or vertical deflection, and the observed period of a signal all have important applications in experimental work.

The simplest type of oscilloscope consists of a cathode-ray tube, a sweep generator circuit, and a vertical input system. The electron beam produces a luminous spot where it strikes the face of the tube. The sweep generator moves the beam from left to right at a uniform rate. When the right-hand limit is reached, the beam is returned to the left-hand limit as rapidly as possible and the cycle is repeated. This provides a uniform time base for any repetitive process. Depending on the instrument, the sweep rate can be adjusted from less than 1 sweep per second to over 10^6 sweeps per second. The signal, amplified if necessary, is fed to the vertical input system, which deflects the beam up or down by an amount proportional to the incoming signal, thereby tracing a graph of the input. If the input signal is repetitive, the sweep can be synchronized with the pulse rate (i.e., the rate at which the input signal repeats). If the pulse rate, and therefore the synchronized sweep rate, are repeated more often than 25 times per second, the trace seems to the eye to be a stationary graph of the variation of the signal with time.

Oscilloscopes are available which display two or more signals, either alternately (chopped) or simultaneously. Two beams are needed for simultaneous dual trace presentation. This makes for an expensive instrument, since much of the electronics has to be duplicated. Modern oscilloscopes often come in modular form. There is a permanent section, or *main frame*, into which interchangeable special purpose units called *plug ins*, are put. This modular construction gives a high degree of flexibility, since the appropriate, relatively inexpensive, plug-in units can quickly and easily enable the scope to be adapted to many different measurements.

The sweep generator circuit may be disconnected and an external signal used to deflect the beam horizontally. With external signals deflecting the beam vertically and horizontally, the trace becomes an X-Y plot, which shows the functional relation between the signals. The Lissajous figures, produced when separate sine waves are connected to the horizontal and vertical inputs, are used to compare phase angles and frequencies. When the X and Y signals have the same frequency, the trace shows the phase relationship between them, as in Figure 15–10a. If the two frequencies are not the same, the curve shows the

FIGURE 15–10 / Phase (a) and frequency (b) relationships (Lissajous figures).

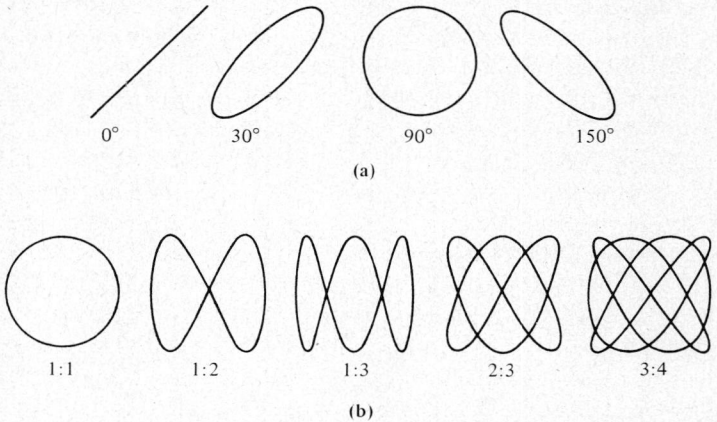

ratio of the frequencies, as in Figure 15-10b. The numbers in Figure 15-10b are the ratio f_X/f_Y of the two input frequencies. These ratios are obtained from the relationship

$$(15\text{-}11) \qquad \frac{f_X}{f_Y} = \frac{n_Y}{n_X}$$

n_Y = number of tangents to a single vertical line
n_X = number of tangents to a single horizontal line

The characteristics of the input amplifiers are critical in determining the usefulness of the oscilloscope. Input impedance is normally 1–10 megohms, in parallel with 10–50 pF capacitance. Sensitivity may be 5–50 mV/cm of deflection, which can normally be reduced all the way down to 100 V/cm with internal voltage dividers. Signal attenuators (i.e., external probes with voltage dividers), can make it possible to have 1000 V signals. These specifications may be improved with an external preamplifier, so that, for example, microvolt pulses may be exhibited on the screen. Only frequency response cannot be improved by a preamplifier. Almost all scopes can accept DC if necessary or can be connected through a capacitor (AC-coupled) to block DC signals. The high frequency response determines the maximum useful sweep rate. Inexpensive oscilloscopes go only to 100 kHz, while more expensive scopes can be used to measure nanosecond (nsec) pulses (1 nsec = 10^{-9} sec).

The trace can be photographed to record the pattern displayed on the oscilloscope. Polaroid film is especially useful, since one gets a permanent record almost immediately. "Memory" oscilloscopes have phosphors with long persistence, so that the trace of even a single sweep remains long enough to be measured directly.

Transient recorders are combined oscilloscopes and analog-to-digital converters with digital memory. The data are stored permanently, in digital form, and can be displayed at any later time or else transferred to a computer.

The strip chart recorder produces an accurate permanent record of transient signals that are slow enough to be measured with mechanical devices. In a typical recorder, chart paper passes beneath a pen at a known constant rate, thus providing a time axis. The pen moves back and forth across the chart, with a deflection that is proportional to the input signal, to draw a continuous graph of the signal. The performance of the recorder depends critically on the pen drive. One widely used design consists of a pen coupled mechanically to a self-balancing potentiometer which measures the amplified input. Various functions of the input can be graphed by using a nonlinear slidewire, such as the logarithmic slidewire often used to record the signal from a spectrophotometer as absorbance. Several signals can be graphed simultaneously with multiple pen recorders. The maximum speed of a signal that can be recorded with negligible distortion depends on the response time of the instrument. Good moderately priced commercial recorders may have a response time of 0.25–0.5 sec for full-scale deflection with 1% error.

In addition to strip chart recorders, which give a record of a measured quantity as a function of time, there are X–Y recorders which draw a graph of the functional relationship between two measured quantities. These have a fixed chart and a pen that is positioned by two independent orthogonal drives, each responding to its own input signal. Besides recording measurements in real time, X–Y recorders can be controlled by computers to draw graphs of stored data or computed functions and to letter the axes of these graphs. Because they need two amplifiers and two drives, X–Y recorders usually cost about twice as much as strip chart recorders.

Often it is necessary to measure and record a small transient change in a large constant signal, for example a 20 mV change in a 200 mV signal. If the measurement is made with an oscilloscope, the input can be AC coupled, eliminating the DC component and letting the relatively high frequency changes pass. Such AC coupling cannot be applied to signals measured with recorders, which generally operate below 1 Hz. Nevertheless, direct recording of the total signal leads to large errors in estimating the change. In the example above, if the measurements were made on a 200 mV scale, a measurement error of 1% of full scale would correspond to 2.0 mV. The error in the total signal would be only 1%, but there would be an error of 10% in the signal of interest. The simplest and most practical solution is to buck out the constant potential by feeding in an opposing constant potential of the same value as the signal to be eliminated. The difference between the opposing signals is the varying signal, whose measurement is desired. In the case of the example above, bucking out the constant potential with an opposed 200 mV potential would enable the result to be read on the 20 mV scale with 1% error.

Digital Instruments

In general, these have important advantages, especially in accuracy and ease of usage. In terms of accuracy, 0.5% of full scale is very good for a metered instrument and requires a mirrored scale behind the needle of the meter, with careful lining up by eye when a reading is made, in order to eliminate parallax. On the other hand, 0.5% of full scale is the *minimum* accuracy that can be expected from a digital voltmeter.

Digital multimeters may measure voltages, current, and resistance to 0.5, 0.05, or even 0.005% of full scale, and the total number of scales may be close to 20. The quantities measured may therefore have a range of seven orders of magnitude. Some voltmeters have even higher accuracy than 0.005%.

Digital counters can be used to determine frequency, to count events, and to determine ratios of frequencies. Such measurements are especially valuable in dielectric contant measurement, radio frequency measurement, and other instrumentation areas. Moderately priced seven digit counters now can go over 100 mHz. As to response time, the present state of the art is several milliseconds, but this will no doubt improve rapidly. In any event, this is more than enough for visual reading, since the unaided eye can make no more than 4–6 readings per second. Laboratory computers, also, can function as digital meters, with microsecond analog-to-digital converter.

References

Applications Manual for Operational Amplifiers. Philbrick Corp., Dedham, Mass., 1968.

Burgess Engineering Manual, Burgess Battery Co. This supplies information needed for battery users.

Malmstadt, H. V., C. G. Enke, and S. R. Crouch, *Electronic Measurements for Scientists*. Benjamin, Menlo Park, Calif., 1973. This is a good overall discussion of topics in this chapter as well as a number of others not covered here. It contains most of the detail needed for practical purposes.

Weissberger, A., and B. W. Rossiter, *Physical Methods of Chemistry*, Part I, A. Wiley-Interscience, New York, 1971.

Chapter 16

Electrochemistry

Electrochemical measurements are used in a wide variety of applications. Under equilibrium conditions *potentiometry*, the measurement of potential, is used for determining activity coefficients and solubility products, determining pH, calculating free energy, entropy, and enthalpy of reaction, determining the oxidation potentials of redox couples, quantitative analysis, and studying the composition of complex ions.

Conductance measurements are used in analysis by conductometric titration, in the determination of the charge and composition of complex ions, in the calculation of ionic velocities, in the determination of equilibrium constants, and in the study of interionic attractions.

Potentiometry under nonequilibrium conditions is used for qualitative and quantitative analysis, for the determination of diffusion coefficients, and for studies of the kinetics and mechanisms of electrode reactions, the adsorption of electroactive species, and the kinetics of formation and dissociation of complex ions.

Amperometry, the measurement of current, is applied in qualitative and quantitative analysis, in determinations of the composition of complex ions, in determinations of reaction mechanisms, and in studies of adsorption and of the reversibility of redox reactions.

Equilibrium Electrode Potentials

An electrode dipped into a solution has, under most conditions, a surface charge and adsorbs ions at its surface, either by specific adsorption (the formation of ion–surface bonds) or by electrostatic attraction. The layer of ions at the

Chapter 16. Electrochemistry

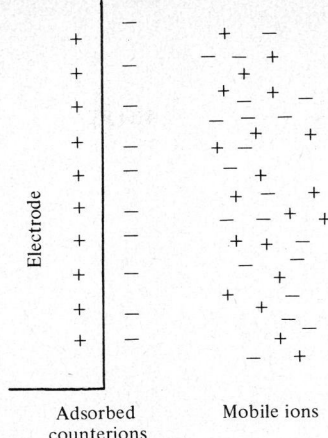

FIGURE 16–1 / Helmholtz and Gouy double layers.

surface induces a layer of opposite charge in the metal. The two layers in effect form a capacitor, called the *adsorbed double layer*, or the *Helmholtz double layer*, since its existence was postulated by Helmholtz. To maintain electroneutrality, the adsorbed layer is blanketed on the solution side by a layer of

FIGURE 16–2 / Potential drop with distance from electrode.

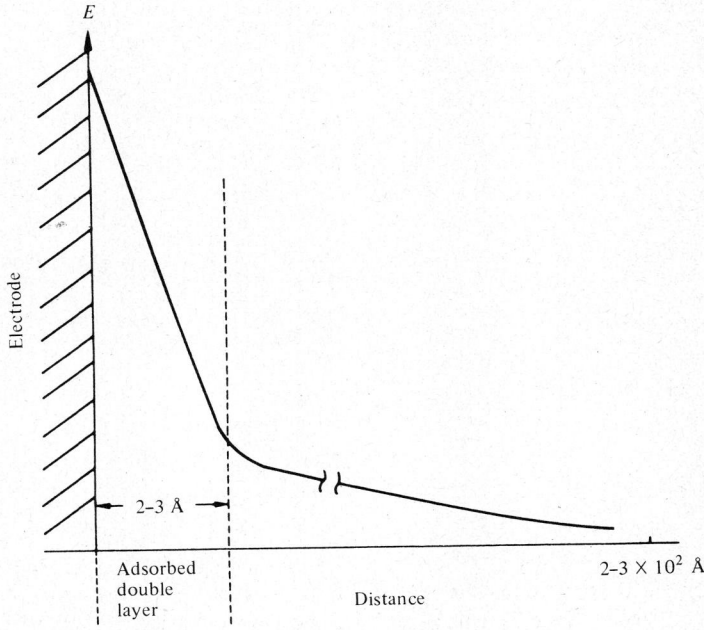

counterions. These two charged layers also form a capacitor, which is called the *diffuse double layer*, or the *Gouy layer*, since Gouy first drew attention to its presence, Figure 16-1 shows the two double layers. There is a potential across each double layer, due to the separation of charges, and the overall potential is the sum of the double-layer potentials. Most of the potential drop between the electrode and the bulk of the solution is across the adsorbed double layer, as shown in Figure 16-2, and after the first few hundred Ångstroms the potential fall is complete. As a result, in making equilibrium potential measurements, the macroscopic distance between electrodes has no effect on potentials.

Since the structure and potentials of the double layer depend on the particular metal of the electrode and on the adsorption of the particular ions at the interface, the potential of the electrode depends on the metal–ion combination and the *activities* of the ionic species (*not* just the concentrations). The potential is also independent of the other electrode. For the electrode reaction

$$x\,A + n\,e \rightleftharpoons y\,B$$
$$e = \text{the electron}$$

the absolute potential is

(16-1) $$E' = K - \frac{RT}{nF} \ln \frac{a_B{}^y}{a_A{}^x}$$

E' = absolute electrode potential
K = absolute standard potential
n = number of electrons
F = the faraday
a = activity
x, y = stoichiometric coefficients

The absolute standard potential K cannot be measured, since any electrical measurement requires that a second electrode be present, thus introducing another unknown K. There is therefore no way to determine K and no way to determine the absolute potential of an electrode.

However, it is not necessary to know absolute potentials. A measurement of emf is actually a measurement of the difference in potential between the two electrodes. Therefore, a set of emf's relative to the same arbitrary reference electrode can be substituted for the absolute potentials. The reference used is the *standard hydrogen electrode*, arbitrarily defined as the electrode of zero potential. The standard hydrogen electrode is a platinized platinum electrode dipping into a HCl solution of unit activity, with hydrogen gas at unit activity (or fugacity) bubbling over the surface. On the hydrogen scale the emf of an electrode is

(16-2) $$E = E° - \frac{RT}{nF} \ln \frac{a_{\text{product}}^x}{a_{\text{reactant}}^y}$$

$E°$ = the potential, vs. hydrogen, at unit activity of reactant and product

Tables of $E°$ for a large number of electrode reactions are to be found in any handbook of physical or analytical chemistry. In using these values it is necessary to be careful that the sign convention is obeyed. Some books list $E°$'s as oxidation potentials, others as reduction potentials. (The accepted international convention is to use reduction potentials, but older American works still list oxidation potentials.) In either case the emf's must be positive for the reaction to be spontaneous. Probably the easiest procedure to follow is simply to take the listed potentials for a reaction that is known to occur, such as

$$2\,\text{Na} + \text{Cl}_2 \rightarrow 2\,\text{NaCl}$$

and see how these must be manipulated to obtain an emf indicating a spontaneous reaction. Then use the same manipulations in all other cases.

Regardless of sign convention, in using tabulated values of $E°$ remember that the $E°$ of an electrode is independent of the other electrode. A $\text{Cu}|\text{Cu}^{2+}$ electrode will develop the same potential regardless of the other electrode present. Therefore, never multiply the $E°$ by any number, but use it as it is. Also, in combining two electrodes there must be only one anode and one cathode. Consequently, in combining the $E°$'s obtained from the tables we must reverse the sign of one $E°$.

Electrodes and Electrode Potential Measurements

Electrodes are arbitrarily divided into three categories. Electrodes of the first kind are metals in equilibrium with their ions:

$$M^{n+} + n\,e \rightleftharpoons M$$

The reaction taking place at the electrode is either the production or the removal of metal ions from the solution. The metal ion therefore determines the potential. The electrode potential is

(16-3) $$E = E° - \frac{RT}{nF} \ln \frac{1}{a_{M^{n+}}}$$

$E°$ = standard reduction potential
$a_{M^{n+}}$ = activity of the metal ion

Examples of this kind of electrode are silver metal dipping into a silver nitrate solution and copper metal dipping into a copper sulfate solution.

An electrode of the second kind is a metal M coated with an insoluble salt $M_x A_y$, dipping into a saturated solution of the salt in which the anion A^{x-} is present in large excess:

$$xy\,e + M_x A_y(s) \rightleftharpoons x\,M + y\,A^{x-}$$

x = valence of the anion
y = valence of the cation

The electrode potential is

(16-4) $$E = E° - \frac{RT}{xyF} \ln a_{A^{x-}}^y$$

Some examples of this type of electrode are silver coated with silver chloride dipping into a concentrated chloride solution, a paste of mercury and mercurous chloride in a KCl solution, and lead coated with lead sulfate in a sulfuric acid solution.

The net reactions taking place at a metal/insoluble salt electrode involve the production or removal of the anion. The anion, rather than the cation, determines the potential. For example, for the silver/silver chloride electrode

$$\begin{aligned} \text{AgCl(s)} &\rightleftharpoons \text{Ag}^+ + \text{Cl}^- \\ \text{Ag}^+ + e &\rightleftharpoons \text{Ag} \\ \hline \text{AgCl(s)} + e &\rightleftharpoons \text{Ag} + \text{Cl}^- \end{aligned}$$

The potential of the electrode is a function of chloride ion activity:

$$E = E° - \frac{RT}{F} \ln a_{Cl^-}$$

Electrodes of the third kind are inert metals dipping into solutions containing the oxidized and reduced forms of the reactant. These electrodes do not themselves react; they simply transfer electrons to and from the reactants and products.

Although the tables of standard electrode potential are based on the hydrogen electrode reference, in most laboratories the hydrogen electrode is rarely used. It is easily poisoned and is a potential fire and explosion hazard. The reference electrodes of choice are the various calomel electrodes and the silver/silver chloride electrode, all of which have well defined values on the hydrogen scale.

The calomel electrode, shown in Figure 16-3, is a metal/insoluble salt electrode consisting of a mixture of mercury, solid mercurous chloride, and a soluble chloride. The reaction is

$$\text{Hg}_2\text{Cl}_2(\text{s}) + 2e \rightleftharpoons 2\,\text{Hg(l)} + 2\,\text{Cl}^-$$

and the emf is

(16-5) $$E = E° - \frac{RT}{F} \ln a_{Cl^-}$$

At any temperature this potential is a function of the activity of the chloride ion only. The chloride solution is usually either saturated KCl, 1.00 M KCl, or 0.100 M KCl. The saturated KCl electrode is easiest to prepare. Excess solid KCl is added to the electrolyte to ensure saturation. Since the activity of the

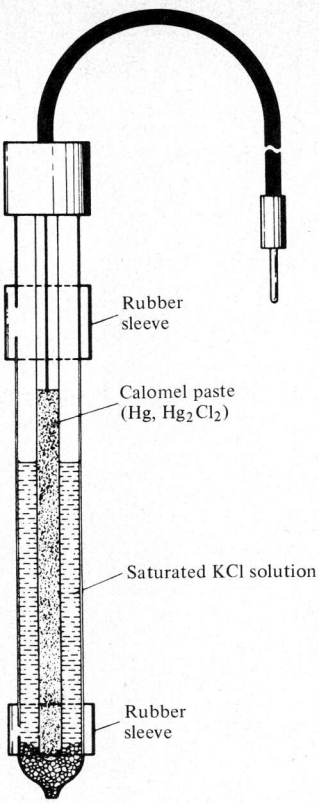

FIGURE 16-3 / Calomel electrode (courtesy of Leeds & Northrup Company).

chloride ion in equilibrium with the solid KCl is constant, the potential of the electrode is kept at a constant value—as long as the temperature does not change. Calomel reference electrodes are usually equipped with salt bridges to separate the calomel from the experimental solution and prevent contamination of the mercury. The salt bridge is flushed with KCl solution after each use to prevent foreign ions from diffusing in and contaminating the mercury and calomel. The liquid junction potential between the KCl solution and the experimental solution, although probably minimized by the KCl, still exists. The calomel electrode should therefore not be used for precise thermodynamic measurements, although it has been so used.

The $Ag\,|\,AgCl(s)$ electrode, like the calomel electrode, is a metal/insoluble salt electrode, but without a salt bridge. It may therefore be used for thermodynamic measurements in solutions containing chloride ions. The overall electrode reaction, as shown previously, is

$$AgCl(s) + e \rightleftharpoons Ag + Cl^-$$

and the emf

(16-6) $$E = E° - \frac{RT}{F} \ln a_{Cl^-}$$

Ag|AgCl(s) electrodes may be prepared by depositing AgCl anodically on a silver electrode of either solid silver or silver plated on plantinum. The potential of the electrode will depend slightly on the conditions of preparation. Each electrode prepared by a student should be calibrated against a standard calomel electrode. Ag|AgCl(s) reference electrodes are available. These have salt bridges similar to those of the calomel electrode, but for thermodynamic measurements the salt bridge may be filled with the experimental solution.

To determine the potential of an electrode, a cell is made by combining the electrode with a reference electrode; the potential is measured, and the reference potential is subtracted from the observed cell potential. Whenever possible the two electrodes should be kept in the same solution to eliminate liquid junctions. In some cases, however, this is not possible, as, for example, when the test solution would react with the reference electrode or when the reacting ions are both either cations or anions. In such cases thermodynamic potentials may be determined with back-to-back reversible electrodes. In effect, this means eliminating liquid junctions by converting the cell into two cells connected back to back. Where liquid junctions are necessary their potentials can be minimized—but not eliminated—by means of salt bridges filled with a concentrated solution of an electrolyte whose ions have approximately equal transference numbers, usually KCl or NH_4NO_3.

FIGURE 16–4 / Weston cell.

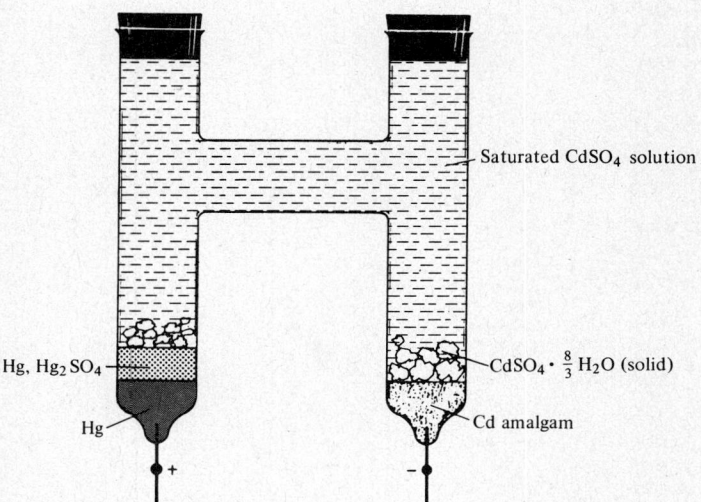

Usually, emf measurements are made potentiometrically (pages 148–49). The standard cell most frequently used in the potentiometer is the Weston cell. As shown in Figure 16-4, this consists of two metal/insoluble salt electrodes whose potentials are each determined by the sulfate ion activity. The electrode reactions are

$$Hg_2SO_4(s) \rightleftharpoons 2Hg(l) + SO_4^{2-} - 2e$$
$$Cd(Hg)(s) + SO_4^{2-} \rightleftharpoons CdSO_4(s) + 2e$$

$$Hg_2SO_4(s) + Cd(Hg)(s) \rightleftharpoons 2Hg(l) + CdSO_4(s)$$

Since all the reactants and products are in their standard states (i.e., solid or pure liquid), the emf is the algebraic sum of the E°'s.

The basic requirements for standard cells are ease of preparation, stability on standing, reproducibility of emf, low temperature coefficient of emf, reversibility of cell reaction, and, most important, ability to handle appreciable currents during balancing. The Weston cells meets most of these requirements. However, care must be taken not to exceed currents of 10^{-4} A, or the concentrations around the electrodes will change. The temperature coefficient of this cell, 4×10^{-5} V/°C, is appreciable. The cell emf must therefore be corrected for temperature change, and care must be taken to avoid abrupt temperature changes immediately before using the cell.

Applications of Potentiometry

Potentiometry is used in chemical analysis by an indirect method, potentiometric titration. Direct analysis, using equation (16-2), is not accurate because, first, the quantity thus calculated is activity, not concentration, and second, since the potential is an exponential function of activity, the error in calculating activity is an exponential of the experimental error in measuring emf.

In a potentiometric titration a standard titrating agent is added to the unknown solution, and potential measurements are made versus a reference after each addition. The resulting curve is shown in Figure 16-5. The endpoint of the titration is the midpoint of the rising portion of the curve. As in most titrations, the endpoint is the region of most rapid change, so it is best located graphically (pages 40–41).

Potentiometric titrations are valuable in determining the composition of a complex ion. For the equilibrium

$$M^{x+} + p\,A^{y-} \rightleftharpoons MA_p^{x-py}$$

the potential of the electrode of metal M dipping into the solution is

(16-7) $$E = E^\circ - \frac{RT}{nF} \ln \frac{1}{a_{M^{x+}}}$$

PRINCIPLES

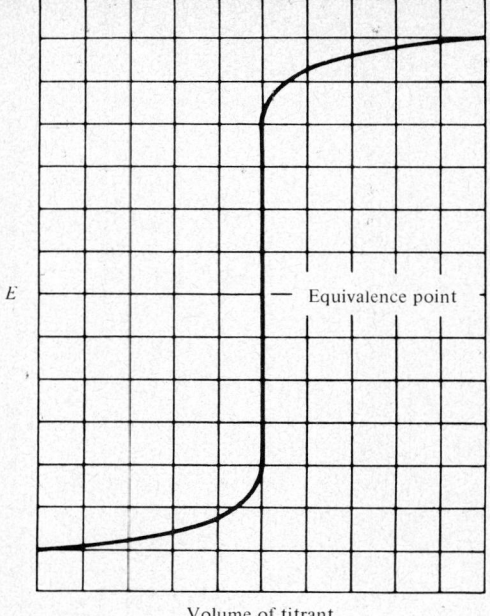

FIGURE 16–5 / Typical curve in a potentiometric titration.

If an excess of A^{y-} is used so that virtually all the metal ion is complexed, the concentration of the complex is almost independent of further changes in the concentration of A^{y-}. Under these conditions

$$a_{M^{x+}} = \frac{k}{(a_{A^{y-}})^p}$$

and the emf is

(16–8) $$E = E° - \frac{RT}{nF} \ln \frac{1}{k} - p \frac{RT}{nF} \ln a_{A^{y-}}$$

A graph of E versus the natural log of the concentration of A^{y-} is a straight line whose slope is $-pRT/nF$.

Emf measurements can be used to determine activities and activity coefficients, but the measurements must be made carefully because of the previously mentioned exponential relationship between activity and emf.

In the determination of activity coefficients, if $E°$'s are known, equation (16–2) may be used. If the individual $E°$'s are unknown, an extrapolation technique must be used to find $E°$ for the cell in order to use equation (16–2) for calculation of activities. The equation is rewritten in the form

(16–9) $$E + \frac{RT}{nF} \ln \frac{c^x_{\text{products}}}{c^y_{\text{reactants}}} = E° - \frac{RT}{nF} \ln \frac{\gamma^x_{\text{products}}}{\gamma^y_{\text{reactants}}}$$

c = concentration
γ = activity coefficient

Values of E are determined at a series of electrolyte concentrations. The left-hand side of equation (16–9) is then plotted against $\sqrt{\mu}$, where μ is the ionic strength, defined as

$$\mu = \tfrac{1}{2} \sum c_i z_i^2$$

c_i = concentration of ionic species i
z_i = charge on ion i

The graph is extrapolated to zero electrolyte concentration. The extrapolated value of the left-hand side equals $E°$, since at zero electrolyte concentration the activity coefficients are unity and the logarithmic term equals zero.

Solubility products of insoluble salts are determined by treating metal/insoluble salt electrodes as metal electrodes, with the activity of the metal determined by the solubility product of the electrolyte. For example, the $Ag|AgCl(s)$ electrode may be considered as an $Ag|Ag^+$ electrode with the Ag^+ activity determined by the Cl^- activity. If a standard $Ag|Ag^+$ electrode is connected through a liquid junction to a standard $Ag|AgCl(s)$ electrode, then, ignoring the liquid junction potential, the standard emf of the combination is a direct measure of the activity product of the saturated silver chloride:

(16–10) $$E° = E°_{Ag|Ag^+} - E°_{Ag|AgCl(s)} = -\frac{RT}{nF} \ln K_a$$

K_a = activity product of $AgCl(s)$
     = $a_{Ag^+} a_{Cl^-}$

For other insoluble salts the appropriate combinations of electrodes can be found and used.

In determining thermodynamic quantities from electrochemical potentials the basic equations are

(16–11) $-\Delta G = nFE$

(16–12) $-\Delta G° = nFE° = RT \ln K_a$

(16–13) $$\Delta H = nF \left[T \left(\frac{\partial E}{\partial T} \right)_P - E \right]$$

(16–14) $$\Delta S = nF \left(\frac{\partial E}{\partial T} \right)_P$$

G = Gibbs free energy
H = enthalpy
S = entropy

The temperature coefficient of the emf of reversible electrochemical cells gives a direct measurement of the entropy of reaction. From the measured emf a skilled experimenter can obtain the Gibbs free energy to five significant figures. The enthalpy may usually be obtained by difference to four figures.

Another application of emf measurements is in the potentiometric determination of pH with a pH sensitive electrode such as a hydrogen electrode, an antimony/antimony oxide electrode, a quinhydrone electrode, or a glass electrode.

The quinhydrone electrode is now only of historical interest. However, it is a good example of a chemical oxidation reduction system sensitive to pH, which can be made into a pH electrode. The material, quinhydrone, $C_6H_4O_2 \cdot C_6H_4(OH)_2$, in aqueous solution dissociates into equal numbers of quinone, $C_6H_4O_2$, and hydroquinone, $C_6H_4(OH)_2$, molecules. The quinone and hydroquinone molecules set up the oxidation-reduction system

$$C_6H_4O_2 + 2\ H_3O^+ + 2\ e \rightleftharpoons C_6H_4(OH)_2 + 2\ H_2O$$

The potential of this half-cell reaction is

$$E = E° - \frac{RT}{2F} \ln \frac{a_{Hq}}{a_Q\, a^2_{H_3O^+}}$$

Q = quinone
Hq = hydroquinone

Since the concentrations of quinone and hydroquinone are the same and the activity coefficients of these nonelectrolytes are approximately equal, the equation reduces to

(16–15) $$E = E° - \frac{RT}{2F} \ln \frac{1}{a^2_{H_3O^+}} = E° - \frac{2.303 RT}{F}\, \text{pH}$$

where pH is defined as the negative logarithm of the hydrogen ion activity. The quinhydrone electrode cannot be used at pH's above about 8 because of oxidation by atmospheric oxygen and because of the ionization of the weakly acidic hydroquinone. The other electrode (i.e., the reference electrode), is usually a calomel electrode with a liquid junction. The thermodynamic potentials are therefore uncertain, and the precise meaning of the measured quantity is not clear. This is an example of the impossibility of measuring an individual ion activity. The liquid junction potential could be eliminated by inserting the calomel electrode directly into the solution containing the quinhydrone electrode. However, the measured potential would then include both the activity of the hydronium ion and that of the chloride ion.

Another pH sensitive electrode is the glass electrode (Figure 16–6), consisting of a thin glass membrane containing a solution of KCl and acetic acid and a silver/silver chloride or calomel electrode. When the glass electrode is

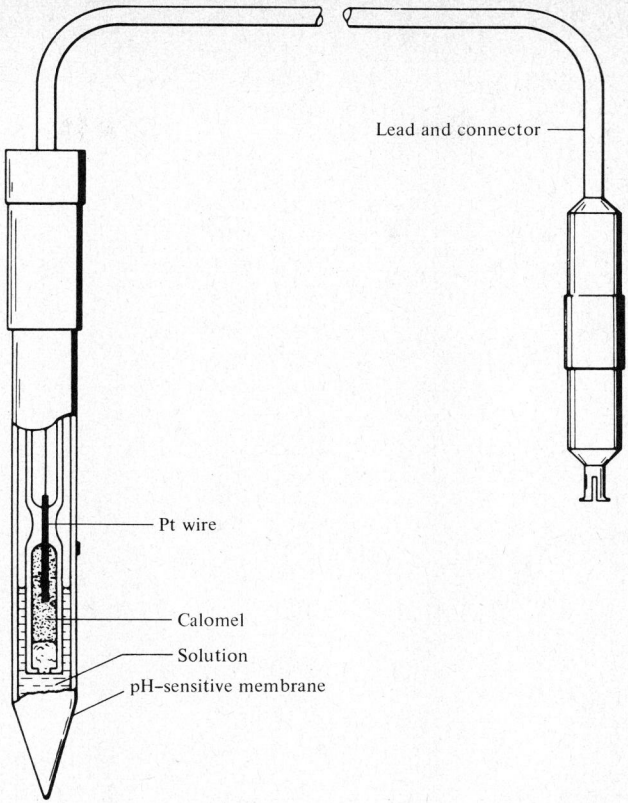

FIGURE 16-6 / Glass electrode (courtesy of Leeds & Northrup Company).

dipped into an acid solution, a potential develops across the glass membrane that is a direct function of the pH of the external solution:

$$(16\text{-}16) \qquad E = K - \frac{RT}{F} \ln \frac{1}{a_{H_3O^+}} = K - \frac{2.303 RT}{F} \text{pH}$$

The constant K is the *asymmetry potential*; it varies from electrode to electrode and depends upon the previous treatment of the particular electrode. It must be determined experimentally or else compensated for whenever the electrode is used. In its way the asymmetry potential is equivalent to the liquid junction potential, precluding the measurement of single ion activities.

The resistance of the glass electrode is so high, of the order of 10–100 megohms, that the potential cannot be measured directly with a potentiometer. Instead, a pH meter, a combination of electronic amplifier and either a milliammeter or a potentiometer, is used. The emf put out by the combination of the glass electrode and the standard calomel electrode is opposed by a standard emf. The resulting potential difference is amplified electronically and operates a meter. In *direct-reading pH meters* the pH is read directly on the meter. In *null point pH meters* an opposing emf is adjusted potentiometrically until the meter gives a null reading.

In addition to measuring pH, glass electrodes are becoming increasingly important in the determination of the activities of specific cations, such as Na^+, K^+, and Ca^{2+}. In all glass electrodes the glass, containing a mixture of oxides, acts as a cation exchange membrane. The potential across the glass membrane develops due to the charge separations when cations diffuse into or out of the glass. The cation specificity of the glass depends on its structure and composition. While pH electrodes have membranes specific for H^+, other electrodes have membranes specific for K^+, and so on. Liquid membrane electrodes and solid state electrodes have been developed for determination of activities of anions as well as cations. There also are gas-sensing electrodes.

Whatever the purpose of the measurement, certain precautions should be taken in measuring emf's. Metal electrodes may be lightly amalgamated to remove surface strains and oxides, provided that one knows what effect, if any, amalgamation will have on the electrode potential. With some metals (zinc, for example), the amalgam electrode is a saturated solution of metal in mercury and has the same electrochemical potential as the unamalgamated electrode. With other metals, such as cadmium and lead, the amalgam is either an alloy or is not in equilibrium with the underlying metal. The potential of such an amalgam electrode is definitely lower than that of the unamalgamated metal. Amalgamation is best accomplished by dipping the electrode into dilute mercuric chloride. The amalgamated electrode should be rubbed vigorously with filter paper to remove any adherent oxide coating. The solutions used should be deaerated just before use, either by flushing with nitrogen or by evacuating the air with an aspirator. (Oxygen diffusing to the electrode surface can react and change the potential especially in dilute solutions.) Sufficient time must be allowed for thermal equilibration of the solution. All binding posts and terminals should be clean and all connections tight. If necessary, corrosion products should be removed with emery paper. Electrodes should be mounted in a fixed position and not allowed to touch each other. Lead wires and electrodes should not be flexed or twisted, or else strains will be set up that will change the potentials. Finally, care should be taken to have the reference electrodes clean and uncontaminated. The salt bridge of the reference electrode should be flushed with a fresh salt solution if necessary. The reference electrodes should be checked against a convenient reliable standard before use.

The cell containing the test solution and electrodes should have a small capacity, perhaps 100 ml at most. The smaller the volume and the thinner the walls, the easier it is to reach thermal equilibrium. The electrodes should be in fixed positions, mounted in glass, if possible. Cork rubber, or plastic stoppers can also be used, but synthetic materials should be checked to be sure that sulfur and other possible contaminents are absent. Even in minute quantities, these can poison electrodes, especially platinum electrodes. There should be provision for passing an inert gas, such as nitrogen, through the solution to remove interfering dissolved gases, when necessary. One simple and useful cell is a small three-neck flask, with the electrodes mounted either in ground glass joints or stoppers and with gas inlet and outlet tubes. Another simple cell, commercially available, is just a lipless beaker with a Teflon or silicone–rubber cap through which holes have been bored for inserting the electrodes and the gas inlet and outlet tubes. For student use, with reactions in which it is not necessary to remove dissolved gases, a simple convenient cell can be made from a 10–20 ml sample vial with a polyethylene cap. Wires are thrust through the cap to make contact. This cell is not only economical of materials and solutions, but it comes to thermal equilibrium very quickly.

For work involving any great degree of accuracy, special equipment is usually required. Students are referred to the references at the end of the chapter.

Steady-State Potentiometry

When current is passed through a solution, charge is transferred to the electrode surface, polarizing or charging up the double layer. If a reactive substance is present, it tends to remove some of the charge, depolarizing the electrode. Such a substance is therefore called a *depolarizer*. Whether or not a depolarizer is present, the surface charge and the composition of the double layer change, thereby changing the potential across the double layer. Consequently, working electrodes do not have the same potentials as equilibrium electrodes.

Part of the potential difference between equilibrium and working electrodes is an artifact of the measuring system. Potentials must be measured against those of reference electrodes. When current is passed, wherever the reference electrode is located in solution, it lies on a current line or line of potential fall. The potential difference measured between the working electrode and the reference electrode therefore contains an ohmic drop. In *amperometry* and *chronopotentiometry* (see below), currents are usually small and solution conductivities high. Ohmic drop under such conditions is negligible. In ordinary potentiometry, however, currents may range up to centiamperes while electrolyte conductances are often low. Under these conditions the ohmic drop may be orders of magnitude greater than the electrode potential, and must either be compensated for or eliminated.

In the *direct method* the reference electrode is connected to the solution by means of a small diameter capillary whose tip is placed very close to the surface of the working electrode, eliminating much of the ohmic drop. For accurate measurements several capillary/reference electrode combinations are used, with each capillary at a different known distance from the surface. A potential measurement is made with each reference electrode, and a graph of potential versus distance from the working electrode surface is plotted and extrapolated to zero distance.

In the *indirect*, or *interrupter, method* an electronic circuit is used to cut the current off periodically for very short time intervals. The current, and therefore the ohmic drop, dies away in fractions of a microsecond, while the electrode potential decays over time periods ranging from microseconds to milliseconds. From the oscilloscopic trace both the ohmic drop and the electrode potential can be observed. The interrupter method has the additional advantage that the shape of the oscilloscope trace indicates the presence or absence of surface contaminents.

Even after the ohmic drop has been eliminated, the potential of the working electrode differs from that of the equilibrium electrode. The difference, or *overpotential*, is the sum of two terms: the polarization due to changes in solution concentration at the electrode surface and the activation overpotential:

(16–17) $\quad \eta = E_w - E_r = \eta_c + \eta_a$

E_w = potential of working electrode
E_r = potential of equilibrium electrode
η_c = concentration polarization
η_a = activation overpotential

The activation overpotential is the increase in potential needed to speed up the electrode reaction, supplying an activation energy. Activation overpotentials have been investigated extensively for many years, and are still the subject of much controversy. Part of the problem in the interpretation of overpotential measurements arises from the difficulty of obtaining reproducible results. Since there is no net reaction at equilibrium electrodes, the impurities in solution do not usually deposit at the surface. When the current is passed, however, impurities deposit at cathodes and oxides, or other films, form on anodes. Since most metal electrodes have about 10^{15}–10^{16} atoms per square centimeter of surface, 10^{-8} mole of a contaminant will cover the electrode surface completely, changing its characteristics. Substances of concentration as low as 10^{-14} mole/liter have been observed to poison completely the working electrode surfaces. Stringent purification procedures are therefore mandatory.

Overpotential measurements can be used to elucidate electrode kinetics, but further discussion is beyond the scope of this introduction.

Conductance

Current is conducted through electrolytic solutions by the movement of the positive and negative ions in opposite directions. The conductance, which is defined as the reciprocal of the resistance, is proportional to the number of ions, the charge on each ion, and the speed with which the ions move. Conductance measurements are used in physical chemistry in determining the ionization constant of weak electrolytes, in determining equilibrium constants and rates of reactions that proceed with the formation or disappearance of ions, in titration of solutions, and in studies of interionic forces.

As the temperature increases, the conductance of a given solution increases because of increased velocity of the ions, decreased viscosity of the solvent, decreased effects of electrostatic interactions among ions, and frequently, in the case of weak electrolytes, increased ionization. A rule of thumb is that the specific conductance of an aqueous solution will increase with increasing temperature by about 2% per degree Celsius. Therefore, accurate conductance measurements require careful temperature control.

Specific conductance, or *conductivity*, is the conductance of a cube of solution contained between two parallel electrodes 1 cm square and 1 cm apart. *Equivalent conductance* is the conductance that would be observed if the volume of solution that contains one equivalent of electrolyte were placed between two sufficiently large parallel electrodes 1 cm apart:

(16–18) $$\Lambda = \kappa v_{eq} = \frac{1000\kappa}{c}$$

Λ = equivalent conductance
κ = specific conductance
v_{eq} = cm³/equivalent
c = concentration, equivalents/liter

Specific conductance decreases as a solution is diluted, eventually reaching the specific conductance of the pure solvent. Equivalent conductance increases on dilution because of decreased interionic attractions and electrophoretic effects and, in the case of weak electrolytes, because of increased dissociation. For uni-univalent electrolytes at concentrations below about 0.01 M, the *Onsager limiting equation* holds:

(16–19) $$\Lambda = \Lambda_0 - (a + b\Lambda_0)\sqrt{c}$$

Λ = equivalent conductance at concentration c
Λ_0 = equivalent conductance at infinite dilution
c = concentration

The constants a and b are calculated from the dielectric constant and viscosity of the solvent, the temperature, the electronic charge, and the valences of the particular ions that make up the electrolyte. Advantage is taken of the linear form of the relationship to determine Λ_0 graphically. Experimental values of Λ are plotted versus $\sqrt{c}$ and the graph is extrapolated back to zero concentration. For solutions of polyvalent ions, deviation from equation (16–19) sets in at much lower concentrations.

For weak electrolytes the Onsager equation takes the form

(16-20) $$\Lambda = \alpha[\Lambda_0 - (a + b\Lambda_0)\sqrt{\alpha c}]$$

α = degree of dissociation

In this case, however, Λ_0 cannot be obtained by the same graphical method as for strong electrolytes because the curvature of the graph is at its greatest in the regions where the experimental precision is at its poorest. Instead, using Kohlrausch's method, Λ_0 is obtained algebraically from the Λ_0 values for three strong electrolytes. For example, if MX is a weak electrolyte and the other three are all strong,

$$\Lambda_{0,\,MX} = \Lambda_{0,\,MCl} + \Lambda_{0,\,NaX} - \Lambda_{0,\,NaCl}$$

The degree of dissociation of a weak electrolyte is approximately the conductance ratio, that is, the ratio of the observed equivalent conductance to that calculated for infinite dilution:

$$\alpha = \frac{\Lambda}{\Lambda_0}$$

This approximation holds more closely as the concentrations are decreased (i.e., in dilute solutions). Conductances are usually calculated from the measured resistances of solutions.

Conductivity cells (Figure 16-7) should be of pyrex, or, better, of quartz, since impurities leach out of soft glass into the solution, markedly increasing conductance at low electrolyte concentration. The filling tubes of the cells should not be too close to the electrode leads, and the leads themselves should

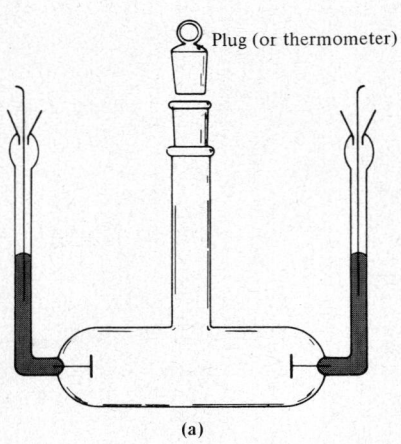

FIGURE 16-7 / Conductivity cells: (a) Jones cell; (b) dip cell.

(a)

(b)

be widely separated to reduce capacitance effects. The electrodes should be of platinum, "platinized" to ensure uniformity of surface and to increase the surface area so that reaction products do not form and accumulate. They should be thoroughly cleaned and rinsed, first with distilled water and then with conductance water.

For research work, aqueous solutions should be made with conductance water, since ordinary distilled water contains CO_2, metal silicates that have leached out of the glass of the condenser, and small amounts of amines that have steam-distilled over. Research grade conductance water is vacuum-distilled from quartz or tin condensers at least five times, starting with ordinary distilled water. For student use a good grade of distilled water is sufficient. Demineralization with an ion exchange resin is also adequate. The specific resistance of the water should be about 10^5 ohm-cm.

In making conductance measurements, the usual experimental procedure is to measure the resistance of a fixed amount of solution contained in a cell. Each cell has a slightly different geometry, and ordinarily the electrodes are not perfectly parallel, nor 1 cm square, nor 1 cm apart, except by coincidence. Therefore, the observed resistance usually is not the specific resistance. However, since the cell geometry is constant, the specific resistance may be calculated using the simple relationship

$$R_{obs} = kR$$

R_{obs} = observed resistance
κ = cell constant
R = specific resistance

Since conductance, the reciprocal of resistance, is the desired property, it is more convenient to use

$$K = \frac{k}{R_{obs}}$$

in order to obtain the specific conductance directly from the experimental measurement. The cell constant k is usually determined by measuring the resistance of a standard calibrating solution, usually a KCl solution, whose specific resistance is known very accurately. The cell constant should be redetermined each time the apparatus is used.

NOTES ON MAKING CONDUCTANCE MEASUREMENTS / All connections must be clean, tight, and free from corrosion. A small spot of dirt or corrosion can cause large errors in a measurement.

The cell should be rinsed at least half a dozen times with distilled water and then twice with conductance water. The cell is not clean until successive samples of conductance water show constant resistivities of at least 10^5 ohm-cm. Always rinse the cell several times with samples of the solution being studied.

When balancing the bridge it is quicker to work from higher resistance values down to smaller ones, following the same general procedure as in weighing an object on an analytical balance. Look for a minimum detector current at the balance point rather than a zero-current.

Do not base your precision on three or four successive measurements of one sample of solution. The factors that limit precision are the preparation of clean solutions and the cleaning and filling of the cell. Make measurements on several portions of each solution and, if time permits, on several samples of each concentration studied.

Always allow time for the solution to reach thermal equilibrium before making the resistance measurement. In case of doubt repeat the resistance measurement after a few minutes to see if there has been any significant drift. To avoid having to wait for thermal equilibrium, solutions should be made up in advance and kept in a thermostat until they are used.

Transference Numbers

Whenever current is passed through an electrolytic solution, each species of ion present carries only a fraction of the total current. This fraction is called the *transference number*. The relationship between transference number, specific ionic conductance, and electrolyte conductance is

(16-21) $$t_i = \frac{\kappa_i}{\kappa}$$

t_i = transference number of species i
κ_i = specific conductance of species i

The transference number of an ion depends on temperature, concentration, and the conductance of the other ions present. In experimental physical chemistry, transference numbers furnish information about hydration and complex ion formation. They are usually determined by a chemical method (the Hittorf method), the moving-boundary method, or the emf method. The Hittorf method is simple in concept but inherently inaccurate. It retains only historical interest and will not be discussed here.

Moving-Boundary Method

In the moving-boundary method (Figure 16-8) the amount of charge carried by an ionic species is calculated from the migration of an interface. Two electrolytic solutions with one common ion are put into a capillary tube, the less

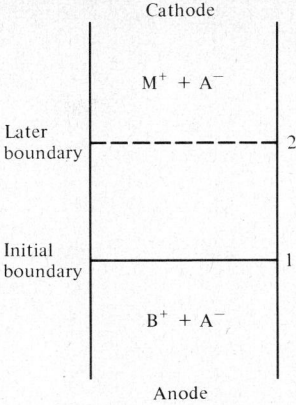

FIGURE 16–8 / Moving boundary.

dense on top, with a sharp boundary between them. Regulated DC is passed in the direction that moves the boundary toward the solution containing the more mobile cation, M^+ in this case. When the boundary has moved a unit distance (e.g., from 1 to 2 in Figure 16–8), a unit volume of solution has been depleted of M^+ ions. The charge (in terms of equivalents) carried by the cation M^+ is the number of equivalents of M^+ originally present in the volume between 1 and 2. The transference number is this number divided by the total charge transferred during the period of current flow. The relationship between the transference number and the measured quantities is

$$(16\text{–}22) \qquad t_+ = \frac{ALc'F}{is}$$

A = cross sectional area of the tube, cm^2
L = distance between positions 1 and 2, cm
c' = equivalents/cm^3
i = current, amp
s = time, sec

The original boundary may be formed either by a mechanical device or electrochemically by filling the tube with one electrolyte and generating the other electrolyte at the anode as the current is passed. Once formed the boundary will remain sharp as long as current is passed if the denser solution is at the bottom and the faster moving ion is out front. The sharpness of the boundary results from the change in the voltage gradient at the interface between the solutions. As the faster moving ion pulls away, the region at the interface develops a higher resistance and therefore a higher ohmic drop. Since the current in the tube is kept constant, the higher ohmic drop at the interface results in a higher voltage gradient, which pulls the slower moving ion along at the same speed as the faster ion travels.

The position of the boundary may be followed by difference in refractive index or color of the two solutions, or by addition of suitable color indicators, especially where the leading ion is H_3O^+. Care must be taken to prevent stirring of the solution by bubbles or thermal convection currents.

The total charge passed may be measured, preferably with an external coulometer, or by measuring the current and the time. The ordinary milliammeter is not sufficiently accurate, so that either a digital meter must be used or else the current must be measured potentiometrically from the ohmic drop across an accurately known resistance in series with the moving boundary cell.

OPERATING INSTRUCTIONS FOR THE MOVING-BOUNDARY APPARATUS

In filling the tube and upper chamber, make sure there are no air bubbles in the capillary. Use a metal/insoluble salt electrode to avoid forming gaseous reaction products, which might stir the solution. Use a machined piece of metal as the anode, keeping the flat surface pressed against the capillary.

If the electrolyte is too concentrated, the boundary will move very slowly. If the solution is too dilute, changes in resistance during the electrolysis may overtax the power supply.

The current must be maintained at a constant value in the range 1–3 mA. If a constant current power supply is not available, a resistance of at least 1 kilohm should be kept in series with the cell to act as a ballast. For precision work the current should be determined from the iR drop across a precision resistor in series with the solution or with a digital milliammeter.

If possible, use a color indicator to locate the boundary. If the leading ion is H_3O^+, use a pH indicator. If there is no appropriate color indicator, locate the boundary by observing the difference in the refractive indices of the two solutions.

EMF Method

The emf method of determining transference numbers is simple and quick but not very accurate. The transference number is obtained from the emf's of cells with and without transference.

The emf of a concentration cell without a liquid junction is

(16–23a) $$E = -\frac{RT}{nF} \ln \frac{a_2}{a_1}$$

n = number of electrons in the electrode reaction
a_1 = activity of more concentrated *electrolyte*
a_2 = activity of more dilute *electrolyte*

The emf of the same cell with transference of ions between the two solutions is

(16–23b) $$E_t = -t\frac{RT}{nF}\ln\frac{a_2}{a_1}$$

t = transference number of the ion *not* taking part in the electrode reaction

Combining the two previous equations,

(16–23c) $$t = \frac{E_t}{E}$$

The transference number may be obtained using equation (16–23c) simply by measuring the emf of a concentration cell, first with back-to-back reversible electrodes and then with a liquid junction. However, since transference numbers vary with concentration, the transference number calculated from equation (16–23c) is an average of the transference numbers of the two different solutions. Consequently, this method is rarely used, except for approximations.

Steady-State Amperometry

In an electrochemical redox reaction, the consecutive steps in the overall process are (1) diffusion to the surface, (2) adsorption, (3) electron transfer, (4) desorption, and (5) diffusion away from the surface. There are often additional complications, such as a surface chemical reaction preceding the electron transfer or following the electron transfer, regeneration of the reactant by a chemical process following the electron transfer, formation of a surface film of product, and so on. However, in all cases the current is some function of the concentration and of the potential, the precise function depending on the type of the electrode reaction. Consequently, amperometry is used in both analysis and in kinetic studies.

Dissolved substances arriving at an electrode are transported to the surface in one or more of three ways:

1. By convection.
2. By migration.
3. By diffusion.

Convection is the movement of masses of material through a solution as a result of stirring and by random vibrations, density differences, and temperature differences. In cases where convection has an effect on the measurement the solution is usually stirred at a steady rate to minimize the effects of random convection.

Migration is the conductance of current through the solution. For electroneutrality, migration must take place at the same rate that current flows. In analysis, in order to prevent concentration changes at the surface from being obscured by migration, a large excess of neutral or indifferent electrolyte is added to the solution. This migrates through the solution, carrying the current, but it does not react at the electrode, so that the ion being studied reaches the electrode by diffusion.

Diffusion of material to the electrode follows *Fick's law*:

$$(16\text{-}24) \qquad \frac{dm}{dt} = -SD_0 \frac{dc}{dx}$$

dm/dt = diffusion flux, moles/s
S = surface area, m^2
D_0 = diffusion constant, m^2/s
dc/dx = concentration gradient, moles/m^4

Converting this into current units, the current due to diffusion is

$$(16\text{-}25) \qquad i_{\text{diff}} = \frac{nFSD_0}{\delta}(c - c_0)$$

n = number of charges per ion
δ = thickness of the diffusion layer
c = concentration in the bulk of solution
c_0 = concentration at the surface

The thickness of the diffusion layer, δ, is defined as the distance that the depolarizer has to diffuse to reach the surface. If the current is increased slowly, the surface concentration c_0 gradually decreases. When the current reaches the value at which c_0 is zero, the depolarizer diffusing in reacts as fast as it arrives at the surface. The current due to the particular depolarizer is therefore at a maximum, called the limiting diffusion current, given the symbol i_{lim}.

The limiting diffusion current is directly proportional to the concentration for a given surface area and stirring speed (the thickness of the diffusion layer falls off with increased stirring speed).

In *amperometric titration* a suitably large potential is applied to the electrodes so that the current reaches its limiting value i_{lim}. The solution is then titrated, the i_{lim} changing linearly with the concentration of the substance being titrated.

In *polarography* the entire current–potential curve is measured, instead of just the limiting diffusion current. The electrode at which the oxidation or reduction takes place is called the *indicator electrode*. The other, or *counterelectrode*, is usually maintained at a constant potential, but its potential is usually not of interest. In a typical polarographic reaction the electrode potential is increased at a slow, fixed rate and the current is determined. Point-by-point potential settings and current measurements may be made, but the usual

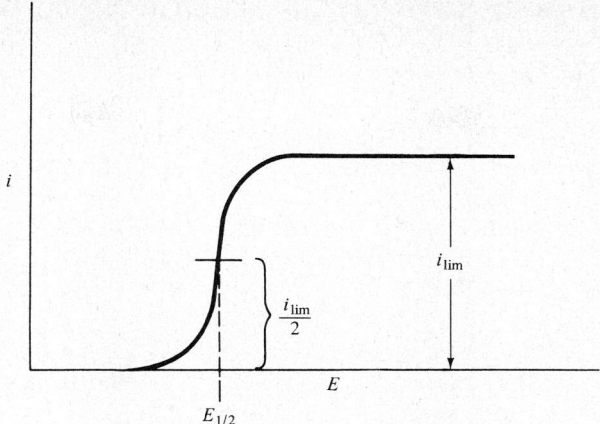

FIGURE 16–9 / Typical current–voltage curve.

method is to record both voltage and current continuously on a strip chart recorder. A typical current–voltage curve is shown in Figure 16–9. The S-shaped curve is really a combination of two curves, one showing an exponential increase in current when the potential reaches a value at which the electrode reaction occurs, and the other showing a rapid leveling off as the current approaches the rate at which the depolarizer diffuses in toward the electrode. At the inflection point on the curve, the diffusion current is equal to one half of the limiting diffusion current (i.e., $i_{diff} - i_0 = \frac{1}{2}i_{lim}$). The potential at this point is called the half-wave potential and is given the symbol $E_{1/2}$. For each redox system, the half-wave potential is a characteristic constant independent of concentration except in special cases.

The polarographic curve of a system gives a complete chemical analysis, $E_{1/2}$ being qualitative and i_{lim} quantitative. If there are several constituents, provided that their half-wave potentials are sufficiently far apart, each constituent will give its own polarographic wave, and one polarogram will suffice for the analysis of the entire multicomponent system.

Applications of Polarography

REVERSIBILITY OF A PROCESS

A reversible electrode reaction is one in which the Nernst equation, equation (16–2), is obeyed. Since the concentrations of reactant and product at the surface of the electrode are directly related to the current, the potential of the indicator electrode is related to the current by

(16-26) $$E = E_{1/2} - \frac{RT}{nF} \ln\left(\frac{i}{i_{\text{lim}} - i}\right)$$

(The derivation of this equation and most of the others in the sections on polarography and non-steady-state potentiometry are beyond the scope of this elementary work. The reader is referred to the general references, especially the invaluable text of Heyrovský and Kůta [1966].)

At 25°C, this equation may be rewritten as

(16-27) $$n\frac{E_{1/2} - E}{0.0592} = \log\left(\frac{i}{i_{\text{lim}} - i}\right)$$

If the graph of $\log[i/(i_{\text{lim}} - i)]$ versus E is a straight line, the electron transfer is reversible and the number of electrons can be calculated from the slope of the line. Another, somewhat quicker, method of determining n is shown in Figure 16-10. The inclined straight line is drawn tangent to the curve. The vertical line is drawn between the extrapolated portions of the two curves. The horizontal segment ΔE measured in millivolts, is $100.7/n$. If n is already known, the agreement of ΔE with $100.7/n$ indicates that the system is reversible.

FIGURE 16-10 / Graphical method for determining, n, the number of electrons in the electrode reaction.

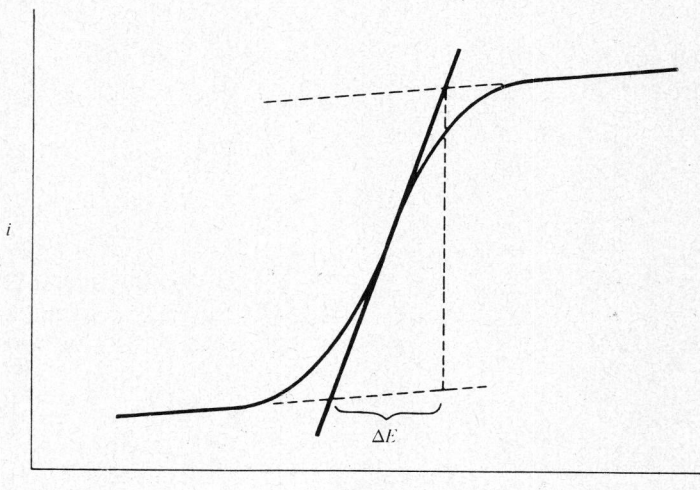

DETERMINATION OF $E°$, THE STANDARD OXIDATION POTENTIAL

For a reversible reaction in dilute solution the half-wave potential is very close to $E°$:

$$(16\text{–}28) \qquad E° = E_{1/2} - \frac{RT}{n\mathrm{F}} \ln \left(\frac{D}{D'} \right)^{1/2}$$

D = diffusion constant of reactant
D' = diffusion constant of product

In most cases the ratio of the diffusion constants is fairly close to unity, and the second term on the right-hand side of equation (16–28) drops out. However, even if the ratio of D/D' was as great as 1.1, the difference between $E°$ and $E_{1/2}$ would be only 1.2 mV for a one-electron process and 0.6 mV for a two-electron process. Polarographic measurements are therefore useful in obtaining quick and fairly accurate values of $E°$, especially so in the case of short-lived intermediates. For accurate thermodynamic values the half-wave potentials should be extrapolated to zero ionic strength.

COMPOSITION AND STABILITY CONSTANTS OF COMPLEX IONS

The half-wave potential of a complex ion is different from that of the free ion because of the energy of formation of the complex. From this difference the stability constant and the number of ligands of the complex can be calculated.

The reaction forming the complex is

$$\mathrm{M}^{x+} + p\,\mathrm{A}^{y-} \rightleftharpoons \mathrm{MA}_p^{x-py}$$

In the presence of an excess of A^{y-} large enough for its concentration at the electrode to be constant and virtually equal to its concentration in the bulk of the solution, the stability constant may be written as

$$K = \frac{[\mathrm{MA}_p]_0^{x-py}}{[\mathrm{M}^{x+}]_0[\mathrm{A}^{y-}]^p}$$

where the subscript 0 refers to the concentration at the surface. The electrode reaction is the reversible reduction of the free ion to metal. The metal atoms are removed by diffusion into the surface of the mercury. The difference between the half-wave potentials of the complex and that of the free ion is

$$(16\text{–}29) \qquad E_{1/2,\,\text{complex}} - E_{1/2,\,\text{ion}} = \frac{RT}{n\mathrm{F}} \left\{ \left(\ln \frac{D_{\text{comp}}}{D_{\text{ion}}} \right)^{1/2} - \ln K - \ln[\mathrm{A}^{y-}]^p \right\}$$

$$\approx -\frac{RT}{n\mathrm{F}} (\ln K + \ln[\mathrm{A}^{y-}]^p)$$

Polarographic waves are obtained at a series of concentrations of A^{y-}, and the half-wave potentials are plotted against $\ln A^{y-}$. The graph is a straight line whose slope is $-pRT/nF$ and whose intercept is $E_{1/2,\,\mathrm{ion}} - (RT/nF) \ln K$.

KINETICS OF CHEMICAL REACTIONS ASSOCIATED WITH ELECTRON TRANSFER

Chemical Reactions Preceding the Electrode Process

$$X \rightleftharpoons A \quad \text{chemical reaction}$$
$$A \pm ne \rightleftharpoons Y \quad \text{electron transfer}$$

If the chemical reaction is fast, the process will be controlled by the rate at which X diffuses to the surface. The maximum current will be the limiting diffusion current due to X. If the chemical reaction is slow, however, the maximum value of the current will be less than the limiting diffusion current of X. Other diagnostic criteria for this type of process are (1) the current is smaller than the diffusion rate, and (2) the temperature coefficient is higher than that for diffusion. (Activation energies for diffusion are of the order of 2–3 kcal, so temperature coefficients are quite small.) One example of this type of process is the reduction of monosaccharides. The limiting currents for these reductions may be two orders of magnitude less than the hypothetical diffusion currents, and the temperature coefficients are an order of magnitude greater than those of diffusion. Equations have been worked out for calculating the rate constants of the kinetic step from the ratios of limiting current to limiting diffusion current.

Chemical Reactions Following the Electron Transfer

$$A \pm ne \rightleftharpoons B \quad \text{reversible electron transfer}$$
$$B \rightleftharpoons X \quad \text{chemical reaction}$$

The chemical step may be considered to be the elimination of B. The polarographic wave should have a reversible form, but the half-wave potential should be shifted, because the surface concentration of product is less than would have been the case without the elimination reaction. Examples of this are the anodic oxidation of ascorbic acid, the anodic oxidation of cadmium amalgam in solutions containing ethylenediaminetetraacetic acid, and reduction of unstable complexes.

Chemical Reactions Following the Electron Transfer, with Regeneration of Reactant

$$A \pm ne \rightleftharpoons B \quad \text{reversible electron transfer}$$
$$B + X \rightleftharpoons A + Y \quad \text{chemical regeneration of reactant}$$

Here the limiting current is larger than that due to the diffusion of A, either as calculated or measured in the absence of X. Various formulas have been derived for calculating the rate constant of the chemical reaction from the ratio of the limiting current to the limiting diffusion current of A. These have been applied to such reactions as the reduction of Ti^{4+}.

Techniques of Polarographic Measurement

The electric circuitry will not be discussed here, as complete packaged instruments are available. These furnish a DC potential that will change at a controlled rate. The current generates a signal that is measured on a strip chart recorder.

Various types of cells are used, either of the beaker type, with all electrodes dipping into the same solution, or of the H type, with a sintered glass disk such as shown in Figure 16-11. The solutions may be separated by plugs of conducting gel (silica gel formed by precipitating sodium silicate with acid or agar saturated with salts).

In polarographic reductions dissolved oxygen must be removed, since it is reducible. This is best accomplished by bubbling nitrogen through the solution before the electrolysis.

The electrode arrangement may be either the one diagrammed in Figure 16-12a or the one in Figure 16-12b. In Figure 16-12a one electrode has a known fixed potential and is used as both the counterelectrode (for the electrolysis) and the reference electrode (for the measurement). The potential difference between the two electrodes is the indicator electrode potential plus a known constant. However, there is also an ohmic drop across the solution that

FIGURE 16-11 / H type of polarography cell with a sintered glass disk.

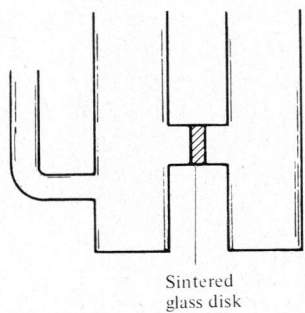

Sintered glass disk

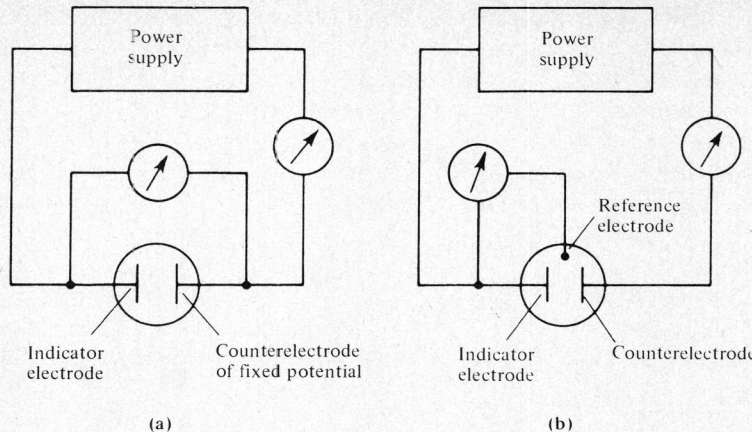

FIGURE 16-12 / Electrode arrangements for polarography.

must either be calculated or compensated for. To avoid ohmic drop, the three-electrode arrangement in Figure 16-12b is used. Here the ohmic drop is negligible, since the reference and indicator electrodes are close to each other.

The most frequently used electrode in polarography is the dropping mercury electrode. It consists of a capillary tube of about 0.03–0.05 mm inner diameter through which mercury flows from a reservoir into the solution, as shown in Figure 16-13. The mercury forms droplets at the tip of the capillary,

FIGURE 16-13 / Dropping mercury electrode.

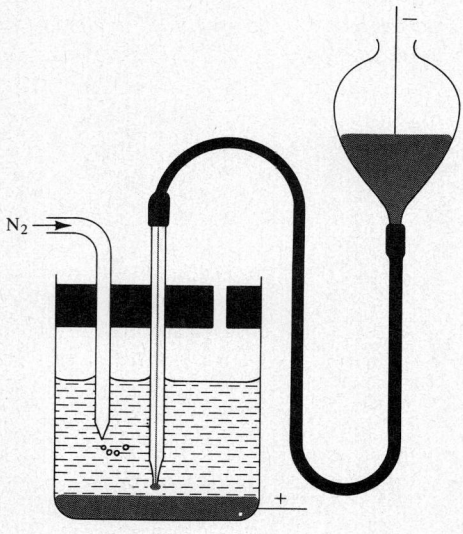

which expand in size and then break off. For most cases the height of the mercury is adjusted so that a new drop forms every 2–6 sec. Electrical contact is made at the mercury reservoir. As the droplet grows in size, its surface area increases steadily. Simultaneously, the current increases to a maximum and then falls off sharply as the drop breaks off and a new drop starts to form. Usually, the mean current is measured. Its relationship to the other experimental quantities was first derived by Ilkovič. At 25°C, the equation is

(16–30) $$i = 607 n c_s D^{1/2} m^{2/3} t^{1/6}$$

i = current, μA
n = number of electrons
c_s = concentration of electrolyte, mmol/L
D = diffusion constant, cm^2/s
m = flow of mercury, mg/s
t = drop time, s

At a solid electrode the passage of current depletes the solution at the surface. Consequently, a concentration gradient rapidly extends out toward the regions where convection starts, and random convection currents can cause irreproducibility. To overcome this, the solution must be stirred or the electrode moved back and forth in the solution. As the drop grows in a dropping-mercury electrode, the surface moves out toward the solution, keeping the diffusion layer narrow. Currents are therefore larger and results more reproducible, even without stirring (except for that caused by the falling droplets).

The dropping mercury electrode has several outstanding advantages, which account for its wide use. Each new droplet formed has a fresh, clean, reproducible surface, so that there is very little chance of the electrode's being poisoned by impurities or reaction products. The surface is smooth, so its real area corresponds to the geometric area. There is no necessity for determining roughness factors as is sometimes necessary with solid electrodes. Currents are larger and results more reproducible, as discussed above. As with other types of mercury electrodes, hydrogen is not evolved at cathodic potentials of less than 1 V, so hydrogen evolution does not interfere with the deposition of metals. One disadvantage, however, is that the anodic potential cannot exceed 0.4 V, at which point the mercury starts to ionize.

Solid electrodes are also used, especially for anodic measurements. Platinum, gold, and carbon are often used, but they have serious disadvantages. Gold and platinum are easily poisoned by traces of impurities, oxides form films, and hydrogen dissolves in the metal and is removed with difficulty. Each electrode has a different roughness factor. Graphite and platinized platinum have especially large ratios of real to apparent surface area. The real surface area may change during the electrolysis or during cleaning procedures. Results are therefore fairly irreproducible and depend greatly on the previous history of the electrode.

As mentioned above, with electrodes other than the dropping mercury electrode, to produce steady state conditions, the solution must be stirred or the electrode rotated or vibrated. The recommended rate of rotation is about 600 rpm. The vibrating electrode is a wire that vibrates in the longitudinal direction about 50–100 times a second, with an amplitude of about 1 mm. Mercury electrodes can also be rotated. There are, in addition, a variety of special purpose electrodes, including the rotating disk electrode, ring and disk electrode, rotating dropping mercury electrode, vibrating dropping mercury electrode, stationary mercury electrode, and many others. Details as to the construction of each of these may be found in the literature.

References

Bates, R. G., *Determination of pH: Theory and Practice.* Wiley, New York, 1964.

Charlot, G., J. Badoz-Lambling, and B. Trémillon, *Electrochemical Reactions.* Elsevier, Amsterdam, 1962.

Delahay, P., *New Instrumental Methods in Electrochemistry.* Wiley–Interscience, New York, 1954.

Heyrovský, J., and J. Kůta, *Principles of Polarography*, Academic Press, New York, 1966.

Ives, D. J. G., and G. L. Janz, *Reference Electrodes, Theory and Practice.* Academic Press, New York, 1961.

Kortüm, G., and J. O'M. Bockris, *Textbook of Electrochemistry*, Elsevier, Amsterdam, 1951.

Meites, L., *Polarographic Techniques.* Wiley–Interscience, New York, 1955.

Nurnberg, H. W., *Electroanalytical Chemistry*, Advances in Analytical Chemistry and Instrumentation, Vol. 10, Wiley, New York, 1974.

Tomilov, A. P., S. G. Mairanovskii, M. Y. Fioshin, and V. A. Smirnov, *Electrochemistry of Organic Compounds.* Halsted Press (Wiley), New York, 1972.

Vetter, K., *Electrochemical Kinetics.* Academic Press, New York, 1967.

Willard, H. H., L. L. Merritt, Jr., and J. A. Dean, *Instrumental Methods of Analysis*, 5th ed. Van Nostrand, New York, 1974.

Weissberger, A., and B. W. Rossiter, *Physical Methods of Chemistry*, Part II, A, chs. 1–5. Wiley-Interscience, New York, 1971.

Chapter 17

Polarimetry

In ordinary (unpolarized) light the electric fields are vibrating at all possible angles in the plane normal to the direction of propagation. In plane-polarized light the electric vibrations are back and forth in one direction in that plane. The direction of vibrations in that plane (the direction of the plane of polarization) depends on the nature and the orientation of the polarizing material.

Polarized light may be produced in a variety of ways. The most widely used method employs prisms made from anisotropic crystals. A beam of ordinary light, on being passed into an anisotropic crystal at an angle neither parallel nor perpendicular to the optic axis, splits into two beams that are plane-polarized at right angles to each other. The splitting results from the interaction between the fields of the light and the spherically asymmetrical fields within the anisotropic crystal.

The most widely used polarizing prism is the *Nicol prism*, a calcite prism cut in two and cemented together with Canada balsam. The angles and refractive indices are such that one of the two polarized beams, the ordinary ray, undergoes total internal reflection, while the other beam, the extraordinary ray, is transmitted. The ray is polarized parallel to the optic axis, and the direction of the plane of polarization therefore depends on the orientation of the Nicol prism. Two Nicol prisms with their optic axes at right angles to each other will transmit no light at all.

Polarized light may also be produced with dichroic crystals such as tourmaline. These do not break the light into two polarized beams, but instead simply absorb all the component polarized in one direction and transmit the residual light as a beam polarized in the other direction. Commercial Polaroid plates consist of many small dichroic crystals aligned in a transparent plastic matrix.

Although polarized light has many uses, to the physical chemist one of its principal functions is in *polarimetry*, the measurement of the angle of rotation, that is, the angle through which the plane of polarization of the light is rotated by passage of the beam through an optically active material. This rotation or change in the plane of vibration of the plane-polarized light depends upon the wavelength of the light, the nature of the material, its density or concentration, and the length of the light path. In solids optical rotation may result from the spatial arrangements of the entities making up the crystal as well as from the structure of the molecules or ions. In gases, liquids, and solutions, optical activity results only from the internal structure of the molecules. Polarimetric measurements on optically active liquids give valuable information about molecular structure, since optical rotation can result from steric hindrance, restricted rotation, or spatial isomerism. Polarimetry is also valuable as a specialized analytical tool.

Split Field Polarimeter

As shown in Figure 17-1, the split field polarimeter consists of two large Nicol prisms—the polarizer and the analyzer—a small Nicol prism to produce the split field, a sample tube, and a lens and eyepiece.

A beam of monochromatic light from a sodium vapor lamp, or sometimes a mercury vapor lamp, is plane-polarized by passage through the polarizing prism. The polarized beam passes through the cell into the analyzing prism, and then to the eyepice. With distilled water or other optically inactive liquid in the tube, the visual field is at maximum brightness with the prisms parallel and at minimum brightness with the prisms crossed at right angles to each other. The usual practice is to work with minimum brightness, so that a small absolute change in light intensity will be a large relative change.

It is easier to match the intensities of two portions of a field than to bring a uniform field to maximum or minimum brightness. Therefore, the visual field is split into two halves, with the third, smaller prism, the splitting prism. This prism is set at a slight angle to the polarizer and rotates the plane of polarization of half the visual field. If the analyzer is then set at a certain angle to the other two prisms, the two halves of the field will be at the sample intensity as long as there is no optically active material in the same tube. However, if there is an optically active material present in the tube, the plane of polarization of

FIGURE 17-1 / Split field polarimeter.

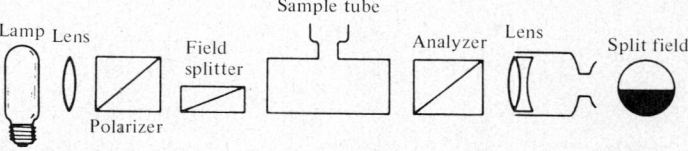

the incident polarized light will be rotated through some angle, depending upon the concentration and the nature of the active material. The plane of polarization of the light in the two halves of the field will then be at different angles to the analyzer, and the two half-fields will have different intensities. Rotating the analyzer through the same angle through which the active material has rotated the plane of polarization of the incident light will restore the original condition of equal intensities in the two halves of the visual field. This angle is the angle of rotation of the optically active material.

OPERATING INSTRUCTIONS

Clean the polarimeter tube and rinse it several times with the liquid being studied. Focus the polarimeter upon a source of monochromatic light about 20 cm distant and adjust the analyzer until both halves of the field are of equal darkness.

Record the setting and repeat the measurement several times, approaching the balance point from both directions. In the calculations use the average value and the appropriate precision measure.

CALCULATION

(17-1) $$\alpha = \frac{[\alpha]_\lambda^t lc}{100}$$

α = observed angle of rotation
$[\alpha]_\lambda^t$ = specific rotation, a constant for the particular system and conditions
t = temperature of the solution, °C
λ = wavelength of the polarized light
l = length of the tube, dm
c = concentration, g solute/100 ml solution

NOTES / The solutions must be clear and the lenses and tube windows clean, with no liquid on the outside and no air bubbles.

Use distilled water to check the calibration of the dial.

References

Crabbé, Pierre, *ORD and CD in Chemistry and Biochemistry: An Introduction.* Academic Press, New York, 1972.

Djerassi, C., *Optical Rotatory Dispersion: Application to Organic Chemistry.* McGraw-Hill, New York, 1960.

Feynman, R. P., R. B. Leighton, and M. Sands, *The Feynman Lectures on Physics*, Vol. I, chs. 32, 33. Addison-Wesley, Reading, Mass., 1963.

Weissberger, A., and B. W. Rossiter, *Physical Methods of Chemistry*, Part III, chs. 1–3 Wiley–Interscience, New York, 1972.

Willard, H. H., L. L. Merritt, Jr., and J. A. Dean, *Instrumental Methods of Analysis*, 5th ed. Van Nostrand, New York, 1974.

Chapter 18

Refractometry

Light consists of electric and magnetic fields vibrating at right angles to each other and to the direction of propagation. In any medium the apparent velocity of light is less than in a vacuum because of the interaction between the fields of the light and the fields associated with the molecules of the medium. When a beam of light passes from a less dense into a more dense medium, its passage is slowed, because of phase shifts. As a result the direction of the light is shifted, or refracted, in a direction toward the normal to the surface. (The effect is analogous to the shift in the direction of a moving automobile when the brakes are applied on one side but not on the other.) As shown in Figure 18-1, when a

FIGURE 18-1 / Critical angle diagram.

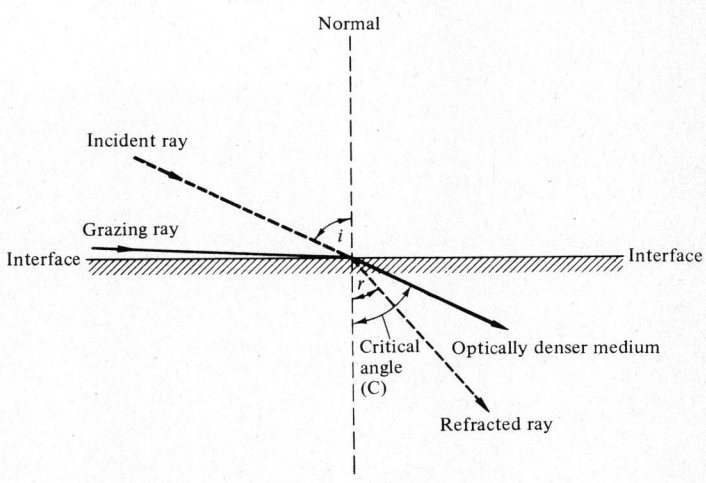

light beam enters a liquid or solid from the air, the angle of incidence i is greater than the angle of refraction r. The ratio of the sine of the angle of incidence to that of the angle of refraction is the *refractive index*, given the symbol η. (Strictly speaking, this definition holds only if the light enters the more dense medium from a vacuum. However, unless the refracting medium is a gas, air at ordinary pressures may be substituted for a vacuum.)

The refractive index depends upon the wavelength of the light and upon the fields in the light path. Experimentally, η increases with decreasing wavelength of light and with increasing pressure and decreasing temperature. Measurements of refractive indices are usually made and values are listed in the literature for 20°C and atmospheric pressure at the wavelength of the convenient and easily reproducible D line of sodium.

Under specified and controlled conditions η is a property of the material and is extremely useful in chemical analysis and process control. In experimental physical chemistry refractive indices furnish valuable information on molecular structure. The *molar refraction* is defined by

$$(18-1) \quad R = \frac{(\eta^2 - 1)M}{(\eta^2 + 2)\rho}$$

R = molar refraction
M = molecular weight
ρ = density

The molar refraction depends upon the number (additive property) and the kinds (constitutive property) of atoms and bonds present. Experimental values of R can be used to distinguish between structural isomers, although the methods of choice are now infrared spectroscopy and nuclear magnetic resonance. More significant to the physical chemist, however, is the fact that the value of molar refraction is approximately equal to that of the *induction polarization*, and that it is of great importance in the calculation of dipole moment.

Liquids, gases, and crystals with a cubic lattice structure are *isotropic*: that is, they have refractive indices that are independent of the direction of the incident light. However, crystals other than cubic are *anisotropic*. In each direction within the crystal there is a different number and arrangement of atoms and therefore different optical, electrical, and mechanical properties. For anisotropic crystals, not only will the refractive index be different for different directions relative to the optic axis, but usually a beam of incident light, unless it is normal to the surface, will split into two polarized beams going in different directions (see Chapter 17 on polarimetry). For such crystals reported refractive indices must specify the direction of the light in addition to the temperature, pressure, and wavelength.

Refractive indices of liquids and solids are usually measured with a refractometer. The most widely used, the Abbe and the Pulfrich refractometers, measure the *critical angle* (see below).

Abbe Refractometer

The *Abbe refractometer* (Figure 18–2) consists essentially of two prisms to refract the light, two additional prisms to converge dispersed light if necessary, lenses to produce and focus a beam, a telescopic eyepiece, and a scale.

A beam of light passes through the "illuminating prism" and is scattered from the frosted glass inner surface, entering the film of liquid at all possible angles. The rays of light pass through the liquid, strike the surface of the dense and highly refractive "refracting prism" at all possible angles, and are refracted toward the normal to the prism surface. The prism–lens system collects all the refracted rays into a beam focused at the eyepiece of the telescope. Rotating the refracting prism relative to the telescope focuses different rays at the center of the visual field.

Measurements are made by rotating the prism relative to the eyepiece until the visual field is split into dark and bright halves, with the line of demarcation located at the cross hairs marking the center. The split field is produced by focusing the telescope on the ray of light that leaves the refracting prism–liquid interface at the angle C, the *critical angle* (Figure 18–1). Since the ray that just grazes the surface (i.e., strikes it at the smallest possible angle) is refracted at angle C, all other rays are refracted at angles less than C. (No ray can strike the prism surface at an angle smaller than the grazing angle.) Therefore, focusing the telescope so as to locate the ray refracted at the critical angle at the center of the cross hairs will produce a split field, half of which is bright with rays of light refracted at angles less than C and half of which is dark because of the absence of light refracted at angles greater than C. From the angle C the refractive index could be calculated using the relationship

(18–2) $\quad \eta = \eta_p \sin C$

η_p = refractive index of the prism

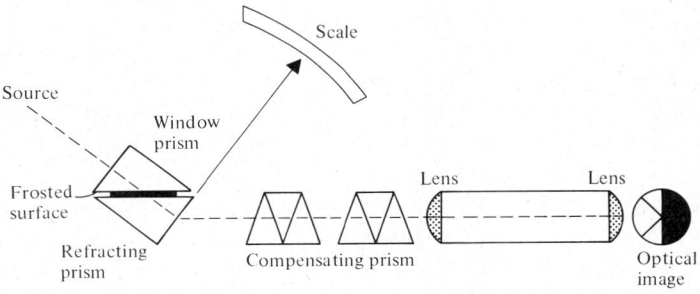

FIGURE 18–2 / Simplified optical system of the Abbe refractometer.

However, it is more convenient to calibrate the scale directly in terms of η.

When polychromatic light is used, the prisms disperse the light into a spectrum, broadening the edge of the split field into a band. Adjustment of the Amici (compensating) prisms converges the band back into a line.

OPERATING INSTRUCTIONS

Adjust the mirror so that light enters the window prism.

Put 2–3 drops of liquid on the lower surface and close and lock the prisms. If the liquid is volatile, close the prisms first and then add the liquid through the groove in the jacket.

Move the prism back and forth by means of the adjustor arm or knob until the visual field is partly bright and partly dark.

Slide the eyepiece up and down until the cross hairs are in sharp focus.

If the light source is not monochromatic, adjust the compensator prism until the edge of the light field is sharp and colorless.

Use the fine adjustment to center the boundary line between the light and dark fields on the cross hairs.

Read the refractive index directly from the scale.

NOTES / Do not use soap or any alkaline material on the soft glass prisms. Clean the prisms by swabbing them gently with cotton or lens paper wet with alcohol. Be careful not to scratch the soft glass.

Circulate water from a thermostat through the prisms if high precision and accuracy are desired or if volatile materials are being used. For each degree of change in temperature the refractive index of most liquids changes by about one or two units in the fourth decimal place.

If the prisms are not thermostatted, keep the light off except when making measurements.

If evaporation occurs, the refractive index may change because the less volatile materials present as solute, solvent, or impurity will become more concentrated. (This is not important for fairly pure materials. To concentrate an impurity from 0.1 to 1%, 90% of the solvent must be removed.)

Distilled water is a handy reference liquid.

EXERCISE

1. Even though polychromatic light is used, the refractive index is measured in terms of the sodium D line. Explain how the compensator prism makes this possible.

Refractive indices of vapors and gases are not usually measured with refractometers. Instead, the instrument used is an interferometer, which measures the interference pattern between two beams of light from the same source that have undergone a relative phase change by being passed through two different materials before arriving at the receiver. From the change in the interference pattern, the refractive index of one material, the length of the light path, and the wavelength of the light the refractive index of the other material may be determined very accurately. This interferometer technique may also be applied to liquids, solids, and solutions, but the apparatus is more expensive and less convenient than the relatively simple refractometers, except for studying very small crystals under a microscope.

References

Feynman, R. P., R. B. Leighton, and M. Sands, *The Feynman Lectures on Physics*, Vol. I, chs. 26, 31. Addison-Wesley, Reading, Mass., 1963.

Weissberger, A., and B. W. Rossiter, *Physical Methods of Chemistry*, Part III,C. Wiley–Interscience, New York, 1972.

Willard, H. H., L. L. Merritt, Jr., and J. A. Dean, *Instrumental Methods of Analysis*, 5th ed. Van Nostrand, New York, 1974.

Chapter 19

Absorption and Emission of Radiation

Phenomena involving the interaction between electromagnetic radiation and matter have a wide range of application, from analysis to structure determination and photochemistry. For clarity and brevity, this chapter is restricted to the absorption and emission of radiation primarily by gases and dilute solutions in the infrared to ultraviolet region of the spectrum.

Theory of Absorption and Emission of Radiation

When radiation is absorbed or emitted, the atom or molecule undergoes a transition from one energy state to another. The absorption of radiation increases internal energy, while emission decreases internal energy. The relationship between the initial and final energy states and the frequency of the radiation is

(19-1) $\Delta E = |E_2 - E_1| = h\nu$

ΔE = energy of photon
E_2 = energy of final state
E_1 = energy of initial state
h = Planck's constant
ν = frequency

Electronic transitions usually require the large amount of energy (on the order of 10^{15} Hz or 4×10^2 kJ/mole) found in ultraviolet light (100–400 nm), although some electronic transitions (e.g., in pigments and dyes) require only the somewhat smaller energies found in visible light (400–700 nm).

Vibrational transitions require less energy than electron transitions (from 10^{13} to 3×10^{14} Hz), and utilize radiation in the near-infrared region (10^3 to 2×10^4 nm). Rotational transitions, except for the very lightest molecules, require only the relatively small amounts of energy (from 10^{12} Hz on down) found in the far-infrared region (10^4 nm and above).

The molecules on any *electronic level* are distributed over the *vibrational levels*, obeying the Boltzmann distribution law. On any *vibrational level* the molecules are distributed over the *rotational levels*, obeying the Boltzmann distribution law. At ordinary temperatures vibrational energy level spacings are greater than kT, so that a relatively small fraction of the molecules are in excited vibrational states. On the other hand, rotational energy level spacings are less than kT, so that rotational levels are more evenly populated. Figure 19-1 is a schematic representation of the various energy levels, with vibrational levels superimposed upon the electronic levels and rotational levels superimposed upon the vibrational levels.

A change in electronic energy changes the charge distribution of the molecules, and therefore changes the energies of the vibrational levels and the equilibrium internuclear distances corresponding to the vibration. According to the Franck–Condon principle, electronic transitions occur so much faster than nuclear motions that the nuclei may be assumed to be stationary during the transition. As a consequence, electronic transitions are accompanied by all possible vibrational transitions, which in turn are accompanied by all possible rotational transitions. Electronic spectra of molecules, therefore, have both vibrational and rotational spectra superimposed on them, and vibrational spectra have superimposed rotational spectra. In the case of electronic spectra, the rotational spectrum is not observed except with very high resolution apparatus in gaseous materials, but vibrational contributions of molecules in the

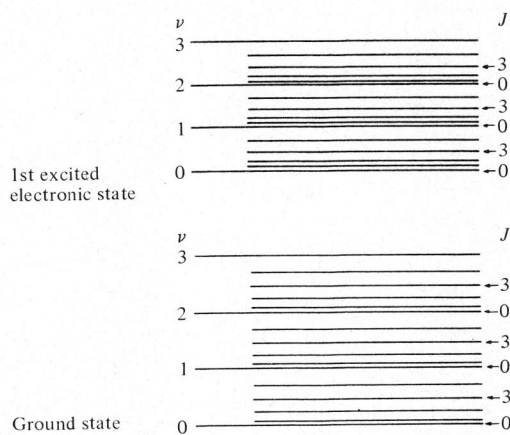

FIGURE 19-1 / Energy levels in a molecule.

gaseous state may be readily observed. In fact, for molecules without dipole moments, it may be necessary to use the electronic spectrum to obtain vibrational spectra.

For studies of the internal structure of molecules, spectra are usually obtained from the gas phase at relatively low pressures. In liquids and solids, intermolecular attractions cause variation in the energy levels and produce *line broadening*. The fine structure disappears and the individual lines coalesce into a broad band. Inelastic collisions in liquids and high pressure gases also change the energies of molecules that are about to emit a photon, thus altering the frequencies of the emitted radiation and broadening the line. In gases this is called *pressure broadening*.

Selection Rules

Experimentally it has long been known that many substances exhibit spectra that seem to be incomplete: sometimes individual lines, corresponding to certain transitions, are missing, and sometimes the spectra for certain types of transitions, such as rotation, are missing. In spectroscopists' terminology, lines that appear are *allowed*, while lines that do not appear at all are *forbidden*. The theoretical explanation of these observations has now been provided by quantum mechanics. Equations have been developed for predicting the probability of transitions. Those with a zero probability are forbidden. The general rules established by solution of the appropriate quantum mechanical equations are called *selection rules*. Table 19-1 lists the selection rules for diatomic molecules.

The restrictions listed in Table 19-1 concerning dipole moments arise from the fact that, as a first approximation for rotational and vibrational transitions, the electromagnetic radiation must interact with an electric dipole. In some polyatomic molecules stretching and bending vibrations produce temporary dipoles that can interact with light whose frequency is the same as that of the molecular vibration. These temporary dipoles do not permit pure rotational transitions, since rotational transitions occur at much lower frequencies than vibrations. Vibrational spectra, however, can be produced when the

TABLE 19-1 / Selection Rules for Diatomic Molecules

Transition Type	Permitted Transitions[a]
Rotation	$\Delta J = \pm 1$
Vibration	$\Delta v = \pm 1, \Delta J = \pm 1$

[a] J, rotational quantum number; v, vibrational quantum number.

molecule has a fluctuating dipole. Diatomic molecules do not have such asymmetrical stretching or bending vibrations, so that for them to exhibit vibrational spectra they must have a permanent dipole.

In addition to the permitted transitions listed in Table 19-1 there are some transitions that occur as a result of interactions between the light and magnetic dipole moments or magnetic or electric quadrupole moments. However, the probability of such a transition is very much less than that of ordinary transitions, and these lines are for the most part rather weak. There are also many transitions that are theoretically forbidden but that actually show up in spectra. In these cases the discrepancy between theory and experiment arises from the fact that transition probability calculations are made using wavefunctions that are known only approximately.

The selection rules in Table 19-1 do *not* apply to electronic transitions in molecules. As a result, all possible rotational and vibrational changes may be superimposed on electronic changes, and some molecules that have no infrared spectrum can be studied in the uv–visible.

Electronic molecular spectra consist of broad bands that are separated into discrete lines only on high resolution. With atoms, however, since there is neither vibration nor rotation, the spectra consist of variably spaced lines in the ultraviolet, visible, and, on occasion, in the near infrared, ranges.

Atomic spectra can be observed by passing continuous radiation through the sample and determining the frequencies that are absorbed. This is possible only with gaseous materials. The more usual method is to excite the electrons by heat or by electric arc discharges. On dropping back to lower energy levels the electron emits radiation in the form of an *atomic bright line spectrum*. One important line spectrum is that of hydrogen excited by an electric voltage. The spectrum consists of several series of bright lines, each of which obeys the formula

(19-2a) $$\bar{\nu} = \frac{1}{\lambda} = R\left(\frac{1}{n_1^2} - \frac{1}{n_2^2}\right)$$

$\bar{\nu}$ = wavenumber, Keysers
λ = wavelength, cm
R = Rydberg constant, 109,677.76 cm^{-1}
n_1 = an integer, the series term, constant for a given series
n_2 = an integer, larger than n_1

For the most familiar series, the Balmer series in the visible and the near infrared, $n_1 = 2$. The spectra of the other hydrogen-like atoms—deuterium, singly ionized helium, doubly ionized lithium, and triply ionized beryllium—all follow a similar formula:

(19-2b) $$\bar{\nu} = RZ^2\left(\frac{1}{n_1^2} - \frac{1}{n_2^2}\right)$$

Z = atomic number

PRINCIPLES

In each of these cases, however, the value of the Rydberg "constant" R changes slightly, since it includes the reduced mass

$$\mu = \frac{m_e m_N}{m_e + m_N}$$

m_e = mass of the electron
m_N = mass of the nucleus

For heavy atoms the value of the Rydberg constant approaches a limiting value, 109,737.4, differing from that for hydrogen by about 6 parts in 10,000.

Atomic bright line spectra represent one type of a general phenomenon known as *fluorescence*, which covers all cases where an electron drops down to a less energetic level by an *allowed* transition. If the transition is *forbidden* by the selection rules but does occur nonetheless, the emission is sometimes called

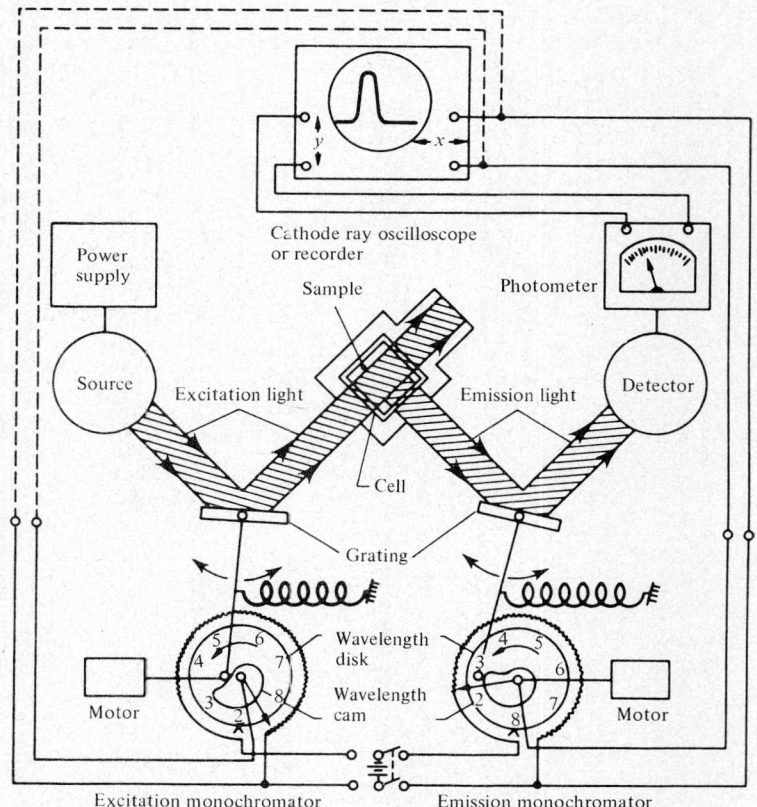

FIGURE 19-2 / Functional diagram of the Aminco–Bowman spectrophotofluorometer (courtesy of American Instrument Company).

phosphorescence, in which case there is a considerable delay before the transition occurs.

The most important use of fluorescence to the chemist is in quantitative chemical analysis. In analysis by spectrometry (see below) the difference in the radiation intensity is measured before and after passage through the absorbing material. Often this represents a small difference between two large quantities. In fluorimetry, at low concentrations emitted radiation is directly proportional to the concentration. Consequently, fluorimetry can be used at concentrations far below those applicable to photometry. From the viewpoint of structure studies and theoretical chemistry, fluorescence measurements furnish much information about the excited state, since the intensity is directly proportional to the number of excited molecules and the wavelengths represent the energy transitions of the excited state.

In fluorimetry the emitted radiation must be separated from the initial exciting radiation, so that the detector of the fluorimeter is set at 90° to the impinging beam (Figure 19–2). Extended discussion of fluorescence is beyond the scope of this text. Further discussion in this chapter is restricted to absorption of radiation.

Applications of Absorption of Radiation to Chemistry

ANALYSIS

The widest application of absorption spectrophotometry is in analytical chemistry. The spectrum is characteristic of the particular material, and so spectrometery is the preferred method of qualitative analysis.

For quantitative analysis the concentration of material is related to the absorption of radiation by the *Beer–Lambert law:*

(19–3a) $\quad P = P_0 \times 10^{-abc}$

P = intensity of transmitted light
P_0 = intensity of incident light
a = absorptivity
b = length of optical path
c = concentration of absorbent

For convenience this is usually expressed in terms of base 10 logarithms:

(19–3b) $\quad A = \log \dfrac{P_0}{P} = abc$

A = absorbance (formerly called optical density)

In many fields, such as biochemistry, spectrophotometry is the preferred method of analysis, and in special areas, such as the kinetics of fast reactions in solution, spectrophotometric methods are indispensable.

MOLECULAR STRUCTURE AND PROPERTIES

In these studies the interest lies in spectrometry, the characteristic frequencies of absorption, rather than in the intensity of the absorption.

The infrared spectrum of a molecule is a composite of the vibration-rotation spectra of the specific groups that make up the molecule, although there are interactions where the spectrum of one group will be affected by the presence of another group. Although for polyatomic molecules the spectra become too complicated to be unraveled by any but the expert, nevertheless the presence or absence of functional groups such as ketone and ether can be established almost by inspection of the spectrum.

From the rotational spectrum, as obtained either in the far-infrared or from the fine structure of the rotation–vibration spectrum, the moments of inertia of the molecule can be determined. Diatomic molecules and linear polyatomic molecules have two moments of inertia, which are equal to each other; that is, they have the same magnitude. Therefore only one value of the moment of inertia shows up in the spectrum. For a diatomic molecule, as a first approximation, good to about 1 part in 10,000,

(19-4) $$v = \frac{2h(J+1)}{8\pi^2 I} = 2B(J+1)$$

v = radiation frequency
J = rotational quantum number of the lower energy state
I = moment of inertia
$B = h/8\pi^2 I$

The frequency difference between two adjacent rotational lines is

(19-5) $$\Delta v = 2B$$

Since the moment of inertia of a diatomic molecule is defined as

$$I = \mu r^2$$

μ = reduced mass
  $= m_1 m_2 / (m_1 + m_2)$
r = internuclear distance

the equilibrium internuclear distance is obtained with good accuracy from the rotational or rotational–vibrational spectrum. For nonlinear polyatomic molecules there are three moments of inertia and there are sometimes internal moments of rotation, and the problem rapidly becomes quite complicated.

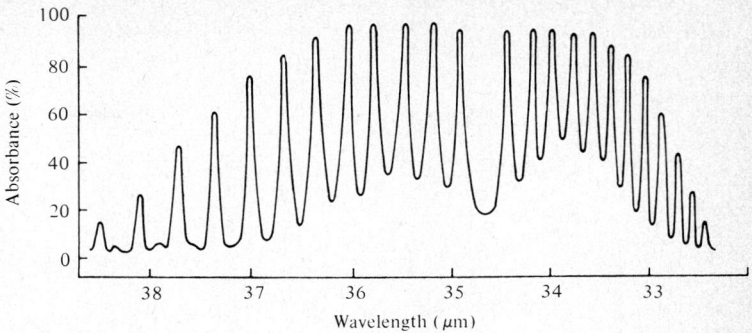

FIGURE 19-3 / Rotational lines in the 3.46 μm vibrational band of gaseous HCl.

Vibrational spectra consist of sets of lines, each of which results from rotational changes superimposed on vibrational changes. The rotation–vibration spectrum of a diatomic gas is shown in Figure 19-3. Note that a line at the center is missing, which corresponds to the forbidden vibrational transition without an accompanying rotational transition, that is, one in which $\Delta J = 0$. The position of the missing line corresponds to the fundamental vibration frequency. The frequency difference between consecutive lines is $2B$, so the moment of inertia can be calculated as described above. The force constant of the vibration is obtained from the (missing) fundamental frequency:

(19-6) $\quad K = 4\pi^2 \bar{v}^2 c^2 \mu$

K = force constant
$\bar{v}$ = wavenumber = wavelength^{-1}
c = speed of light

From the vibration frequency and the moments of inertia the partition functions can be calculated, and from these all the thermodynamic properties of the substance can be determined:

(19-7) $\quad Z_{vib} = \dfrac{e^{-hc\bar{v}/2kT}}{1 - e^{-hc\bar{v}/2kT}}$

Z_{vib} = vibrational partition function

(19-8) $\quad E_{vib} = RT^2 \dfrac{d \ln Z_{vib}}{dT}$

(19-9) $\quad C_{vib} = \dfrac{R}{T^2} \dfrac{d^2 \ln Z_{vib}}{d(1/T)^2}$

For a diatomic gas at 300°K, equation (19–8) becomes

(19–10) $$E_{vib} = \frac{Nhc\bar{v}}{e^{hc\bar{v}/kT} - 1} + \frac{Nhc\bar{v}}{2}$$

The first term on the right is the vibrational energy in excess of the zero-point energy per mole. The second term is the zero-point vibrational energy.

For a diatomic gas, equation (19–9) becomes

(19–11) $$C_{vib} = \frac{Re^{hc\bar{v}/kT}(hc\bar{v}/kT)^2}{(e^{hc\bar{v}/kT} - 1)^2}$$

In electronic spectra the superimposed rotational energies are too small to be observed without very high resolution, but the vibrational lines show up. If the vibrating atoms formed a harmonic oscillator, the vibrational lines would be equally spaced. Since the molecule is not a harmonic oscillator, the lines become closer to each other at high vibrational quantum numbers. Ultimately, the vibrational lines converge in a continuum. The frequency at which the continuum starts is called the *convergence limit*. Above the convergence limit radiation absorption is continuous because the absorbing molecule has dissociated into fragments of variable kinetic energies. Convergence limits are especially valuable in determining bond energies. Figure 19–4a and c shows the relationships for a diatomic molecule, between the frequency at the convergence limit, the atomic excitation energy, and the bond energy. Figure 19–4b and d shows schematic spectra corresponding to the transitions in Figure 19–4a and c, respectively.

The example shown in Figure 19–4a is that of a transition in which the internuclear distance in the first excited state is much greater than that in the ground state, while in Figure 19–4c the internuclear distances are about the same in both the ground state and the first excited state.

V_c is the energy required for the overall process

$$X_2 + hv \rightarrow X + X^*$$

where X^* is an atom in the first excited electronic state. It is calculated from

$$V_c = hv_c$$
v_c = frequency at the convergence limit

V_d is the dissociation or bond energy and corresponds to the process

$$X_2 \rightarrow 2X$$

and V_e is the excitation energy corresponding to the process

$$X \rightarrow X^*$$

If the value of V_e is known from atomic line spectroscopy, determination of V_c furnishes the information for the calculation of accurate bond dissociation

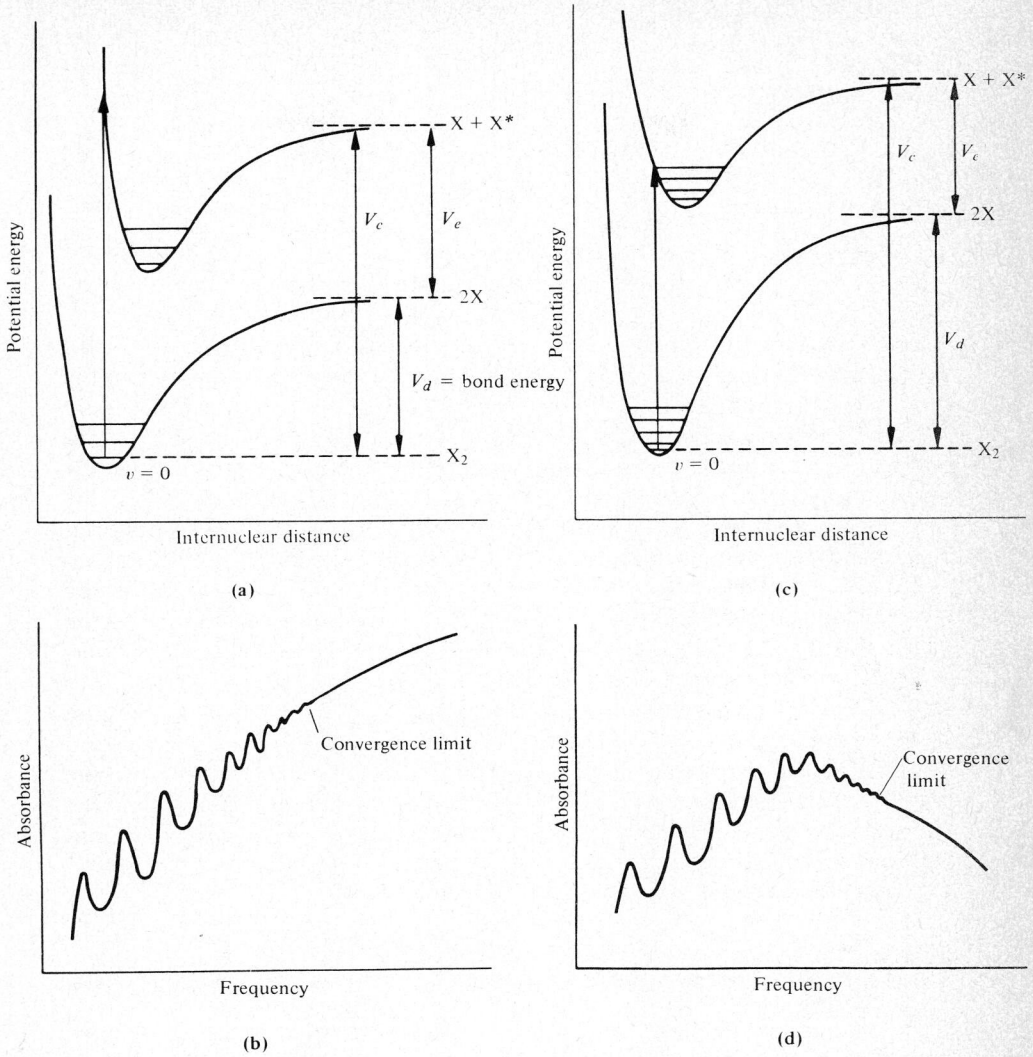

FIGURE 19-4 / Transitions and corresponding spectra in a diatomic molecule. The schematic spectra in (b) and (d) correspond to the transitions in (a) and (c), respectively.

energies. However, V_c is not always easy to determine experimentally, as can be seen from Figure 19-4d.

In the ground vibrational state the molecule has an internuclear distance somewhere near the center of the horizontal line, while in the excited vibrational states the molecule is most probably at a distance represented by either extreme of the horizontal line. On electronic excitation, according to the Franck–Condon principle, the internuclear distance does not change. The transition is therefore represented by a vertical line from the central portion of the ground state vibrational line to the end of an excited-state vibrational line in the excited electronic state. In Figure 19–4a, because of the shift of the excited electronic state to large internuclear distances (i.e., to the right), the most probable transitions end at vibrational levels above the dissociation energy. Therefore the wavelength of maximum absorption is in the continuum, and the convergence limit can be determined by inspection. In other words, the increase in internuclear distance increases the probability that an electronic transition will cause a dissociation of the molecule. On the other hand, if the two electronic states have almost the same internuclear distance, as in Figure 19–4c, the most probable transition will be to a vibrational state that will not dissociate. In this case the probability of dissociation is slight, and to locate the convergence limit an extrapolation technique is required.

Each line in the spectrum in Figure 19–4d represents a transition starting at the ground vibrational state with an initial vibrational quantum number of zero. The vibrating atoms form an anharmonic oscillator, whose energy, neglecting all but the first anharmonic term, is

(19–12) $\quad E_v = h\nu_e(v + \tfrac{1}{2}) - h\nu_e x_e(v + \tfrac{1}{2})^2$

$v =$ vibrational quantum number
$\nu_e =$ fundamental vibration frequency
$x_e =$ anharmonicity constant

The energy of any transition from the ground vibrational state and the ground electronic level to a higher vibrational state on the first excited electronic level is

(19–13) $\quad E_v - E_0 = \Delta E_{el} + h\nu_e(v) - h\nu_e x_e(v^2 + v)$

$\Delta E_{el} =$ energy of the pure electronic transition

The energy difference between any two adjacent spectral lines is therefore

(19–14) $\quad \Delta E = E_2 - E_1 = h\nu_e - h\nu_e x_e(2v + 2)$

Since $E = hc\bar{\nu}$, writing ΔE in terms of $\Delta \bar{\nu}$ gives

(19–15) $\quad \Delta \bar{\nu} = \bar{\nu}_2 - \bar{\nu}_1 = \dfrac{\nu_e - 2\nu_e x_e(v + 1)}{c} = a - bv$

$\bar{\nu} =$ wavenumber of observed line
$a = (\nu_e - 2\nu_e x_e)/c$
$b = 2\nu_e x_e/c$

Equation (19–15) shows that $\Delta \bar{\nu}$ is a straight line function of v, going to zero at the convergence limit. To locate the convergence limit, the wavenumber of each of a large number of lines is measured and the difference in wavenumber between each adjacent pair is calculated. Each line is assigned an arbitrary value of v, assigning the value zero to the line of lowest wavenumber. Equation (19-15) is then graphed and the line extrapolated to $\Delta\bar{v} = 0$. The value of the arbitrary quantum number at this point is multiplied by $\frac{1}{2}\Delta\bar{v}_0$, where $\Delta\bar{v}$ is the initial graphed value of $\Delta\bar{v}$. This product is then added to the wavenumber of the line corresponding to the start of the graph, to give the wavenumber of the convergence limit.

PHOTOCHEMISTRY

Photochemistry is defined as the study of the chemical reactions of atoms or molecules excited by radiation. The radiation used in photochemistry is in the uv–visible range, corresponding to electron transitions. The energy per *einstein* of this radiation (1 einstein = Avogadro's number of photons) is well above the activation energy of most reactions and even above many bond energies. It is several orders of magnitude above thermal energies. There is therefore a fundamental difference between ordinary thermal chemical reactions and photochemical reactions. The latter involve high energy intermediates. One interesting consequence is the small temperature coefficient of most photochemical reactions.

The primary process in photochemistry is the formation of an electronically excited state. It may react chemically or it may dissociate into excited fragments that themselves react chemically. Figure 19-5 shows the four possible cases. In the case shown in Figure 19-5a the ground state and the excited state have almost the same internuclear distances, and the probability of a dissociation is very small. Photochemical reactions therefore occur only by the excited molecule reacting. In the case shown in Figure 19-5c the excited state is unstable and dissociates immediately. All reactions therefore involve reactive fragments. In the case shown in Figure 19-5b some transitions cause dissociation and others do not. The path of the photochemical process will depend on the radiation frequency. At low frequencies the excited molecule reacts, and at high frequencies the fragments react. Figure 19-5d shows an interesting case in which there are two excited states whose energies are very close to each other. One of these is stable and the other unstable. Transition occurs to the stable state, and at low frequencies dissociation does not occur. At high energies the molecule changes over from the stable excited state to the unstable excited state and dissociates. This phenomenon is called *predissociation*. The spectrum of the absorbing molecule must be carefully studied to determine the stability of the excited state in order to decide if the excited molecule reacts before dissociation or not.

In photochemistry it is necessary to determine the number of molecules of product that result from the absorption of each photon. This quantity is called the quantum efficiency or the *quantum yield*, and is expressed as either the

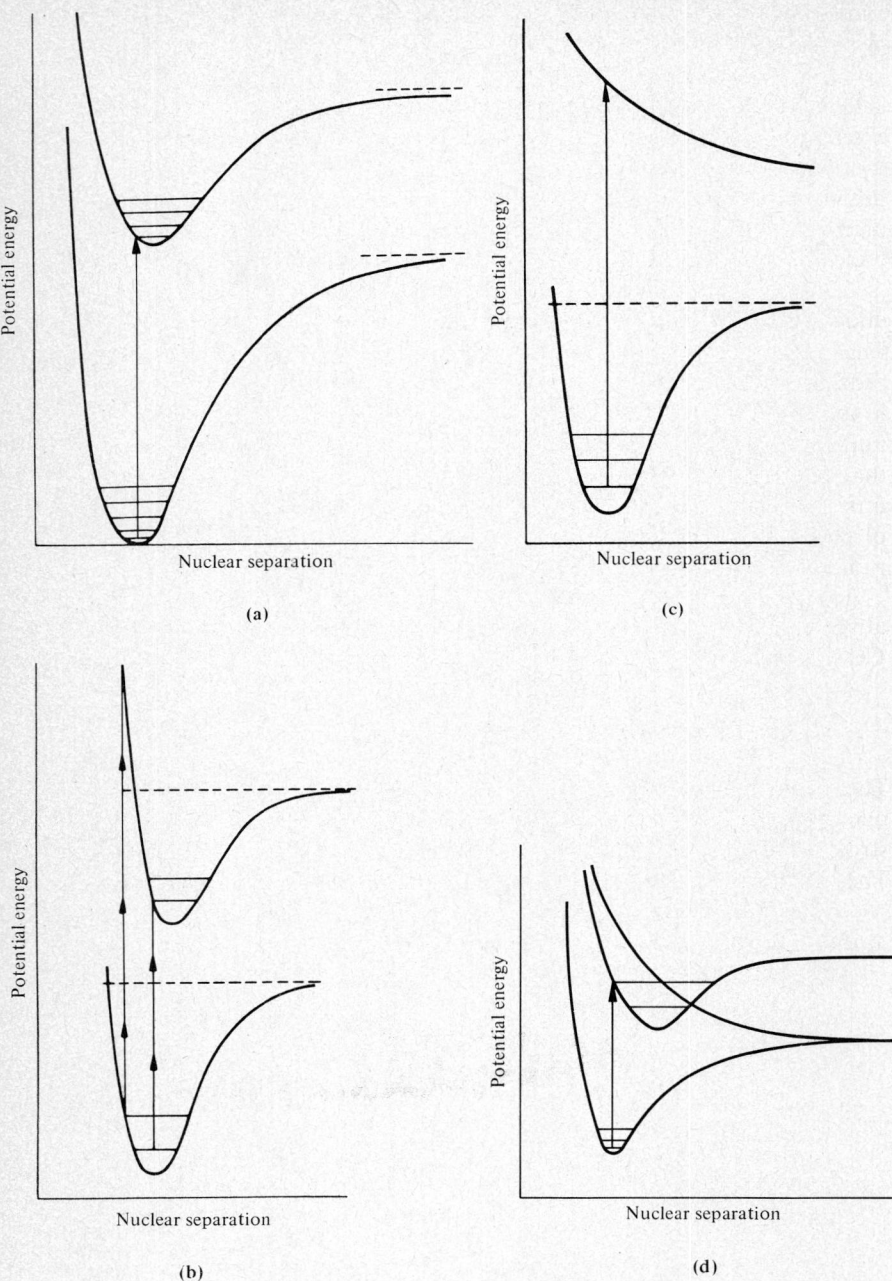

FIGURE 19-5 / Possible photochemical reactions.

number of molecules of reactant per absorbed photon or the number of moles of product per einstein of radiation.

To determine quantum yields, the intensity of incident radiation must be determined, the number of photons absorbed must be calculated, and the extent of the reaction must be measured by any of the usual analytical methods.

To determine the intensity of radiation, the light source must be calibrated. Calibration by chemical means is termed *actinometry*. Physical measurements are more accurate than chemical ones, but the latter are usually preferred as being more comparable with the reacting system under study.

A chemical actinometer is simply a reproducible chemical system that undergoes a measurable photochemical reaction of known quantum efficiency. The actinometer solution is put into the reaction vessel and exposed to the radiation for a measured time. From the extent of the photochemical reaction in this reference system the number of photons absorbed in the reaction cell per unit time can be calculated. The great advantage of the chemical actinometer is that the solution conforms to the shape of the reaction vessel and responds only to the radiation being absorbed by the light sensitive material. The extent of the reaction should, of course, be linear with the quantity of absorbed radiation.

One standard actinometer solution is the ferrioxalate system. This absorbs light in the range 300–500 nm, the ferrioxalate ion decomposing to CO_2, Fe^{2+}, and oxalate ions:

$$2\ Fe(C_2O_4)_3^{3-} \xrightarrow{h\nu} 2\ Fe^{2+} + 2\ CO_2 + 5\ C_2O_4^{2-}$$

The extent of this reaction is determined photometrically. A solution of 1,10-phenanthroline is added, forming a complex with Fe^{2+} that absorbs light strongly at 510 nm, the molar absorptivity being 1.11×10^4 liter/mole-cm. The quantum yield is 1.04 moles/einstein, at 510 nm and 25°C.

Another standard solution is uranyl oxalate. The uranyl ions absorb light in the range 320–440 nm and react catalytically with the oxalic acid:

$$UO_2^{2+} + H_2C_2O_4 \xrightarrow{h\nu} UO_2^{2+} + H_2O + CO + CO_2$$

The amount of oxalic acid decomposed is determined by titration with permanganate. The quantum efficiency is 0.57 mole/einstein.

Once the number of photons absorbed by the actinometer solution is known, the intensity of the light source can be calculated. The intensity absorbed, P_{abs}, is usually only a fraction of the total light incident on the solution, P_{inc}. This fraction is

(19-16a) $$\frac{P_{abs}}{P_{inc}} = 1 - 10^{-A}$$

The fraction of light not absorbed (i.e., transmitted) is

$$(19\text{-}16\text{b}) \qquad \frac{P_{tr}}{P_{inc}} = 10^{-A}$$

After the intensity of radiation is calculated from the extent of the actinometer reaction and the absorbency of the actinometer solution using equation (19-16a), the calculation must be reversed and the reacting solution investigated. The absorbency at the wavelength of irradiation is determined, and from this and the calculated intensity of radiation, equation (19-16a) is used to obtain the number of photons absorbed by the reacting system.

There are many subtleties involved in the measurement and interpretation of photochemical data. Light is a most difficult "reagent" to work with. Light sources sometimes change characteristics upon aging, power sources fluctuate, reflections of cells change, and so forth. Furthermore, photochemical results are often complex, with competing side reactions, parallel thermal reactions, and so on. As a result, quantum yields may depend on temperature, concentration, and light intensity, and vary during the course of a reaction. The quantum yield may actually be an average, as measured experimentally. Another problem, especially in cases where the quantum yield varies with concentration, is the inhomogeneity of light absorption by the solution. More light is absorbed, per unit volume, at the front window than elsewhere in the cell. There is therefore a concentration gradient of the excited species throughout the cell unless the solutions are stirred, or the cells are thin and the irradiation is stopped before the reaction has proceeded very far. Such a concentration gradient would produce an average quantum number that may be misleading when used to interpret a reaction mechanism.

In photochemical work procedures vary widely from laboratory to laboratory. However, there are several general rules that apply:

1. Always use blanks, to eliminate thermal (nonphotochemical) reactions. The blank should consist of a sample of the same solution as that being irradiated and should be kept in a similar cell. The blank should undergo all the same procedures as the sample except that it should be kept in an opaque container or a black wrapper during the irradiation period.
2. Do not calibrate a radiation source one day and run a series of reactions at some other time. It is necessary to make a calibration run each time there is a radiation run to guard against fluctuations or variations in the light source.
3. All photosensitive materials should be stored in opaque containers or shielded from light in some fashion.
4. Since intensity of radiation obeys the inverse square law, the distance between the cell and the light source should be kept fixed or else determined accurately for consistent reproduction.

5. All direct and reflected ultraviolet light should be screened with absorbing shielding to protect the eyes of laboratory personnel.

Instrumentation

Spectrometry is concerned with the determination of the emission and absorption spectra of elements and compounds. *Photometry* is a quantitative measurement of the magnitude of the absorbed radiation. Most modern instruments are equipped for both, and are called *spectrophotometers*. The basic components of a spectrophotometer are the light source, a dispersing element for breaking the light up into its component frequencies, a scanning device for changing the frequencies being used, a sample cell, a detector, and the various optical components needed to produce and focus beams.

LASER LIGHT SOURCES

In recent years, *lasers* (laser is an acronym for light amplification by stimulated emission of radiation) have altered the technology of many fields. Lasers are devices for producing a beam of essentially monochromatic, unidirectional, coherent (in phase) radiation in the infrared, visible, or ultraviolet region. Their unusual properties permit uses in which they cannot be replaced by ordinary sources of radiation. Because it is unidirectional, the intensity of a laser beam, unlike the intensity of a beam of ordinary radiation, does not decrease with the inverse square of the distance. Therefore, laser beams can be and have been bounced off the moon. Because the radiation remains in phase for a second or more, beat frequencies between two different lasers can be detected. Because the radiation is monochromatic, prism or grating monochromaters are unnecessary, and they provide an intense light source at their emission frequency.

In lasers photons induce electrons in a high-energy state to drop back to a lower energy state, emitting radiation in the process. Stimulated emission of radiation was first predicted by Einstein in 1916. He showed that when an incident photon interacts with an electron in an atom or molecule, it can either be absorbed and raise the electron to a higher energy state, or cause the electron to drop to a lower energy state, if one is available. If an electron is in an excited state, it can return to a lower energy state in one of three ways: by the stimulated emission of radiation, by spontaneous emission of radiation, or by a nonradiative transition in which the energy is dissipated as heat. In spontaneous emission the emitted photon has no particular phase or direction in relation to any other photon because the events are independent and random. In stimulated emission (i.e., lasers) the emitted photon has the same direction and is in phase with the photon that induced the transition.

According to the Boltzmann distribution law, at equilibrium

$$\frac{n_2}{n_1} = \frac{g_2}{g_1} e^{-hv/kT}$$

n_1, n_2 = populations of electrons in two energy states
g_1, g_2 = degeneracies of the two states
v = frequency corresponding to the transition between the two states
T = absolute temperature

Because the energy difference between electronic states is considerable, $n_1 \gg n_2$. Consequently, the probability of a photon interacting with an electron in an excited state is so small that normally we observe only absorption at ordinary temperatures. However, in a laser, the normal population ratio is inverted. The population of a higher energy state is "pumped" to exceed the population of a lower energy state. Therefore, there is a high probability of a photon stimulating the emission of radiation. In addition, there is a "cascade" effect. The emitted photons stimulate further emission all in the same phase and the same direction. The cascade effect is often enhanced by the construction of the laser. Lasers are often in the shape of a cylinder with mirrors on the end faces. These mirrors reflect the light back and forth until it finally leaves through a port in one mirror, as an intense unidirectional one-phase beam.

The population inversion is made possible by the existence of a higher energy metastable state from which unstimulated transitions to lower energy states are very slow (see Figure 19-6). Atoms in the ground state are "pumped" up to a very high energy state (state 3) by light or by an electric current or a

FIGURE 19-6 / Laser energy levels.

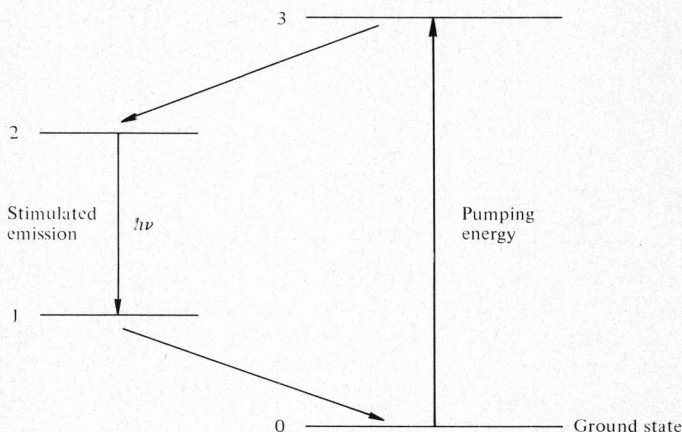

plasma discharge. From this state they drop to the metastable state (state 2), where they are trapped. If pumping is more rapid than spontaneous deexcitation, the population of the metastable state can become very large. In that case stimulation by incident radiation, which induces the transition from state 2 to state 1, produces an intense beam. (The first devices employing this principle were masers, microwave amplifiers.)

There are a number of types of lasers. Lasers may be of gas or of semiconductors or of solids such as the ruby. Some lasers give continuous radiation while others produce pulses. The pulsed radiation is of high *power*, since the total energy is produced over a short time. Ruby lasers can produce nanosecond or even picosecond pulses, used in flash photolysis kinetic studies (Chapter 24). Continuous lasers range from milliwatts up to kilowatts in power.

One of the most widely used lasers is the helium–neon laser, an inexpensive laser whose radiation is at 632.8 nm and ranges to 50–75 mW. It may be obtained in single mode (polarized) or unpolarized models. The argon laser is a gas laser typically used at 488.0 nm, although sometimes used to produce other wavelengths in the range 350–520 nm; Ar^+ ions are involved in the discharge. This laser is much more powerful than the helium–neon laser. The CO_2 laser uses molecular vibration levels at 10.6 nm (infrared). This laser is exceedingly powerful, up to kilowatts being possible, and quite efficient. There are also semiconductor lasers, such as neodymium, garnet, and gallium arsenide, and a variety of dye lasers, which can be tuned to select a frequency.

Obviously, a huge variety of applications exist in chemistry, as in light scattering, and spectroscopy, especially Raman spectroscopy. There is also a great number of applications outside chemistry, such as in the communications field. It should be noted that because laser radiation is coherent, the effective "noisiness," which limits the sensitivity of many experiments, is more than an order of magnitude better for lasers than for incoherent light sources.

One extremely important note on handling lasers. *Never* look directly into a laser beam or even a sharp reflection of one. It is possible to burn out an eye faster than one can blink. The more powerful lasers are especially dangerous to the eyes, but so are ir and uv lasers, which are the more insidious for being invisible.

LIGHT SOURCES OTHER THAN LASERS

For absorption spectrophotometry continuous radiation is required. There is no single source that emits sufficient radiation at all wavelengths. In the visible range, from the near infrared (about 3000 nm) to the near ultraviolet (about 350 nm), a tungsten filament lamp contained in a pyrex envelope can be used. The beam emitted is too weak to be useful at wavelengths shorter than 350 nm because the intensity of the tungsten light drops sharply and the glass of the light bulb absorbs large amounts of the radiation. For ultraviolet work the light source must be either in quartz envelopes or in pyrex envelopes with quartz windows. The radiation source for the ultraviolet is usually the low pressure (on the order of a few torr) hydrogen or deuterium lamp. The excited deuterium or hydrogen is molecular and produces wide bands that are continuous in the ultraviolet region usually investigated. Commerically available hydrogen lamps give a spectrum that is continuous in the region from 160 to 360 nm, and the commercial deuterium lamp gives a continuous spectrum between 185 and 375 nm. For infrared radiation the usual sources are tungsten filament lamps, Nernst glowers (filaments of zirconium and yttrium oxides), and Globars (silicon carbon rods). All of these emit continuous radiation when heated.

PRISMS AND GRATINGS

In prism instruments the light to be analyzed passes through a slit and is collimated into a beam by a lens. The beam is then passed through a prism and is dispersed, that is, spread out into a broad spectrum. The *linear dispersion* is the distance between lines on the dispersed spectrum, as seen with a spectroscope or spectrograph, expressed in ångström units per millimeter of spectrum length. The resolution is the minimum wavelength difference that can be detected between two adjacent lines, and is the more important quantity. In grating instruments a collimating lens is not used. The grating is itself a mirror, focusing the beam. Prisms give good resolution, especially in terms of difference in wave number, at shorter wavelengths, but give increasingly poor resolution at longer wavelengths. In a prism instrument the scale is nonlinear. Grating instruments give good resolution, in terms of difference in wave number, at longer wavelengths, but increasingly poor resolution at shorter wavelengths. The scale is linear in terms of wavelength.

SLITS

The light that strikes the prism, or grating, is first collimated into a beam by a slit-and-lens combination. The slit width is critical. Too wide a slit leads to deviation from Beer's law, loss of fine structure, and distortion of the spectrum by the rounding off of peaks, the decrease of maxima, and the increase of minima, even with relatively broad bands. Too narrow a slit lowers the intensity of radiation. The slit must be sufficiently wide so that diffraction from the slit edges can be ignored. Fortunately, for most applications the slit width of a Beckman DB, a Cary spectrophotometer, or similar instruments can be set or programmed without seriously affecting resolution.

DETERMINATION OF WAVELENGTH

In spectroscopy and spectrography the wavelengths of the lines or the bands are determined from their positions in the visual field or on the film. The

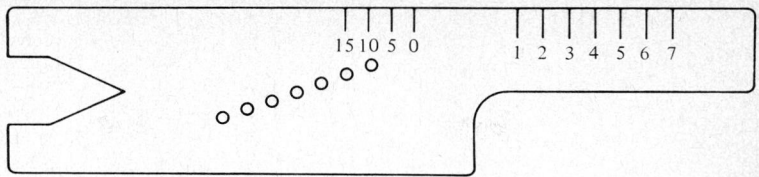

FIGURE 19-7 / Hartmann diaphragm.

spectroscope has a rough scale, to be used for approximate location only. In spectrographs devices such as the Hartmann diaphragm (Figure 19-7) are used to photograph the unknown spectrum with one or more standards on the same film. Each hole in the diaphragm exposes a different region of the entry slit so that a different portion of the film is exposed. Seven different exposures may be taken with one film. Several standards should be included on the same film with the spectrum being studied, both for calibrating the instrument and for

TABLE 19-2 / Wavelengths of Reference Lines (in nm)

Sodium	Mercury	
589.6	690.8	
589.0	623.435	
330.3	612.3	
330.1	607.2	
	579.066	Very strong
	576.960	
	546.072	Very strong
	491.6	
	435.834	Strong
	434.750	
	433.921	
	410.807	
	407.78	
	404.677	Strong
	390.644	Very weak
	366.327	Very strong
	366.288	
	365.483	
	365.015	Strong
	313.184	
	313.156	
	302.348	Weak
	302.150	

adjusting for possible film shrinkage during development. In determining the wavelengths of the unknown lines, each line should be bracketed with a pair of known lines. Interpolation may be assumed to be linear, within a few Ångström units, or perhaps more, depending on the degree to which approximation is acceptable. In cases where standard lines do not bracket the unknown line, various interpolation formulas can be used, such as the *Hartmann formula*:

$$(19\text{-}17) \qquad \lambda = \lambda_0 + \frac{c}{d_0 - d}$$

λ = wavelength of line
λ_0, c, d = empirical constants
d_0 = position of line, on the film or plate

The three empirical constants are obtained from any three standard lines. Table 19-2 lists some convenient reference lines in the sodium and mercury spectra.

MATERIALS

Scanning spectrometers require lenses or mirrors for focusing the dispersed radiation into beams and cells with windows for admitting radiation. Spectrometers, including spectrographs and spectroscopes, employ collimating lenses or mirrors. Care should be taken that these materials do not themselves absorb the radiation being studied. For ultraviolet and visible light, lenses, prisms, and cells should be made of quartz, although glass can be used for visible light. For infrared work NaCl, KBr, CaF_2, AgCl, and similar salts are used.

Liquids and gases may be studied directly (gases require special cells), but solids must sometimes be dissolved, suspended in Nujol, or embedded in a matrix such as KBr. The solvent must also be transparent in the range used (in spectroscopic terminology, it must have large window areas), and it must not interact with the absorbing materials in the range used. Carbon disulfide, carbon tetrachloride, and chloroform are the most common solvents for the infrared, while for visible and ultraviolet light carefully purified water, cyclohexane, ether, and alcohol may be used.

RADIATION DETECTORS

Radiation detectors include the eye, photographic emulsions, photocells, photovoltaic cells, photoconductors, bolometers, and thermopiles. Generally, photocells and photovoltaic cells are used with visible and ultraviolet radiation, and thermopiles and bolometers are used for infrared cells. Photocells are composed of materials, such as cesium oxide, that emit electrons on being struck with photons of sufficient energy (i.e., visible or ultraviolet radiation).

Photovoltaic, or *barrier layer*, *cells* are semiconductor devices in which illumination with light of the appropriate frequency causes electrons to flow from the semiconductor to a metal electrode in contact with it. Photoconductors are devices whose electrical resistance decreases on irradiation. Infrared light does not have enough energy to activate photocells and photovoltaic cells; thus bolometers and thermopiles are used for infrared measurements. *Bolometers* are essentially sensitive resistance thermometers. *Thermopiles* are detectors composed of several very small thermocouples in series with each other. The mass of the thermopile or bolometer is kept small so that the incident infrared radiation will produce a relatively large temperature change in the thermoelement. Both bolometers and thermopiles can be used for detection of radiation at all wavelengths, but due to their low sensitivity, the intricacies of the required techniques, and, to a lesser extent, the cost, their use is usually restricted to the infrared. Solid state photodiodes are steadily replacing bolometers and thermopiles as infrared detectors. The materials PbS and PbSe have been used for a number of years. Commercial InSb detectors, cooled with liquid nitrogen, are sensitive from the visible down to 5.5 nm. Recently, PbSnTe detectors cooled with liquid nitrogen have become available with an optimum range from 5 to 13 nm, as have helium-cooled PbSnTe detectors with an optimum range from 6.6 to 18 nm.

SINGLE AND DOUBLE BEAM OPTICAL PATHS

In all precise methods, the absorbing substance is compared to a reference, either in the same beam or in a beam as nearly equivalent as possible. In single beam photometry, the beam is first passed through the reference and the instrument is then adjusted to read zero absorbence, or 100% transmittance. The reference is then removed, the sample inserted in the beam, and the reading noted. Photometry requires either closely matched cells or cells whose relative dimensions have been determined with a reference liquid. (In spectrometry, where frequency is the desired quantity, cell matching is not necessary. The beam intensity need only be kept constant for the time needed to make measurements.)

In double beam photometry, the light is formed into two identical beams by prisms, rotating sectors, rotating or vibrating mirrors, or other mechanical or optical devices. One beam passes through the sample and the other through a reference material or blank.

In single beam operation, the signal impinging upon the detector produces an electric current that is either suitably amplified and read or that produces an electric signal that is in turn bucked with a potentiometric circuit, producing an electric null reading. Since detection of an electric null point is more sensitive than the reading of a meter, especially in the lower range of the meter scale, the null method is preferred for accurate work.

In double beam operation the two signals being compared may be used to produce either an electric null or an optical null. An optical null results when

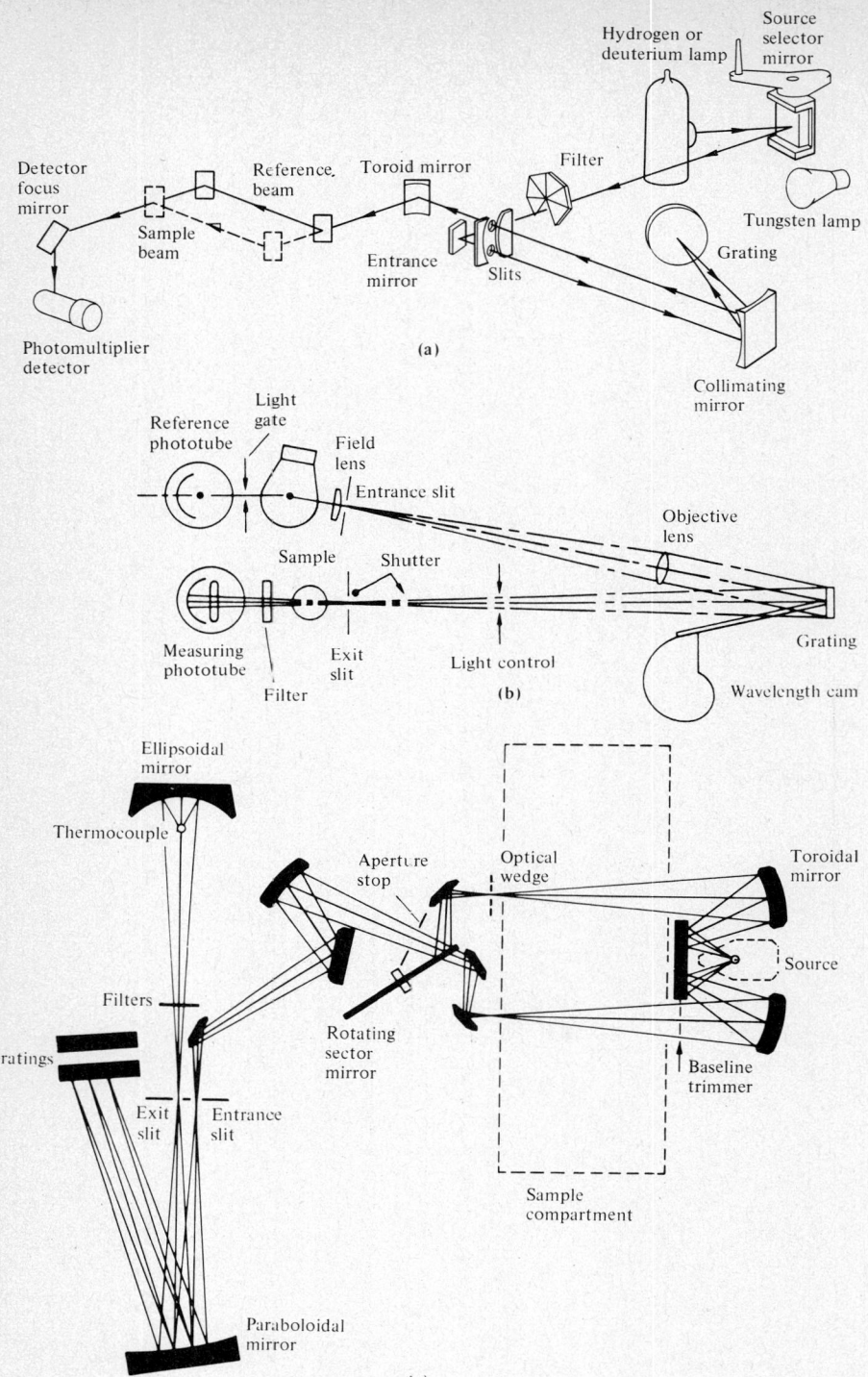

both the sample and the reference beams have the same intensity. The two beams alternately strike a detector. If the sample absorbs light, during one half of the cycle the intensity of the light reaching the detector is lowered. As a result the detector puts out an unbalanced signal, smaller during one half of the cycle than during the other. This unbalanced signal is in effect an AC superimposed on a DC. The AC component is amplified and activates a servomotor, which moves an opaque object into the reference beam, attenuating it. As the reference beam is weakened, the imbalance signal becomes smaller and smaller. Finally, when the two beams are of equal intensity the detector puts out essentially DC, which is not amplified, and the servomotor stops. The extent of absorption is indicated by the distances the optical attenuator has moved, as shown by the synchronized motion of a pen over a chart paper.

Since the output of the radiation source, the resolution of the dispersing element, the absorption of the cell windows, and the response of the detector per incident photon all vary with the frequency, it becomes necessary to maintain the incident beam at constant energy. This is usually accomplished by having the beam pass through slits that are programmed to open or close as the instrument scans through the wavelength range.

Figure 19-8 shows some typical optical systems for spectrophotometers. In all cases the instrument is furnished with a set of detailed instructions that vary with the make and model of the instrument. Unfortunately, no set of instructions is universally applicable.

References

Calder, A. B., *Photometric Methods of Analysis.* American Elsevier, New York, 1969.

Calvert, J. G., and J. N. Pitts, Jr., *Photochemistry.* Wiley, New York, 1966.

Jaffe, H. H., and M. Orchin, *Theory and Application of Ultraviolet Spectroscopy.* Wiley, New York, 1962.

Lambert, J. B., et al., *Organic Structural Analysis*, Parts Two and Three. Macmillan, New York, 1976.

Martin, A. E., *Infrared Instrumentation and Techniques.* Elsevier, Amsterdam and New York, 1966.

Stearns, E., *The Practice of Absorption Spectrophotometry.* Wiley-Interscience, New York, 1969.

FIGURE 19-8 / Typical optical systems for spectrophotometers. (a) Optical diagram of Beckman 24/25 (courtesy of Beckman Instruments Inc.). (b) Optical system of Bausch and Lomb Spectronic 20 Spectrophotometer (courtesy of Bausch & Lomb). (c) Optical layout of Perkin-Elmer Model 281/283 Spectrophotometer (courtesy of the Perkins-Elmer Corporation).

Weissberger, A., and B. W. Rossiter, *Physical Methods of Chemistry*, Part III,D. Wiley–Interscience, New York, 1972.

Willard, H. H., L. L. Merritt, Jr., and J. A. Dean, *Instrumental Methods of Analysis*, 5th ed. Van Nostrand, New York, 1974.

Chapter 20

X-Ray Diffraction

X rays are high-energy electromagnetic radiation with wavelengths in the range 10^{-1}–10^2 Å, although for diffraction studies the range is usually 0.5–2.5 Å. They are produced by the bombardment of a metal target with high-energy electrons. The bombarding electrons remove electrons from the K and L shells of the target atoms, leaving vacant low-energy orbitals. Higher energy electrons from the other shells drop into these vacancies, emitting the energy difference in the form of X rays. X rays can also be produced by bombarding the target with high speed particles accelerated in a cyclotron or produced in radioactive decay. These, however, are methods requiring special facilities.

Production of X Rays

The beam of bombarding electrons is accelerated by a field of 25–50 kV. To prevent deceleration of these electrons by random collisions with air molecules, the bombardment takes place in an evacuated tube (Figure 20-1). The radiation produced by the bombardment is emitted in all directions, and most of it is absorbed by the metallic housing of the tube. The portion that strikes the windows in the tube passes through them, is collimated into beams, and is aimed at the sample. (The windows are usually made of aluminum, beryllium, or mica.) The coils for circulating water are necessary to conduct away the heat generated by the bombardment, which constitutes 99% of the energy of the electron beam. Only about 1% of the beam energy goes into emitted radiation.

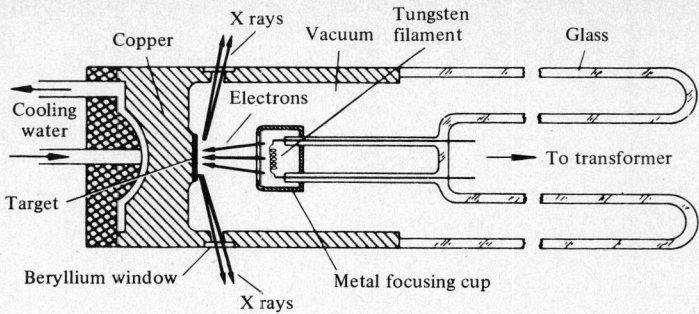

FIGURE 20-1 / Schematic diagram of the cross section of a sealed-off filament X-ray tube (after Cullity [1966]).

Monochromatic X-Ray Beams

The radiation emitted by the target consists of a broad band of *continuous*, or *white*, radiation, which is produced at all accelerating voltages, and some relatively sharp peaks called *characteristic lines*, emitted only above a critical or excitation voltage. The white radiation results from bombarding electrons striking the target without knocking target electrons out of orbit. The kinetic energy of most electrons is converted into heat, but some electrons emit energy in the form of radiation, according to

(20-1) $\quad \frac{1}{2}mv^2 = h\nu = \frac{hc}{\lambda}$

ν = frequency
c = speed of light
λ = wavelength

If all the kinetic energy of a bombarding electron is converted into radiation, the wavelength, in Ångströms, of the emitted radiation is

(20-2) $\quad \lambda = \frac{hc}{eV} = \frac{12,400}{V}$

e = electronic charge
V = accelerating potential

This is the shortest wavelength that can be emitted by an electron beam accelerated by a given voltage; it is called the short-wave limit or the short-wave cutoff. Since most of the incoming electrons do not lose all their kinetic energy at once but are instead decelerated in a series of random impacts, most of the radiation will be emitted as a continuous series of wavelengths longer

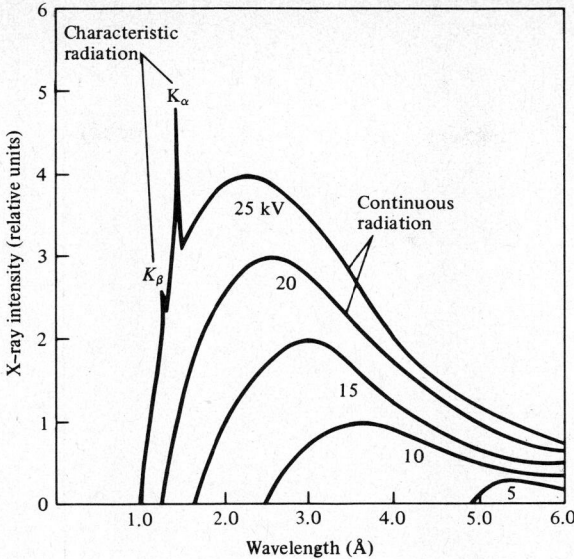

FIGURE 20–2 / X-ray spectrum of molybdenum as a function of applied voltage. The linewidths are not to scale.

than the short-wave cutoff. An increase in the accelerating potential decreases the short-wave limit and shifts the bulk of the white radiation to shorter wavelengths, as shown in Figure 20–2.

The *characteristic lines* of the X-ray spectrum are those emitted when a vacancy is created in an inner shell and an electron from an outer shell drops into it. Since the energies of the initial and final states of the electron are well defined, the emitted photon has a definite wavelength characteristic of the target element. The excitation voltage for the characteristic lines is the minimum potential needed to remove the target electron, which depends only on the atomic number of the target atom and on the particular shell losing the electron.

Lines resulting from the removal of an electron from the K shell are called K lines; L and M lines result from the removal of L and M electrons. There are two series of K lines, α and β, the first resulting from an L electron dropping into the K-shell vacancy and the second from an M electron falling into the K-shell vacancy. Since L electrons are closer and of lower energy, the K_α line is more intense and of lower energy (longer wavelength) than the K_β line. In X-ray studies the K_α line is the one most often used.

Optical techniques of focusing with lenses and prisms cannot be used because the wavelengths are too short. The X rays would go straight through the lens or prism almost as if it were not there. Instead, all rays of unwanted wavelengths, or whose direction is wrong, are absorbed, leaving a greatly attenuated beam of rays with the right direction and wavelength.

The X rays produced at the target have wavelengths over a large range, as shown in Figure 20–2, and are at 180° of solid angle relative to the target. As previously mentioned, only that portion striking the small windows gets out of the tube. This radiation is then passed through a filter that absorbs almost all rays whose wavelength is shorter than that of the K_α line (which is the desired wavelength). After going through the filter, the now largely monochromatic radiation enters a collimator, a long narrow brass tube with a slit window at each end. Any radiation that strikes the brass wall is absorbed leaving only those rays that are in line with both of the slits (i.e., a beam). Obviously, only a minute fraction of the radiation generated can get out of the tube, through the filter and through the collimator, so that the efficiency of the overall process is very low, in terms of energy.

The filter absorbs the unwanted wavelengths, using them to produce photoelectrons. Photons of the unwanted K_β radiation strike the K or L electrons of the filter metal and furnish the energy for ionization, that is, produce photoelectrons. The energy of a K_β photon from a copper target is just the energy difference between the M and K shells of the copper atom, which is less than the ionization energy of a K or L electron from a copper atom. Therefore, a copper filter could not absorb K_β photons from a copper target. Only if the filter atoms had a smaller nuclear charge than copper atoms would their K and L electrons have ionization energies so low that the K_β photons from copper could remove them. In other words, to filter out the K_β photons from a copper target, the filter would have to be of a metal with a smaller atomic number than copper. On the other hand, if the filter were made of a metal whose nuclear charge was much smaller than that of copper, the K_α radiation from the copper target would also be absorbed. So, a reasonable compromise is reached. The filter metal is made of the atom whose atomic number is one unit less than that of the target metal. For example, nickel (atomic number 28) is used to filter radiation from a copper target (atomic number 29).

Each metal is effective for wavelengths of higher energy (lower wavelength) than a minimum, called the *absorption edge*. Since the energy of the white radiation shifts in the direction of shorter wavelengths, with increasing accelerating potential (Figure 20–2), it becomes relatively simple to remove it. The accelerating potential is simply increased until the bulk of the white radiation is below the wavelength of the absorption edge. For copper targets and nickel filters, the accelerating potential would be at 30–35 kV.

Unfortunately, no filter will absorb all the β lines without also absorbing some of the α lines. A reasonable compromise must be made. Too thick a filter will attenuate the beam, even though virtually all the unwanted radiation is removed. Too thin a filter results in a strong beam but with much unwanted radiation. Usually, filters are between 0.01 and 0.1 mm in thickness, depending on the particular target metal. For copper radiation, a 0.18 mm thick nickel

Chapter 20. X-Ray Diffraction

filter reduces the incident K_α line to half its original intensity and the K_β line to $\frac{1}{85}$ of its initial intensity. Since in the unfiltered beam the α/β ratio is 6 : 1, the resultant beam has an α/β ratio of 250 : 1.

X-Ray Detection and Intensity Measurement

The diffracted X rays are usually detected photographically. The record obtained is permanent and shows a large number of relationships to the trained eye that are not readily apparent if the results are expressed as a column of figures. Relative line intensities are easily estimated from a film in a semiquantitative fashion. Furthermore, photographic equipment is relatively inexpensive. However, there are certain disadvantages inherent in the photographic method. Since film tends to warp on developing, for accurate measurement the film shrinkage factor must always be taken into account. Intensities cannot be quantitatively determined unless photodensitometers are used, and these are relatively expensive and slow. (Semiquantitative relative measurements can be made by using film strips with standard calibrated exposures and by comparing the observed lines to these.) Most troublesome is the low sensitivity and nonlinearity of films in comparison with ionization detectors and counters.

For very accurate measurements the intensities of the diffracted X rays are measured with an electronic radiation detector, which feeds the data into a recorder, giving accurate and rapid measurements both of diffraction angles and of integrated intensities of diffracted radiation.

GENERAL OPERATING INSTRUCTIONS FOR X-RAY GENERATORS

Each manufacturer furnishes an instruction manual for the particular instrument used. It is absolutely essential that with equipment as complicated, potentially hazardous, and expensive as X-ray generators, the operator be familiar with the particular instruction manual. However, some general procedures are applicable to most equipment.

A safety check must be made constantly. X radiation is extremely dangerous, and its effects are cumulative. All equipment should be checked from time to time to make sure that there is no radiation leakage around the instrument, that the cameras are equipped with radiation stops, that safety devices are in operating order, that the amount of air scattering is negligible, and that an emergency cutoff switch is located in an easily accessible spot. Warning signs should be placed around the equipment, indicating that it is a radiation area. Operators should always stay out of the direct path of the beam, in case the shielding does not work or a leak has developed.

The water circulating through the cooling coils should be filtered to prevent internal clogging of the coils. There should be safety relays to cut off the tube if the water pressure drops below a preset minimum or if the temperature rises above a fixed level.

The apparatus should be warmed up for at least $\frac{1}{2}$ hr before an exposure is made.

Beam alignments should be checked periodically.

A photograph of the source should be taken from each window as soon as a new tube is installed. Only a small percentage of tubes have defective targets, but the high cost of tubes makes such a routine check advisable.

Tubes should be used at least a few hours each month and should not remain unused for more than a month. Unused tubes accumulate gas in the filaments, lowing the beam intensity. Spare tubes should be changed every few weeks. If an X-ray supply house is close by, spares should not be kept on hand.

In starting up the generator, the cooling system should first be started, and then the tube turned on and allowed to stabilize. To avoid power surges, the voltage and current controls should be turned to minimum positions before switching on the tube.

To prolong tube life, it is better to keep the filament voltage and current somewhat below the manufacturer's ratings, increasing exposure times if necessary.

X-Ray Diffraction

Atoms are electrical systems, the electrons oscillating with various frequencies and amplitudes in the central field of the nucleus and the other electrons. When electromagnetic radiation impinges upon the atoms, the system is disturbed, the electrons are displaced from their unperturbed positions and they oscillate at the frequency of the perturbing radiation, absorbing energy from the impinging beam. This radiation is reemitted in all directions as the electrons return to their original positions during the oscillation. The process is called *scattering*.

Crystals are composed of large numbers of atoms arranged in a geometric pattern. When radiation is scattered from such a large number of atoms, most of the scattered radiation is out of phase and cancels out. Only in certain directions will scattering be in phase, producing detectable radiation. For the scattered radiation to be in phase, it is necessary that the radiation scattered by one atom travel an integral number of wavelengths farther than that scattered by another atom; that is, the light paths differ by an integral number of wavelengths. Since in X-ray crystallography we deal with beams instead of individual photons, the scattering is by planes of atoms rather than by individual atoms. In-phase scattering is called *diffraction* and occurs only at certain angles, depending on the wavelength of the radiation and the geometry of the crystal. The basic relationship is the *Bragg equation*,

(20-3) $\quad n\lambda = 2d \sin \theta$

$n =$ an integer
$d =$ distance between planes
$\theta =$ angle of reflection, diffraction, or incidence

The sets of planes that produce the scattering are characterized by the *Miller indices*, which are defined as the reciprocals of the intercepts of the plane upon the unit crystallographic axes. Alternatively, the Miller indices may be considered to be the coefficients of the equation for the plane in three dimensions:

$$hx + ky + lz = 1$$

(Hexagonal systems may be characterized by four Miller indices.)

In X-ray crystallography the quantities of interest are the angles at which the X rays are diffracted and the intensities of the refracted beams. Measurement of the angles at which the diffraction occurs furnishes the information for calculating the interplanar distances and ultimately the shape and size of the unit cell. However, it is important to note that except for the simplest systems, such angle measurement tells nothing about the arrangement of atoms within the planes. The positions of the X-ray reflections from a crystal depend only on the size and type of the lattice. Two materials with the same type and size of unit cell will give X-ray reflections at the same angles even if the chemical structure and composition are completely different. For example, SnI_4 reflects X rays at the same angles as $RbAl(SO_4)_3 \cdot 12H_2O$ because the unit cells are of the same shape and size. However, the relative intensities of the various reflected rays are completely different. Consequently, for a complete structure determination the relative intensities of the various reflections must be determined.

TABLE 20-1 / Some Bragg Relationships

Cell Type	$\sin^2 \theta$
Cubic	$\dfrac{\lambda^2}{4a^2}(h^2 + k^2 + l^2)$
Tetragonal	$\dfrac{\lambda^2}{4}\left(\dfrac{h^2 + k^2}{a^2} + \dfrac{l^2}{c^2}\right)$
Hexagonal	$\dfrac{\lambda^2}{4}\left[\dfrac{\frac{4}{3}(h^2 + kh + k^2)}{a^2} + \dfrac{l^2}{c^2}\right]$
Orthorhombic	$\dfrac{\lambda^2}{4}\left(\dfrac{h^2}{a^2} + \dfrac{k^2}{b^2} + \dfrac{l^2}{c^2}\right)$

[a] h, k, and l are the Miller indices on the x, y, z axes; a, b, and c are the cell parameters.

The mathematics involved in such a structure determination is beyond the scope of this introductory text, and will not be discussed here. Interested readers are invited to refer to the References at the end of the chapter. The rest of this discussion will be concerned with determination of the lattice geometry.

Table 20-1 shows the relationship between the measured diffraction angles, the lattice parameters, and the Miller indices of the four simplest crystal systems. For other cell types—the monoclinic, triclinic, and rhombohedral—the equations are too complicated to be of use to the nonspecialist, and have therefore been omitted from the table.

In determining the cell type and lattice dimensions from the diffraction angles, all the various analytical and graphical methods amount to trial-and-error procedures or educated guesses. Given a structure, the diffraction pattern can be calculated relatively easily, but the reverse is much more difficult. Much labor can be saved in attempting to find lattice types and parameters if a preliminary study can be made with a microscope, especially a polarizing microscope, to observe the crystal form. The internal structure does not always coincide with the macroscopic form, but usually the faces that show up in the macroscopic crystal correspond to the most important planes in the unit cell. The appearance of right angles will suggest a cubic, tetragonal, or orthorhombic lattice. A polarizing microscope will distinguish the isotropic cubic lattice from all others.

If nothing is known about the crystal system, the best procedure is to assume the simplest case: cubic. The various lines are assigned arbitrary integral values of h, k, and l to see if they fit into the cubic pattern. If not, the next simplest system is tried, and then the next, and so on. In most cases only the very simplest systems can be determined from diffraction angle data alone.

If the powder method is used, diffraction angles and intensities are determined, but again indexing is difficult except for the simplest systems. If the material is studied by single crystal methods, much more information is obtained. The rotating- and oscillating-crystal methods yield data from which the

TABLE 20-2 / Some Characteristic Extinctions

Cell Type	Missing Lines	Cell Designation
Only corners of the cell are occupied	None	Primitive
Centers of all faces are occupied	h, k, l mixed (lines appear only when h, k, l are all odd or all even)	Face-centered
Center of cell is occupied	$h + k + l$ odd	Body-centered
Centers of top and of bottom are occupied	$h + k$ odd	Base-centered

TABLE 20-3 / Lines Characteristic of Cubical Cells

				Values of $h^2 + k^2 + l^2$	
h	k	l	Simple Cubic	Face-Centered	Body-Centered
1	0	0	1		
1	1	0	2		2
1	1	1	3	3	
2	0	0	4	4	4
2	1	0	5		
2	1	1	6		6
2	2	0	8	8	8
2	2	1	9		
3	0	0	9		
3	1	0	10		10
3	1	1	11	11	
2	2	2	12	12	12
3	2	0	13		
3	2	1	14		14
4	0	0	16	16	16
4	1	0	17		
4	1	1	18		18
3	3	1	19	19	

lattice parameters may be quickly calculated. The Weissenberg and the precession methods not only furnish data for calculating lattice parameters and interaxial angles but make it possible for reflections to be immediately indexed. These methods, in fact, permit determination of crystal symmetries almost by inspection of the films.

Whatever the experimental method used, not all the possible mathematical combinations of the various Miller indices need appear in experimental measurements. Certain lattice types have systematic absences, as shown in Table 20-2. All substances have, in addition, nonsystematic absences characteristic of the structure and not of the crystal system. Table 20-3 shows the lines characteristic of cubical cells.

In a face-centered cubic lattice the 100, 110, and 201 reflections are missing, as are all the others that have some indices even and some odd. A body-centered cube, for example, would have the 100 and the 111 reflections missing, as well as all other reflections whose indices add up to an odd number.

Single Crystal Methods

Historically, the first diffraction technique was the *Laue method*. Polychromatic radiation was diffracted through a single crystal, producing patterns on a film.

The points of reinforcement on the exposed film corresponded to the intersections on diffracted ellipses. Today the Laue method is most often used for aligning crystals.

The *rotating-crystal method* and its variation, the *oscillating-crystal method*, produce reflections much easier to index than the Laue or the powder pattern and in addition furnish valuable information on lattice parameters and give some information on symmetry.

In the rotating-crystal method the crystal is mounted, as shown in Figure 20–3a, with one of its rational axes perpendicular to the X-ray beam and is then rotated. The internal planes in the crystal have a fixed orientation relative to the axis or rotation. The X rays striking the sample generate cones of diffracted radiation coaxial with the axis of rotation of the crystal and therefore at fixed angles relative to the internal planes. If the sample is surrounded by a coaxial cylindrical film, the cones of diffracted radiation produce reflections that lie on lines on the film called *layer lines*, as shown in Figure 20–4. The layer lines correspond to reflection from the families of planes with a fixed orientation relative to the rotation axis. If the crystal is mounted

FIGURE 20–3 / Rotating crystal method of X-ray diffraction (after Buerger [1942]).

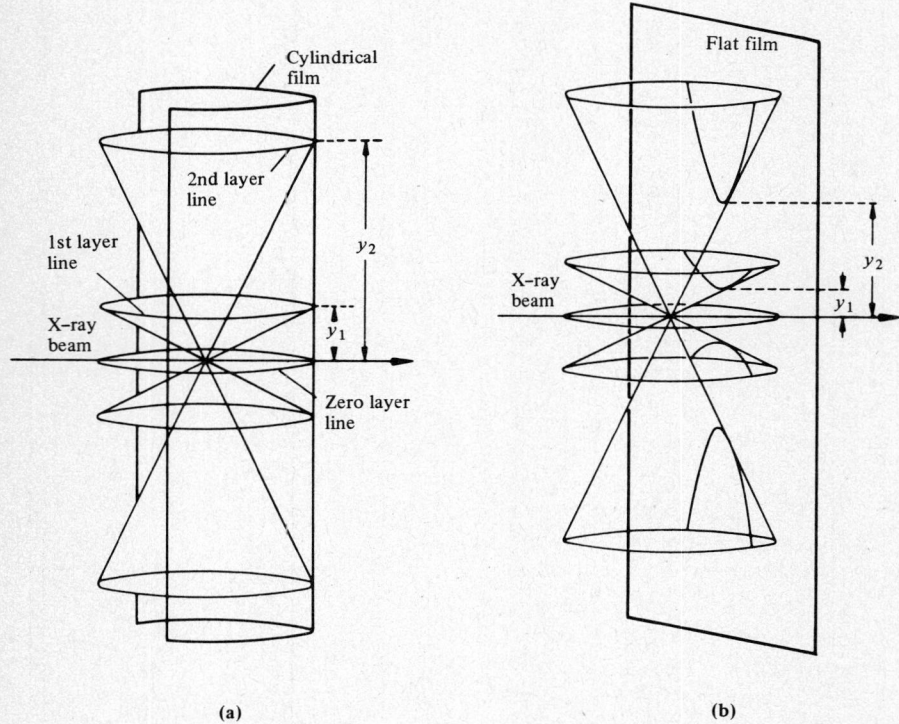

(a)   (b)

Chapter 20. X-Ray Diffraction

FIGURE 20–4 / Schematic representation of a section of rotating-crystal photograph taken with cylindrical camera.

parallel to the c axis, each line has a fixed value of the Miller index l. The spots on the equatorial line have indices of $h, k, 0$; on the first lines above and below the equator the indices are $h, k, 1$; on the second lines the indices are $h, k, 2$; and so on. If the crystal is mounted parallel to the b axis, the indices on the equator would be $h, 0, l$, and so forth. The indexing procedure is therefore simplified in that for each line one index is always known. There are, however, additional complications that make indexing a tedious and often difficult job.

Before indexing a rotation photograph, a valuable piece of information may be obtained, almost by inspection. The *repeat distance* along the axis of rotation is a simple function of the radius of the camera and the distance of the layer line from the equatorial line:

$$(20\text{-}4) \qquad a = \frac{n\lambda}{\sin \arctan(y_n/R)}$$

a = repeat distance
n = number of layer lines
y_n = distance from equator to nth layer line
R = radius of the camera

If the axis has been properly selected, the repeat distance is the edge of the unit cell. Although in practice the selection of the proper axis can be quite difficult, in principle the lattice constants and the cell volume can be determined on the basis of three photographs. Although for the crystallographer this is the start of the location of each atom, for the chemist it is often the solution of the problem, which is the determination of the structure of the molecule. Once the volume of the cell is known, the number of molecules per unit cell can be calculated from the density of the material. Molecular models can be made to scale and fitted together in an attempt to duplicate the known cell density. Very often this is sufficient to decide between spatial and structural isomers.

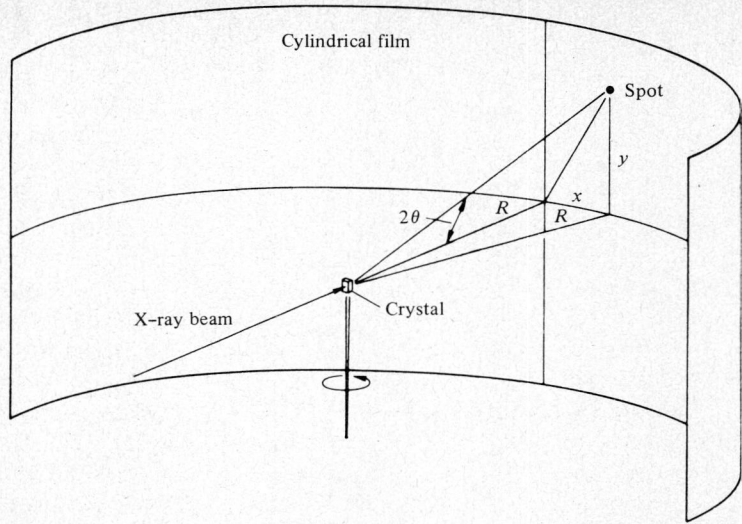

FIGURE 20-5 / Calculation of diffraction angle θ for indexing reflections in a rotating crystal photograph (after Buerger [1942]).

To index each reflection of a rotating-crystal photograph, the diffraction angle θ must be calculated from the measurements shown in Figure 20-5.

(20-5) $$\cos 2\theta = \cos \frac{x}{R} \cos \arctan \frac{y}{R}$$

or

(20-6) $$d = \frac{n\lambda}{2 \sin\{\frac{1}{2} \arccos[\cos(x/R) \cos \arctan(y/R)]\}}$$

y = height of spot above equator
x = distance of spot from the direct beam, measured along the equator
R = camera radius

Once θ is obtained, the Miller indices are calculated by trial and error as discussed above. The major problem is the large number of planes that reflect onto the same spot or onto spots very close to one another, making it difficult to decide which of a large number of planes has produced a given reflection. For example, in a cubic crystal, reflections from plane 511 always coincide with those from plane 333.

Many substances cannot readily be prepared in the form of single crystals, even though X-ray studies are desirable. Such substances include long chain organic and biochemical polymers, such as nylon, silk, and hair (especially polymers that have been stretch-pulled), and fibrous minerals such as asbestos. It is, however, possible to conduct X-ray studies on fibers of such materials, which very often consist of crystals with one axis parallel to the fiber axis. An X-ray photograph of such a fiber, mounted at right angles to the incident beam, is equivalent to a rotation photograph, even if the sample is not rotated. The large number of small crystals, all with the same axis perpendicular to the beam, produces the effect of one crystal being rotated around an axis perpendicular to the beam. There are, of course, differences. First, the orientation of the crystals in the fiber is not exact and the spots in the single-crystal photograph become arcs in the fiber photograph. Second, the crystals are very small and the reflections therefore less sharply defined. Third, there is less crystallinity in fibers than in single crystals, and consequently the pattern is often not so well developed.

In the oscillating crystal variation of the rotating crystal method, the crystal is oscillated through a narrow angular range, perhaps 15–30°. This cuts down the number of reflections and permits discrimination between possible reflecting planes. The oscillation method also furnishes information concerning crystal symmetry, which is masked in the rotating crystal photograph. If, for example, the top half of the photograph is identical with the bottom half, the crystal is mounted along an orthogonal axis. If the pattern on the left hand side is duplicated on the right hand side of the photograph, there is at least a twofold rotational symmetry. In most cases more than one photograph is necessary to establish rotational symmetry, since the pattern will not repeat itself during oscillation through a narrow angular range. If twofold symmetry is expected, a photograph is taken, the crystal is then rotated 180°, and another photograph is taken. A similar procedure is adopted for threefold symmetry, fourfold symmetry, and so on.

Moving Film Methods

The entire problem of indexing rotating and oscillating crystal photographs is essentially that of solving two equations for three unknowns. There are three indices (at least) for each reflecting plane. A film has only two coordinates, and so it is not possible to index all reflections without additional information. In the rotating- and oscillating-crystal methods, one index is fixed for each layer line, but even so the one remaining film coordinate, the distance along the layer line, must be solved for two unknowns. In the moving-film methods, one more coordinate is added. A screen with a slit in it is placed between the film and the crystal in such a fashion that only one layer line is photographed. As a result the photograph corresponds to all reflections from planes with one predetermined Miller index. The diffraction spots along the layer line are then spread out along the film in such fashion that the two coordinates of the film correspond to a function of the two remaining Miller indices of the diffracting planes.

In the *Weissenberg method*, as the crystal is rotated, the film is moved at a prefixed rate at an angle to the direction of the incident X-ray beam and parallel to the direction of the rotation axis. The diffraction spots are spread out, and from their positions a model of the crystal lattice may be constructed.

In the *precession method*, which is easier to interpret, the crystal is *not* rotated around an axis perpendicular to the X-ray beam, as in the rotating and oscillating crystal and Weissenberg methods. Instead, the crystal is rotated around an axis that is itself rotating around the X-ray beam at a preset rate with a preselected angle of inclination. The crystal is therefore precessing around the beam. The film is coupled to the crystal and exactly duplicates each crystal motion, so that as the crystal precesses, the film precesses (for the geometry of the method, see Buerger [1964]). The great advantages of the method are that it gives at a glance the symmetry of the direction parallel to the X-ray beam and that the computations of the lattice parameters and interplanar distances become almost trivial.

To understand the calculations, it is first necessary to understand the concept of the *reciprocal lattice*. From the Bragg equation, $\sin \theta = n\lambda/2d$, it is apparent that the distances one sees on a film, corresponding as they do to the $\sin \theta$, are *not* the interplanar distances but are proportional to the *reciprocals* of the distances. If one starts at any point in the crystal lattice and from that origin draws a normal to each plane, a collection of vectors is obtained whose directions are the normals to the planes and whose lengths are the interplanar distances. At some arbitrary origin new vectors are drawn parallel to the original vectors but with lengths proportional to the *reciprocals* of the lengths of the original lattice vectors. By putting a point at the end of each of these new vectors, a new lattice is obtained whose planes are *normal* to those in the crystal and whose interplanar distances are the reciprocals of the interplanar distances in the crystal. This is the reciprocal lattice. If the procedure is reversed, the direct crystal lattice is obtained from the reciprocal lattice.

The great importance of the reciprocal lattice is that the X-ray photographs correspond (in most methods with some distortion) to the reciprocal lattice, *not* to the direct lattice. The symmetry of the reciprocal lattice is the same as that of the direct lattice, but the smallest interplanar distance in the reciprocal lattice is the largest in the direct lattice (i.e., the edge of the cell). The advantage of the precession camera method over all other methods is that it gives an undistorted picture of the reciprocal lattice, and therefore the direct lattice can be obtained with a minimum of calculation.

Debye–Hull–Scherrer Powder Method

The *Debye–Hull–Scherrer powder method* is, technically, the simplest and easiest method, although it provides the least information about structure. It is

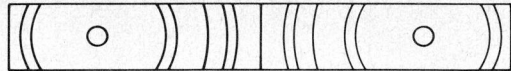

FIGURE 20–6 / Typical pattern appearing on a filmstrip in the powder method.

probably of most value in the identification of unknown solids. In structural analysis, the method is applicable mostly to the determination of cubic, hexagonal, and tetragonal lattice constants.

Unlike the specimens used in the methods previously discussed, the sample is *not* a single crystal that must be oriented carefully. Instead, the sample is a fine, uniform powder, about 200–300 mesh, formed into a fine rod 0.5–1.0 mm in diameter. The sample is shaped either by inserting the powder into a fine, thin-walled glass capillary or by wetting a borosilicate glass fiber or a nylon fiber with petroleum jelly, mucilage, or shellac and dipping the wet fiber into the powder. The powder sample is placed at the center of a cylindrical filmstrip and rotated while being exposed to X rays. The powder is actually a large number of microscopic crystals randomly arranged. The rotation of the sample ensures that crystals at all possible orientations will be exposed to the X rays. Most of the radiation diffracted from most of the lattice planes is canceled by out-of-phase radiation. However, there will always be some in-phase radiation from planes oriented at the angles for which the Bragg relation holds, as found from Table 20–1. Figure 20–6 shows a typical pattern.

The distance in millimeters between the center of the direct beam and any line is $k\theta$. For most commercially available powder cameras k is either 1 (as in the 57.3 mm diameter camera) or 2 (as in the 114.6 mm diameter camera). For identification of solids the film is compared with those of known materials. The American Society for Testing and Materials has published lists of the chief characteristic lines of a very large number of organic and inorganic solids, with detailed instructions for rapid and easy location of appropriate spectra.

For structural work, if the crystal form is not known, Miller indices are assigned to the lines on a trial-and-error basis, as outlined above. In general it is not possible to use the powder method for crystals other than cubic, hexagonal, or tetragonal. The difficulty lies in attempting to assign three or four constants on the basis of one experimental quantity, the value of $\sin \theta$.

In addition to information about the structure of the crystal, the powder pattern furnishes information concerning the size of the particles making up the powder. If the lines of the pattern are very broad and diffuse, the substance is a very finely ground powder, possibly smaller than 100 Å. If the lines are spotty, the crystals are quite large, possibly larger than 0.001 cm in diameter.

Lattice constants within 0.5% error can be determined by the powder method without too much difficulty. For higher precision, single crystal

FIGURE 20-7 / Philips–Norelco powder camera (courtesy of Philips Electronic Instruments, Inc.).

methods are preferred. The greatest single source of error is film shrinkage. This can be minimized by using the *Straumanis technique* of comparing the length of the developed film with a standard length. In most accurately machined cameras, the diameter is either 57.3 or 114.6 mm. The distance between the centers of the two holes punched to receive the collimating tube and the beam stop tube should therefore be 180 and 360 mm, respectively. If the film has shrunk uniformly during the developing process, all distances are multiplied by the shrinkage factor, the ratio of either 180 or 360 mm to the measured distance (in millimeters) between the centers of the two punched holes in the film.

The powder camera is a flat, hollow cylinder, with the film held firmly against the inner wall and the sample in the center. Figure 20-7 shows the Philips–Norelco powder camera. (Other companies have cameras that differ slightly in detail.) In this model of the camera the film is held in place by means of the two pins shown at the top. The pin on the right slides back and forth and is fastened with the set screw on the outside. In operation the film is put into the camera with the cut ends flush against the pins. Then the sliding pin is forced as far to the right as it goes, clamping the film tight against the wall. The brass tubes are for focussing the entering X-ray beam upon the sample and for absorbing the excess radiation which has passed through the sample. The tubes are removed while the sample and the film are being inserted. They are of soft brass and must be handled with care, since even a slight impact will bend the brass and knock the beam out of alignment. The beam absorber tube has a fluorescent screen to aid in focusing the beam and aligning the camera and track. The collimator tube has a pin that fits into a hole machined in the camera case, so it may be mounted in only one position, that which aligns and focuses the beam.

The sample is mounted in plasticine or some other semisoft material held in a brass pin in the sample holder. During the exposure the sample is rotated by means of a rubber belt connected to two pulleys, one on the camera and one on an electric motor. The sample must be carefully centered on the axis of rotation. This is done with an adjustable sample holder that slides over the flat surface of the rotor. The clutch screw on top of the camera operates a rod that pushes the sample holder downward along the rotating plate. To center the sample, the specimen is rotated until it is displaced upward as far as it goes. It is then centered by pushing it down with the rod. The rod is retracted, and the rotating and pushing are repeated again and again until the sample is just over the axis of rotation and no displacement is observed when it is rotated. For the final check, the rotation should be observed through a lens mounted on the beam collimating tube, to ensure that there is no lateral motion. The sample must, of course, be coaxial with the sample holder, or no amount of adjustment of the holder will keep the sample centered during rotation.

INSTRUCTIONS FOR TAKING POWDER PHOTOGRAPHS WITH THE NORELCO CAMERA

Preparation of the Sample. Wet the end of a 1–2 cm nylon or borosilicate glass fiber with mucilage, holding it with forceps. Then dip the wet end of the fiber into about 50–100 mg of powder and rotate it so that the crystals are deposited uniformly.

Mounting and Aligning the Sample. Fill the brass insert plug in the sample holder with putty or plasticine. Insert the fiber into the putty with its axis parallel to that of the insert plug. Center the sample over the axis of rotation of the rotor plate. If the sample is centered, there is no visible lateral displacement when the rotation is observed with a viewing lens mounted on the inlet collimator tube. (Note that the two brass tubes must be removed in order to insert the sample.)

Loading the Camera. In the darkroom, using a safety light if necessary, remove the cover and collimator tubes. It is best to have a holder for the tubes so that they may be easily handled in the dark. Notch one corner of the film for identification purposes, then slide the film into the camera along the inner circumference and fasten it in place. Insert the tubes in their appropriate positions and put the lid back on the camera.

Photographing the Powder Pattern. Start the generator and let it warm up for $\frac{1}{2}$ hr. Fasten the camera at the desired exit port with the beam inlet flush against the port, making sure the proper filter is in place. Adjust the voltage and current according to the manufacturer's instructions.

Indexing the Lines by the Straumanis Method. The forward reflections are centered around the beam exit tube, while the back reflections are centered around the beam entry tube. Fasten the film to the viewer. There usually are more forward than backward reflections, and so there are more lines around the beam exit tube. Determine the position of the center of each line. Then, from pairs of symmetrical lines locate the center of each hole. Calculate the distance in millimeters between the centers of the holes. The ratio of one half of the camera circumference to this distance is the shrinkage factor. Determine the distance of each line in millimeters from the center of the direct beam and multiply this by the shrinkage factor. The corrected distance in millimeters corresponds to the value of θ in degrees for the 57.3 mm camera and twice the value of θ in degrees for the 114.6 mm camera.

References

Azaroff, L. V., and M. J. Buerger, *The Powder Method in X-Ray Cystallography.* McGraw-Hill, New York, 1958.

Buerger, M. J., *The Precession Method in X-Ray Crystallography.* Wiley, New York, 1964.

Buerger, M. J., *X-Ray Crystallography.* Wiley, New York, 1942.

Cullity, B. D., *Elements of X-Ray Diffraction.* Addison-Wesley, Reading, Mass., 1966.

Lambert, J. B., et al., *Organic Structural Analysis*, Part Five. Macmillan, New York, 1976.

Marton, L., in *Methods in Experimental Physics*, Vol. 6. Academic Press, New York, 1959.

Weissberger, A., and B. W. Rossiter, *Physical Methods of Chemistry*, Part III, D, ch 1. Wiley-Interscience, New York, 1972.

Woolfson, M. M., *An Introduction to X-Ray Crystallography.* Cambridge Univ. Press, London and New York, 1970.

Chapter 21

Magnetochemistry

The magnetic phenomena associated with nuclear and electronic spin and electronic orbital motion cause paramagnetism and diamagnetism and yield extremely valuable information about molecular, radical, ionic, and crystal structure. Static measurements of paramagnetism and diamagnetism furnish information about the overall magnetism of the atom, ion, or molecule. Resonance methods, such as electron spin resonance (esr) and nuclear magnetic resonance (nmr), furnish detailed information about the structures concerned, including such things as internal charge distribution and location of equivalent atoms.

A charged moving particle, such as an electron in an orbital or a spinning electron or nucleus, is, in effect, a current and generates a magnetic field. If the particle moves in a circular path, its angular velocity in radians per unit time is $v/2\pi r$. The current circulating is

$$i = \frac{qv}{2\pi r}$$

q = charge of the particle, coulomb
v = velocity, m/sec
r = radius of path, m

If the radius of the path is small compared to its distance from the observer, the magnetic field is a magnetic dipole whose moment is

(21-1) $$\mu = iA = \frac{q(mvr)}{2m} = \frac{qP_\phi}{2m} \equiv \gamma P_\phi$$

μ = magnetic dipole moment, Am^2
A = area
   = πr^2
m = mass of the particle
P_ϕ = angular momentum
γ = gyromagnetic ratio

If the angular momentum is known, the magnitude of the magnetic moment may be calculated from equation (21-1). When the particle is an electron, the orbital angular momentum is given by

(21-2a) $\quad P_\phi = [l(l+1)]^{1/2} \dfrac{h}{2\pi} = [l(l+1)]^{1/2} \hbar$

l = orbital quantum number
$\hbar \equiv h/2\pi$

The electron spin angular momentum is given by

(21-2b) $\quad P_\phi = [s(s+1)]^{1/2} \hbar$

s = electron spin quantum number
$= \frac{1}{2}$

The equation for nuclear spin angular momentum is

(21-2c) $\quad P_\phi = [I(I+1)]^{1/2} \hbar$

I = nuclear spin quantum number

In magnetochemistry the quantities of interest are usually not the total magnetic moments but the maximum observable components in some direction. Usually, the direction along which the external magnetic field is applied is considered to be the z axis, so the quantities measured and calculated are the (arbitrary) z components. Most textbooks now use the symbol m_l in place of l_z, m_s in place of s_z, and m_I in place of I_z, and we will conform to that usage.

For the *angular momenta* the maximum components in the z direction are, for the orbiting electron,

(21-3a) $\quad P_{\phi_z} = l_z \hbar \equiv m_l \hbar$

for electron spin,

(21-3b) $\quad P_{\phi_z} = s_z \hbar \equiv m_s \hbar$

and for the nuclear spin,

(21-3c) $\quad P_{\phi_z} = I_z \hbar \equiv m_I \hbar$

The components of magnetic moment along the z axis are calculated by the appropriate combinations of equations (21-3a), (21-3b), and (21-3c) with (21-1). For the orbiting electron,

(21-4a) $\quad \mu_z = B m_l$

B = Bohr magneton,
$= 0.927 \times 10^{-24}$ J/T

TABLE 21-1 / Nuclear Spins and g Factors

Nucleus	g_N	I	Nucleus	g_N	I	Nucleus	g_N	I
^{1}H	5.585	$\frac{1}{2}$	^{15}N	−0.567	$\frac{1}{2}$	^{31}P	2.2283	$\frac{1}{2}$
^{2}H(D)	0.857	1	^{17}O	−0.757	$\frac{5}{2}$	^{33}S	0.429	$\frac{3}{2}$
^{7}Li	2.171	$\frac{3}{2}$	^{19}F	5.257	$\frac{1}{2}$	^{35}Cl	0.548	$\frac{3}{2}$
^{13}C	1.405	$\frac{1}{2}$	^{23}Na	1.478	$\frac{3}{2}$	^{39}K	0.261	$\frac{3}{2}$
^{14}N	0.403	1						

For both the electronic spin and the nuclear spin, the expressions are analogous, but differ from equation (21–4a) in that the Landé g factor, a dimensionless constant, is included. For the spinning electron,

(21–4b) $\quad \mu_z = -gBm_s$
$\quad\quad\quad g =$ electronic spin g factor $= 2.00232$

and for the spinning nucleus,

(21–4c) $\quad \mu_z = g_N B_N m_I$
$\quad\quad\quad g_N =$ nuclear spin g factor
$\quad\quad\quad B_N =$ nuclear Bohr magneton for the proton
$\quad\quad\quad\quad = 0.505 \times 10^{-27}$ J/T

(The magneton includes terms for the mass of the particle and the charge of the particle.) For convenience the Bohr magneton is defined as being positive, and since the electronic charge is negative, there is a negative sign in equation (21–4b). The difference between the masses of the proton and of the electron has important consequences in the study of nmr and esr, especially in the instrumentation. In this chapter B is used for the magneton, instead of μ_B, to avoid confusion with μ_z.

Table 21–1 is a compilation of nuclear I and g factors for the most commonly studied nuclides.

Paramagnetic and Diamagnetic Susceptibilities

Paramagnetism, positive magnetic susceptibility, is observed experimentally as an attraction toward a magnetic field. A paramagnetic substance suspended at the edges of a magnetic field is pulled into the field, with the field lines traveling preferentially through the substance. Paramagnetism is demonstrated by all materials with observable magnetic dipole moments, the interaction between the field and the dipole producing the magnetic attraction.

The electronic motion generating the magnetic dipole moment may be electron spin or orbital motion or both. Paired electrons, which according to the Pauli principle, must have opposing spins, do not contribute to magnetic dipole moments. Only unpaired electrons contribute to observed magnetic attractions. All orbiting electrons produce a magnetic dipole moment, but not all orbital magnetic dipole moments can interact with an applied field. For interactions to occur the orbital moment must be able to align itself in the direction of the field. When the internal fields are much stronger than the external field the interactions are "quenched," and the orbital moments do not then contribute to the experimentally observed magnetism. This quenching effect usually takes place in molecules, because internuclear fields are stronger than external fields. It takes place in atoms only when the orbitals in question are in interior shells, where the field of the atom is more intense than the external field.

Diamagnetism, negative magnetic susceptibility, is observed when a material is repelled from a magnetic field. A field is generated within the atom which opposes the effect of external field. It produces a force that pushes the sample in the direction of lower external field strength. Diamagnetic effects are two or three orders of magnitude weaker than paramagnetic effects. All substances exhibit diamagnetism, even paramagnetic materials, in which the paramagnetism masks the diamagnetism.

The relationship between the paramagnetism and the calculated magnetic dipole moment is

(21-5) $$\chi_m = N\left(\alpha + \frac{\mu^2}{3kT}\right)$$

χ_m = molar magnetic susceptibility
N = Avogadro's number
α = induced magnetic dipole moment
k = Boltzmann's constant
T = Kelvin temperature

The molar magnetic susceptibility is analogous to the molar polarization (page 446). It is calculated from the observed paramagnetic effects using

(21-6) $$\chi_m = \chi_{sp} M$$

χ_{sp} = specific magnetic susceptibility
   = magnetism per gram
M = molecular weight

The magnetic dipole moment induced by the applied field is small enough at ordinary temperatures so that it may be neglected for most purposes, but for accurate work it must be measured and taken into account. It is obtained from susceptibility determinations at a series of temperatures.

In molecules, where orbital moments are quenched, the magnetic dipole moments depend only on the number of unpaired spins, as shown by

(21-7) $$\mu = [n(n+2)]^{1/2}B$$

n = number of unpaired spins

In polyelectronic atoms, however, orbital moments must be taken into account. There are interactions between orbital magnetic moment vectors and spin moment vectors of the various electrons. The quantity obtained from the molar magnetic susceptibility is the resultant of all the various vectors involved.

The moments of interest in susceptibility measurements, in contrast to resonance measurements, are the total magnetic dipole moments, rather than the components. For each electron the magnitude of the orbital magnetic moment is $[l(l+1)]^{1/2}B$ and the magnitude of the spin magnetic moment is $[s(s+1)]^{1/2}B$. If there is one electron, the two magnetic moments couple only by algebraic addition; that is, their magnitudes either add or subtract because the vectors are collinear. The resultant vector is the j vector, where j is called the *inner quantum number*. However, if there are many electrons, two different types of vector coupling are possible. In heavier atoms the orbital and spin vectors of each electron combine to form a j vector, and the j vectors of all the electrons then add algebraically to give a resultant J vector for the atom. This J vector determines the magnetic dipole moment. This type of coupling is known as j–j coupling. In lighter atoms all the individual l vectors combine to form a total orbital moment vector, characterized by a resultant quantum number L. The L value is taken as the sum of the m_l vectors, where m_l is the magnetic quantum number and has values defined by $l \geq m_l \geq -l$. All the individual s vectors also combine, to form a resultant S vector, which is the sum of the m_s vectors, over all electrons, where $m_s = \pm s$. The L and S vectors then combine to form a resultant J vector. This is known as *Russell–Saunders coupling*.

In Russell–Saunders coupling the values of L, S, and J in the ground state can be calculated from Hund's three rules, based on the Pauli principle and the repulsion of electrons from each other:

1. Other things being equal, the state of highest multiplicity is the most stable. Since multiplicity is $2S + 1$, the most stable state has the highest S value; that is, the spins are parallel.
2. The most stable state has the largest orbital angular momentum; that is, the L value is at a maximum integral value. Since in the presence of a magnetic field the value of L is determined by the m_l's, the magnetic quantum numbers (i.e., $L = \sum m_l$), this is equivalent to requiring that the sum of the m_l values of the individual electrons be a maximum.
3. For shells more than half filled the most stable state has the largest J value. For those less than half filled, the most stable state has the smallest J value.

For an example of the use of these rules, let us find the ground state of chromium(III). This ion has three $3d$ electrons in the valence shell. Hund's first rule tells us that we need maximum unpaired spins, so $S = \sum m_s = \frac{1}{2} + \frac{1}{2} + \frac{1}{2} = \frac{3}{2}$; the second rule, maximum orbital angular momentum, gives $L = \sum m_l = 2 + 1 + 0 = 3$ (note that the Pauli principle forbids $2 + 2 + 1$); the third rule, for a less-than-half-filled shell, requires that $J = L - S = \frac{3}{2}$.

When the coupling is of the Russell–Saunders type, the total angular momentum in an atom is directly proportional to $[J(J + 1)]^{1/2}$. The magnetic dipole moment is more complicated because the magnitude of the unit spin magnetic vector is twice that of the unit orbital magnetic vector. This is taken into account by the Landé g factor,

$$g = 1 + \frac{J(J+1) + S(S+1) - L(L+1)}{2J(J+1)}$$

Note that for pure orbital motion, $S = 0$, $J = L$, and the g factor would be unity. This is the explanation for the fact that there is no g factor in equation (21-4a), while it is included in (21-4b) and (21-4c). Note also that for quenched orbital motion $L = 0$, $J = S$, and the g factor is 2. The actual g factor for the free spinning electron is 2.00232, when a small relativistic correction is included. Taking the g factor into account, the final equation for the angular magnetic moment is an atom where Russell–Saunders coupling is followed is

(21-8) $\qquad \mu = g[J(J+1)]^{1/2} B$

Experimental Determination of Magnetic Susceptibilities

To determine the magnetic susceptibility of a substance, a sample is suspended in an inhomogeneous magnetic field and the force exerted by the field upon the sample is measured, usually by means of an analytical balance. The Faraday version of this method employs small samples and large field gradients. The Gouy version employs large samples and small field gradient.

If the sample is suspended in air, and if the direction of the field gradient is downward,

(21-9) $\qquad dF_s = \chi dv H \dfrac{dH}{ds}$

$\qquad dF_s$ = force exerted on a small sample at a given point; the differential force
$\qquad \chi$ = volume magnetic susceptibility of the sample
$\qquad dv$ = differential sample volume
$\qquad dH/ds$ = field gradient in s direction
$\qquad H$ = field strength
$\qquad s$ = distance in vertical direction

If the sample is large and passes through a finite region of the field, equation (21-9) is integrated and mg' is substituted for force, giving

$$(21\text{-}10) \quad \Delta m = (m - m_0) = \tfrac{1}{2}\left(\chi \frac{a}{g'}\right)(H_2^2 - H_1^2)$$

m = apparent mass of sample in field
m_0 = sample mass in absence of field
a = cross sectional area
g' = gravitational constant
H_2 = field strength at high-field end of sample
H_1 = field strength at low-field end

Equation (21-10) is valid only for cases where the field varies uniformly with distance. In practice, it is easier to use a reference substance of known susceptibility than to calibrate the field at both ends of the sample. If equal volumes of reference material and the substance being studied are employed, the sample tubes are of the same dimensions, and the samples are suspended at the same point, then

$$(21\text{-}11) \quad \Delta m = k'\chi$$

k' = apparatus constant

To calculate molar susceptibility,

$$\chi_m = \chi_{sp} M = \frac{\chi M}{\rho}$$

ρ = density

In making paramagnetic susceptibility measurements it is always necessary to use magnetically dilute samples to prevent interactions between dipoles that would result in gross errors in interpretation. Magnetic dilution is accomplished for solids by using double salts with many molecules of water of hydration. so that each paramagnetic ion is separated from the nearest paramagnetic neighbor by many diamagnetic atoms. Solutions may also be employed, of concentration 1–0.1 M, depending on the material, sample size, and field strength. With solutions the volume susceptibility is calculated assuming that the solute is dispersed uniformly through the volume.

Resonance Methods

In the absence of an external magnetic field the electron or proton spins randomly with no preferred orientation. Although there are two spin states, the energies of both states are the same, and the system is said to be *doubly degenerate*. The application of an external magnetic field removes the degeneracy, as shown in Figures 21–1 and 21–2. Figure 21–2 is the more frequently used

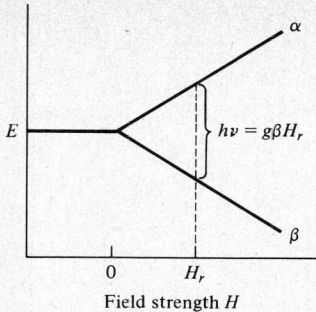

FIGURE 21-1 / Zeeman splitting.

form of energy level diagram. Diagrams of this type show the energy level splittings caused by a magnetic field of fixed strength H. This is the *Zeeman effect*. The energy difference between the ground state and the excited state is a direct function of the field strength, and transitions between the two spin-energy levels will take place when electromagnetic radiation of the appropriate energy, $h\nu$, is supplied. However, the main object of resonance studies is not the Zeeman energies themselves, but the perturbations due to internal fields. In nmr studies one observes the effect of neighboring nuclei upon the Zeeman interaction (nuclear spin-spin coupling) and the screening of the nucleus by the surrounding atoms (chemical shift). In esr one observes the effect of neighboring nuclei upon the Zeeman interaction (hyperfine splitting). Occasionally, one observes electron-electron interactions. These give rise to fine structure with rather large splitting compared to hyperfine splitting.

The Zeeman energy is the energy of interaction between the applied magnetic field of strength H and the spin of a charged particle. Its magnitude is

FIGURE 21-2 / Energy level diagram for the hydrogen atom. Resonance conditions: (a) $h\nu = g\beta h$; (b) $h\nu = g\beta H + \frac{1}{2}a$; (c) $h\nu = g\beta H - \frac{1}{2}a$.

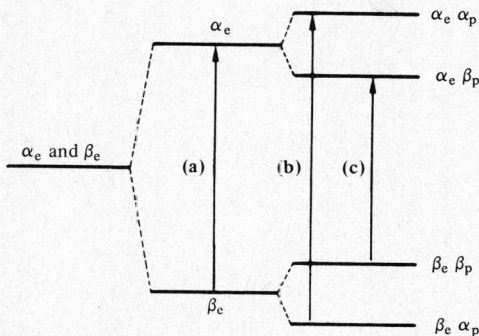

the negative of the vector dot product of field strength and magnetic moment:

(21-12) $\quad E = -\mathbf{\mu} \cdot \mathbf{H}$

Substituting equations (21-4b) and (21-4c) in (21-12) gives the Zeeman energy for electron spin.

(21-13a) $\quad E = gBHm_s$

and for a spinning nucleus,

(21-13b) $\quad E = -g_N B_N H m_I$

For both the electron and the proton the spin quantum numbers are $\pm\frac{1}{2}$. Energies corresponding to spin $+\frac{1}{2}$ are given the symbol α and those corresponding to spin $-\frac{1}{2}$ are given the symbol β. No nucleus other than the proton will be considered here. For a more complete discussion see the References. Electrons of β spin are of lower Zeeman energy than those of α spin, while the opposite is true of protons. The inversion of ground state energies results from the opposite sign of the charges carried by the two particles.

For both protons and electrons the difference in Zeeman energy between particles of opposite spin is

(21-14) $\quad \Delta E = gBH\, \Delta m_s$

Δm_s = difference in spin quantum number

For the electron the α state is the excited state, the β state is the ground state, and since the m_s values are $\pm\frac{1}{2}$,

(21-15a) $\quad \Delta E_e = gBH[\frac{1}{2} - (-\frac{1}{2})] = gBH$

For the proton the excited state is the β state, which has spin of $-\frac{1}{2}$, and so

(21-15b) $\quad \Delta E_p = -g_N B_N H[-\frac{1}{2} - (\frac{1}{2})] = g_N B_N H$

Transitions between energy levels can take place when energy is supplied to the system in the form of electromagnetic radiation whose energy $h\nu$ is equal to the ΔE of the transition. The equations above give the magnitude of the transition energies resulting from the primary interactions between the applied external field and the spinning particles, without taking into account possible perturbations due to interactions of spinning particles with each other. However, these secondary interactions do cause small perturbations in the resonance requirements and result in further splitting of the energy levels, the hyperfine splitting of the esr spectrum previously mentioned, and the spin-spin coupling and chemical shift in the nmr spectrum. Since these secondary effects depend on the environment of the spinning particle, their study furnishes valuable information to the chemist who is interested in molecular structure.

Both esr and nmr originate from the same fundamental processes, spin transitions in an externally imposed field. They differ only in the magnitude of the interactions, and therefore in the field strengths, radiation frequencies, and instrumentation involved. The rest of this discussion will be devoted to esr only.

Electron Spin Resonance Spectra

When the relationship between the field strength and the frequency of the microwave radiation is such that the energy of a photon corresponds to the energy of a transition, radiation is absorbed, as indicated by a detector. In practice, the radiation frequency is fixed and the field strength is slowly varied until resonance absorption occurs. The instrumentation of most apparatus feeds out the first derivative of the absorption curve, rather than the absorption curve itself (see pages 284–287).

The presence of neighboring protons, with their small but significant magnetic moments, changes the energy requirements for electron spin transitions. According to equation (21-15a), the single energy expected to produce a spin transition at a given field strength is split into two, three, or more energies, each of which produces a transition at a different field strength. The number of such different transitions depends upon the number of protons interacting with the electron and on whether or not their interactions are equivalent. The simplest case is the hydrogen atom, with one proton and one electron in the system. Figure 21-2 shows the electron–proton interactions in the hydrogen atom. The magnetic moments of the proton and the electron couple, splitting the energy levels of the electron. Because of the opposite charges on the two particles, the α_p proton lowers the energy of the β_e electron but raises that of the α_e electron, while the β_p proton raises the energy of the β_e electron and lowers that of the α_e electron. The magnitude of this energy change depends on a, the coupling constant. (The units of a may be joules, millihertz, centimeters^{-1}, or tesla. When expressed in tesla, a is referred to as the hyperfine splitting or coupling constant. When expressed in joules, millihertz, or centimeters^{-1}, it is referred to as the hyperfine coupling energy. In practice, the constant is most frequently expressed in tesla.) The energy needed to raise an electron from the

FIGURE 21-3 / Energy difference between two transitions in the hydrogen atom.

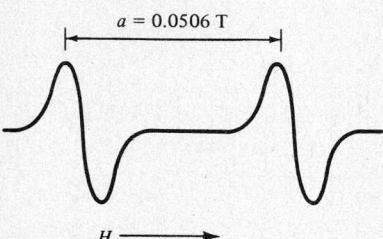

$a = 0.0506$ T

$H \longrightarrow$

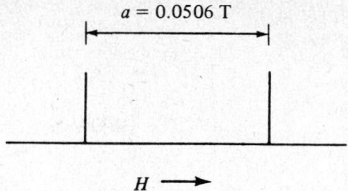

FIGURE 21-4 / Stick plot for showing position of absorption, number of peaks, and relative intensities of peaks.

β_e spin state to the α_e spin state, including the hyperfine coupling energy, is therefore

(21-16) $\quad \Delta E = gBH + am_I$

$$m_I = +\tfrac{1}{2} \text{ for } \alpha_p \text{ spin state}$$
$$ = -\tfrac{1}{2} \text{ for } \beta_p \text{ spin state}$$

Therefore, for protons of α_p spin,

(21-17a) $\quad \Delta E = hv = gBH + \tfrac{1}{2}a$

and for protons of β_p spin,

(21-17b) $\quad \Delta E = hv = gBH - \tfrac{1}{2}a$

The two transitions shown in Figure 21-2 would require energies that differed from each other by 9.28×10^{-25} J or 0.0506 T, as shown in Figure 21-3, the derivative graph corresponding to the transitions of Figure 21-2. Implicit in

FIGURE 21-5 / Energy level diagram for a $CH_3\cdot$ radical.

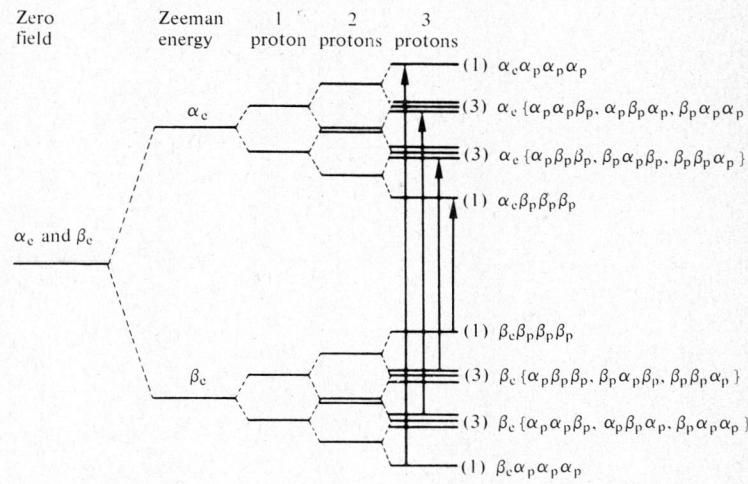

Figure 21-2 shows an important rule that is generally obeyed: For electron spin transitions, where coupling exists the spin quantum number of the proton or of any other nucleus is unchanged. It should be noted that if there are two or more protons in a compound, those that are equivalent in terms of symmetry have the same value of the coupling constant, while nonequivalent protons have different coupling constants.

The stick plot shown in Figure 21-4 is a convenient and simple way of showing the position of absorption (i.e., the field strength at which absorption occurs), the number of peaks, and the relative intensities of each peak (their relative heights on the stick plot).

When there are two or more protons the spectrum becomes considerably more complex. However, it is often relatively simple to interpret quite complex diagrams. Figure 21-5 shows the energy level diagram for a $CH_3 \cdot$ radical, and Figure 21-6 shows the stick plot. In these diagrams the Zeeman splitting is shown first, then the interaction with the first proton, then the second proton interaction, and finally the third proton interaction. The line heights in the stick plots show the relative absorption intensities, and are indicated in parentheses. These intensities can be calculated from the energy-level diagram simply by counting the number of different ways the same total spin can be obtained. The different combinations of total spin are shown in the parentheses. In the $CH_3 \cdot$ radical the protons are equivalent, and the hyperfine splitting is the same for each. The Zeeman energy levels are therefore perturbed to the same extent by each proton. The three protons produce four lines whose intensities are in the ratio $1:3:3:1$. An easy way to remember the number and relative intensities is to note that the number of lines is $p + 1$, where p is the

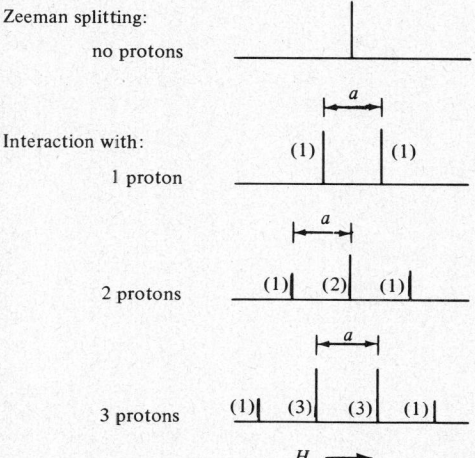

FIGURE 21-6 / Stick plot corresponding to energy level diagram of Figure 21-5.

Number of protons											Number of lines observed
0					1						1
1				1		1					2
2				1	2	1					3
3			1	3		3	1				4
4			1	4	6	4	1				5
5		1	5	10		10	5	1			6
6		1	6	15	20	15	6	1			7
7	1	7	21	35		35	21	7	1		8
8	1	8	28	56	70	56	28	8	1		9

FIGURE 21-7 / Pascal's triangle.

number of equivalent protons, and that the relative intensity of a line is proportional to the number of different ways to reach the level, starting from the origin of the diagram, at the left. For example, there is only one path to reach the topmost level at the right of Figure 21-5. At each bifurcation† the upward path must be taken. On the other hand, to reach the second level there are three paths: the effect of the first proton can be positive, the second positive, and the third negative; or the first can be positive, the second negative, and the third positive; or the first can be negative, with the second and third both positive. Alternatively, one may say that the three proton spins can be $\alpha\alpha\beta$, $\alpha\beta\alpha$, and $\beta\alpha\alpha$.

A simple diagram that can be applied to any number of equivalent protons for computing the number and relative intensity of the lines is the Pascal triangle shown in Figure 21-7. The esr spectrum of n equivalent protons will be represented by the nth row in the triangle, with the apex being the zeroth row. For example, the spectrum of a compound with five equivalent protons is represented by the fifth row. There are therefore six lines, of relative intensity 1 : 5 : 10 : 10 : 5 : 1. To construct the triangle, begin and end each new row with the number 1. Then simply add each adjacent pair of numbers in the previous

FIGURE 21-8 / Nonequivalent protons in biphenylene (a) and naphthalene (b) radicals. For protons 1, 4, 5, and 8, $a_I = 2.1 \times 10^{-5}$ T in biphenylene and 4.90×10^{-4} T in naphthalene. For protons 2, 3, 6, and 7, $a_{II} = 2.86 \times 10^{-4}$ T in biphenylene and 1.83×10^{-4} T in naphthalene.

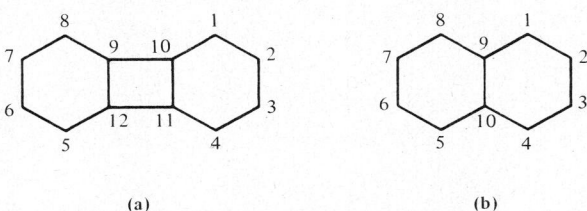

(a) (b)

† The number of energy levels obtained when a nucleus of spin quantum number I interacts with an electron is $(2I + 1)$. The proton $(I = \frac{1}{2})$ therefore splits each energy level in two, as shown in Figure 21-5.

row and place its sum just below them in the new row so as to form an equilateral triangle; for example,

from the row   1 2 1

we obtain   1 3 3 1

and so on.

If the protons are nonequivalent the spectrum is more complex. Radicals such as biphenylene (Figure 21-8a) and naphthalene (Figure 21-8b) contain two sets of nonequivalent protons, each set consisting of four equivalent protons. Each set of equivalent protons has its own value of the hyperfine splitting constant, which may differ considerably from that of the other set. The greater the relative difference, the easier it is to interpret the esr spectrum. Figure 21-9 shows the stick-plot diagrams for the biphenylene and naphthalene free radical anions. For the biphenylene anion the constants are $a_I = 2.1 \times 10^{-5}$ T and $a_{II} = 2.86 \times 10^{-4}$ T. For naphthalene anion, $a_I = 4.90 \times 10^{-4}$ T and $a_{II} = 1.83 \times 10^{-4}$ T. Both spectra may be viewed as originating from the hyperfine splitting by one set of equivalent protons, forming lines that are in turn split by the other set of protons. Since each set contains four protons, there is a splitting into five lines of relative intensity $1:4:6:4:1$ by the first set, with

FIGURE 21-9 / Stick plot diagram for the biphenylene (a) and naphthalene (b) free radical anions.

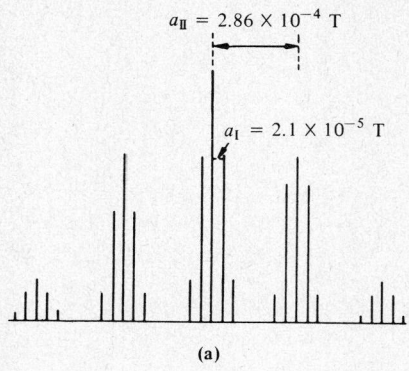

(a)

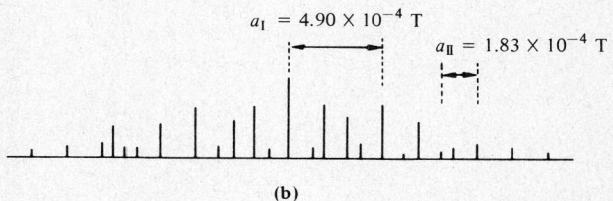

(b)

each of these lines then being split into lines by the second set, also of intensity 1 : 4 : 6 : 4 : 1. In Figure 21–9a the lines are easy to interpret because of the very large relative difference between a_I and a_{II}. There are five groups of lines, each group consisting of five lines of the characteristic ratios. The line separation within any group is a_I and that between corresponding lines in two groups is a_{II}. In the naphthalene anion spectrum the overlapping of the lines makes interpretation much more difficult.

INSTRUMENTATION FOR ESR STUDIES

The Zeeman splitting of the spin states is studied by esr spectrometry. This technique is so different from optical spectroscopy that it requires separate consideration. The magnitude of the splitting is gBH; for a free electron the gB product is 1.85×10^{-24} J/T. At a field of 10^{-2} T, $E = h\nu = 1.85 \times 10^{-25}$ J, or the frequency is 2.8×10^8 Hz and the wavelength is 107 cm. This is in the high radio frequency region. In a field of 1 T the frequency is 3×10^{10} Hz in the microwave region.

The relative population of the two spin states is

$$\frac{n_{\alpha_e}}{n_{\beta_e}} = e^{-\Delta E/kT}$$

At room temperature kT is 4.4×10^{-21} J, so the Boltzmann factor, $e^{-\Delta E/kT} = e^{-gBH/kT} \approx 1-5 \times 10^{-5}$ for a 10^{-2} T field and $1-5 \times 10^{-3}$ for a 1 T field. Therefore, the population of the upper level will be only very slightly less than that of the lower level. Since the radiation will stimulate downward transitions with the same probability that it stimulates upward transitions, power will be absorbed only because of the slight excess of spins in the lower level. In this region of the spectrum, stimulated downward transitions are more important than are spontaneous downward transitions (see Moore [1972] for a discussion of the two mechanisms). Obviously, one of the chief problems in esr spectrometry is the accurate detection of very small changes in a radiation field.

Spectrometer components consist of a magnet, a radiation source, a radiation detector, and the electronics for amplifying and recording the output signal.

HIGH FIELD SPECTROMETRY

The most convenient field range is of the order of a few tenths of a tesla for most spectrometers. It is occasionally necessary to scan from zero field strength to a field several times greater than the normal free electron resonance field strength. In such scanning there are fairly stringent stability requirements. In esr, lines can be less than 10^{-5} T wide. The background noise or jitter must therefore be much less than this. In general, a stability of 1 part in 10^5 is adequate (in nmr, for comparison, the stability requirement is 1 part in 10^8).

The radiation is in the microwave range and is produced by an oscillator, usually either a klystron or a traveling wave tube. These devices are somewhat like modified vacuum tubes, producing oscillations at well defined frequencies. The frequency of a given oscillator can be varied over only a very narrow range, so that to scan for energy absorption, the magnetic field must be varied.

The microwaves put out by the oscillator must be conducted by waveguides, which are hollow highly reflecting metal tubes of precisely defined dimensions that allow the waves to propagate in only certain modes as they move to the sample chamber. Essentially, the waveguides are transmission lines, and much the same analysis can be applied to them as to conventional power transmission lines.

The sample chamber is a metallic, highly reflecting cavity of dimensions matched within fairly narrow tolerances to the particular frequency and waveguide, although in some applications slightly "detuned" cavities are useful. The sample itself should be at a point in the cavity where the magnetic field of the electromagnetic radiation is perpendicular to the steady magnetic field (to allow resonance), the varying magnetic field is at a maximum, and the varying electric field is at a minimum. (These last conditions provide the best signal/noise ratio.) This combination of requirements means that the sample size must be appreciably smaller than λ, where λ is the wavelength. A good cavity should have a Q factor of at least several thousand, where Q is the ratio of energy stored in the cavity to energy lost per cycle. For this reason it is not so easy to study aqueous solutions with microwaves, because the water will absorb the microwaves. For aqueous solutions a low-field frequency spectrometer is useful. To sample the energy, that is, to detect it, it is necessary to take out energy without adding reflections or disturbing the resonance conditions in the cavity. The output waveguides must be carefully matched to the inputs.

The detector is a transducer that converts the microwave energy into electric voltages that can be handled by standard electronics. Crystal diodes, bolometers, and masers may be used as detectors. Usually, cost, convenience, signal-to-noise ratio, and frequency characteristics govern the choice of the particular system.

It is possible to measure the absorption of power from the microwave field directly, but the change is so small, relative to the original microwave field intensity, that accuracy is not possible except for the most intense lines. Direct absorption is used only for demonstration purposes. Instead, the magnetic field is *modulated*, producing a periodic variation in the microwave power. This variation is then used to obtain the *change* in microwave power during absorption, that is, the *derivative* of the original absorption curve.

The amplifying system operates at the modulating frequency, which may be at audio frequency or slightly above. Phase-sensitive detectors are used to eliminate noise at different frequencies. Some noise does get through, but it is greatly decreased in amplitude.

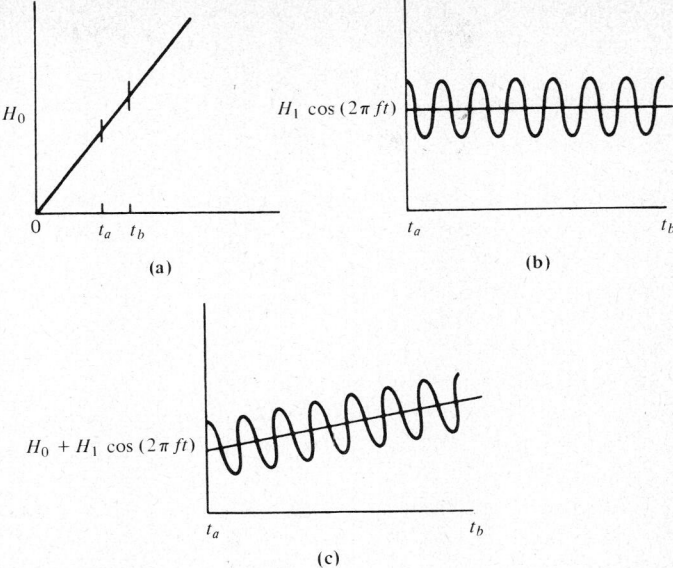

FIGURE 21–10 / (a) H_0 as a function of time. (b) The modulating field, $H_1 \cos(2\pi ft)$. (c) The total field as a function of time. Note that the field decreases during a large part of the cycle of the oscillating field. Actually, the scale is distorted: H_1 is shown too large with respect to H_0, and f is too small for the rate of increase shown for H_0 (also note the expansion of the time scale; the straight line represents H_0 from t_a to t_b).

To modulate the magnetic field, a pair of auxiliary coils produce a small magnetic field that oscillates at a preselected frequency. The two magnetic fields superimpose to form a resultant magnetic field whose intensity varies with time, as shown in Figure 21–10. The equation for the resultant field is

(21–18) $\qquad H = H_0 + H_1 \cos(2\pi ft)$

$\qquad H$ = resultant field
$\qquad H_0$ = slowly changing field of magnet
$\qquad H_1 \cos(2\pi ft)$ = modulating field; field of the coil
$\qquad f$ = frequency
$\qquad t$ = time

In the region where the resultant field H is large enough to cause microwave absorption, the oscillations in H produce an oscillation in the microwave intensity at the modulating frequency (i.e., the frequency of the coil). Figure 21–11a shows power absorption in an unmodulated field and Figure 21–11b shows the power absorption seen with a modulated field, such as that in Figure 21–10c.

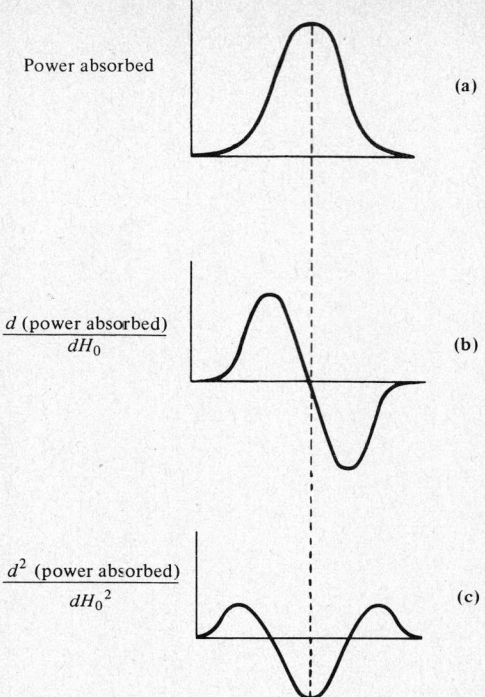

FIGURE 21-11 / Power absorption in unmodulated (a) and modulated (b) fields; (c) is a second derivative curve.

Note that in Figure 21-11b the curve is zero where microwave absorption does not take place, zero where the absorption is at a maximum, that is, at the peak in Figure 21-11a, and at a maximum at the points where the slope of the absorption curve is at its greatest. In other words, Figure 21-11b is proportional to the first derivative of the absorption curve shown in Figure 25-11a.

To measure the power absorption at the modulating frequency, the signal is picked up with a phase sensitive detector that responds only to the total power at the modulating frequency. The detector compares the signal with the original generated signal. The difference between the two signals, which corresponds to the change in the microwave field intensity, is sent to an oscilloscope or a recorder. Almost all noise at different frequencies will be out of phase as much as in phase, and so will cancel.

The power-absorption curve is not linear with magnetic field. Therefore higher harmonics are produced as well as the initial changes at the modulating frequency. With phase-sensitive detectors the amplifier can be tuned to give either the first or the second derivative. These are shown in Figure 21-11c

Chapter 21. Magnetochemistry

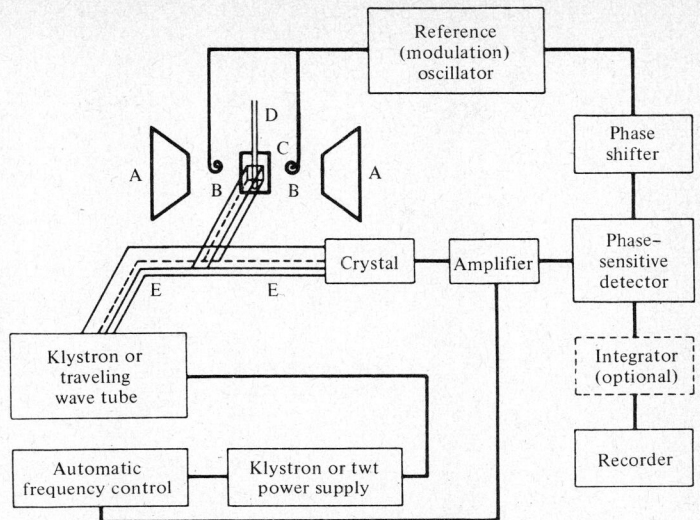

FIGURE 21-12 / Simplified block diagram of a "typical" esr spectrophotometer. The automatic frequency control keeps the microwave frequency correct. Most of the microwave system has been omitted, as have the oscilloscope, frequency meters, and other items needed to calibrate the instrument and to check that the phase balance is correct. A = magnet, B = modulating (Helmholtz) coil, C = cavity, D = sample, E = waveguide.

for a single peak. The second derivative gives an additional measure of the peak slope and is sometimes used to work out details of structure.

In the absorption curve and in the first derivative curve the linewidth is given in terms of H, the change in applied field. If H is reasonably large, greater than perhaps 10^{-5} T, 100 kHz modulation will cause negligible distortion. The amplitude of the modulation must also be appreciably less than H and the sweep time must be long compared to $1/f$ where f is the modulation frequency. These conditions generally can be met without too great difficulty.

The block diagram in Figure 21-12 shows the elements of a microwave spectrometer as discussed up to this point.

LOW FIELD SPECTROMETERS

If H is small, ν, the frequency of the electromagnetic radiation, can be small, too. It is therefore possible to have a spectrometer operating at low fields and frequencies. For example, if the magnetic field is 10^{-2} T, ν is 280 MHz when $g = 2$. This frequency is not even in the microwave range—it is in the radio frequency range. The field modulation and phase detection techniques are not seriously affected. However, the microwave connections, the cavity, and so on, can be dispensed with and the klystron can be replaced by an rf oscillator. The sensitivity, however, is seriously affected. Sensitivity is proportional to v^2. Dropping the frequency by a factor of 10 lowers the sensitivity by a factor of 100. To compensate for this, the sample size can be increased. Unfortunately, this too has a drawback, in that absorption of energy by the solvent becomes important.

Low-frequency spectrometers have the advantage of low cost and can be used with aqueous solutions, but the disadvantages are such that they are used only infrequently.

References

Alger, R. S., *Electron Paramagnetic Resonance: Techniques and Applications*. Wiley-Interscience, New York, 1968.

Carrington, A., and A. D. McLachlan, *Introduction to Magnetic Resonance*. Harper & Row, New York, 1967.

Kauzmann, W., *Quantum Mechanics*. Academic Press, New York, 1957.

McLauchlan, K., *Magnetic Resonance*. Clarendon Press, Oxford, 1972.

Moore, W. J., *Physical Chemistry*, 4th ed. Prentice-Hall, Englewood Cliffs, N.J., 1972.

Selwood, P. W., *Magnetochemistry*. Wiley–Interscience, New York, 1943.

Weissberger, A., and B. W. Rossiter, *Physical Methods of Chemistry*, Part III,A, chs. 6, 7, and Part IV, ch. 7. Wiley–Interscience, New York, 1972.

Chapter 22

Mass Spectrometry

Mass spectrometry is an extremely powerful and versatile method for the identification and structural analysis of materials. It is unique in three respects: First, a complete analysis may be run on a sample smaller than a microgram, and constituents may be detected in the 10^{-8} g range. Since the analysis is conducted on gas molecules at pressures below 10^{-5} torr, the method is applicable to all gases and liquids and all solids that can be melted or that on heating develop vapor pressures of 10^{-8} torr (or even less). Second, the results furnish a complete analysis of each constituent present, including all *isotopic species*. Almost all other analytical methods are based on the *average* properties of a mixture of isotopic species. For example, Cl_2 consists of a mixture of $^{35}Cl^{35}Cl$, $^{35}Cl^{37}Cl$, and $^{37}Cl^{37}Cl$. In mass spectrometry each of these shows up as a separate constituent, but in all other analytical methods they all show up only as Cl_2, with an average molecular weight of 71. Third, the method of analysis breaks down the compounds into fragments that show the detailed structure of the compound, including the location of the various functional groups. In addition, from the energies required to produce these fragments, bond energies can be calculated and much information can be obtained concerning electronic energy levels within the molecule.

Instrumentation

In mass spectrometry a sample of substance is converted into gaseous ions. The ions are then sorted out according to their ratios of mass to charge and the relative abundance of each ion species is determined. Instruments employing photographic detection are called *mass spectrographs* and are employed in analyzing for elements. Instruments that use electronic detection are called *mass spectrometers* and are applicable to studies of molecules that fragment and where accurate determination of small differences is required.

The essential elements of a mass spectrometer are the sample inlet, the ion source, the ion sorter, the detector, and a vacuum system to keep the pressure in the mass spectrometer below 10^{-5} torr, even in the presence of the sample. It is necessary to keep the pressure low so that the mean free path of the ions is large compared to the internal dimensions of the spectrometer. Otherwise, the ions would collide with each other and with residual gas molecules, reducing the resolving power of the instrument and even causing secondary reactions that confuse the results.

SAMPLE INTRODUCTION

Samples are usually present at atmospheric pressure, while the analysis takes place at high vacuum. If the sample is a gaseous or volatile substance, it is introduced into the system through a *leak*—a sintered glass disk a ceramic, a pinhole in a metal foil, or a precision needle valve. If the sample is not volatile, it may be heated in a furnace and the vapors admitted through a leak or it may be placed inside the mass spectrometer and heated there electrically.

For the rate of passage of gas through the leak to be small enough to prevent the system pressure from rising above 10^{-6} torr the pressure on the outside of the leak must be kept fairly small, typically about 1 torr. If gas at a pressure approaching atmospheric is to be monitored continuously, most of it must be pumped away and discarded, with only a small portion entering through the leak. A less wasteful procedure used for analysis of static samples is to expand a small portion of gas to a large volume, thus decreasing the pressure to the 1 torr range. At these low pressures the gas passes through the leak by molecular flow, and the inlet is called a *molecular leak*. In another type of sample inlet system, the gas chamber is connected to the inlet by a long fine capillary. If the capillary is long enough and fine enough, there is a considerable pressure drop along its length, and the pressure at the pinhole is sufficiently low. The gas flows through the capillary mainly by viscous flow, and this type of inlet is called a *viscous leak*.

In the analysis of a mixture of gases under molecular flow conditions the lighter components diffuse into the system more rapidly. The rate of diffusion is

(22-1) $$N = kp\sqrt{\frac{T}{M}}$$

N = number of molecules entering in unit time
k = constant dependent on the system, not the compound
p = pressure on outside of leak
T = Kelvin temperature
M = molecular weight

Although the gas mixture is fractionated by diffusion into the mass spectrometer, since the lighter components also diffuse more rapidly out of the mass spectrometer into the vacuum pump, following a similar law, the composition in the mass spectrometer chamber is the same as that of the original sample unless, of course, so much of the gas has been removed that the sample composition has itself changed. Usually, the system is designed so that the reservoir loses no more than 1% during the analysis.

With a viscous leak a steady state is rapidly established in which the gas enters the mass spectrometer chamber with the same composition as that of the original sample. In this case, since the lighter gas leaves the chamber at a greater relative rate than it entered, the mixture is fractionated within the sample chamber. To compensate for this fractionation, a simple correction factor is used:

$$(22\text{-}2) \qquad \left(\frac{A}{B}\right)_{\text{init}} = \left(\frac{A}{B}\right)_{\text{obs}} \sqrt{\frac{M_B}{M_A}}$$

IONIZATION

In the most widely used ionization method the molecules of gas that enter the ionization chamber are bombarded with high energy electrons. These electrons knock electrons out of the sample molecules and in addition may break chemical bonds, producing a variety of positive ions. For example, C_2H_6 breaks into $C_2H_5^+$, CH_3^+, and H^+; N_2 becomes N_2^+, N^+, and N^{2+}. The ion formed and the extent of the ionization will depend on the energy and intensity of the electron beam, which may be varied. For analysis of gas mixtures the beam energy is usually set at about 70–100 eV, sufficient to ionize whatever gas particles the mixture is expected to contain. In the literature 70 eV is used for standard tables of mass spectra. In studies of current efficiencies the energy of the beam is varied over the range from 0 to 20–30 eV, depending on the particular study. The intensity of the electron current may also be varied, depending on the particular gas, the objective in mind, and the instrument being used. For detector currents in the 10^{-13} A range, the ionizing currents are in the range 10^{-5}–10^{-6} A. Obviously, only a very small fraction of the sample molecules are ionized and reach the detector. The detector efficiency is high, but the ionization efficiency low.

Chemical ionization (CI) has recently been developed to supplement electron impact ionization (EI). The reagent gas, usually methane or isobutane, is ionized by EI. These ions then react with the sample molecules, ionizing them. There is less fractionation than in EI and more chance of observing the parent molecular ion.

ION SORTING

There are many ways of sorting ions. In one widely used method, the ionized molecules are formed into a beam by passage through a slit, accelerated to a high velocity by an electric field, and passed into a magnetic field. The magnetic field accelerates each ion into a circular path whose radius at any combination of electric and magnetic field strengths is directly proportional to the mass/charge ratio of the ion. The beam is broken up into a number of smaller beams, each of which is characterized by a definite mass/charge ratio. In effect, the combination of magnetic and electric fields sorts the beam into its constituents.

The acceleration by the electric field transforms the potential energy of each ion in the field into kinetic energy according to

$$(22-3) \qquad eV = \tfrac{1}{2}mu^2$$

e = charge on the ion
V = accelerating potential, V
m = mass of the particle
u = velocity

In the magnetic field the acceleration is the centripetal force per unit mass, as given by

$$(22-4) \qquad \frac{u^2}{R} = \frac{Heu}{m}$$

R = radius of curvature of ion path
H = magnetic field strength

The combination of equations (22–3) and (22–4) gives

$$(22-5) \qquad \frac{m}{e} = \frac{R^2 H^2}{2V}$$

There are a number of different designs for combinations of electric and magnetic fields. Figure 22–1 shows a simple Dempster-type apparatus, the Picker-REI MS 10. In this instrument the ions, once formed, are propelled by a small positive voltage (up to 10 V), through a slit, into the intense accelerating electric field of 20–2000 V. The magnetic field is provided by a permanent magnet.

FOCUSING AND DETECTING THE ION BEAMS

In mass spectrometers the detector is placed at a fixed point. Each particle of definite mass/charge ratio will reach the detector only if the combination

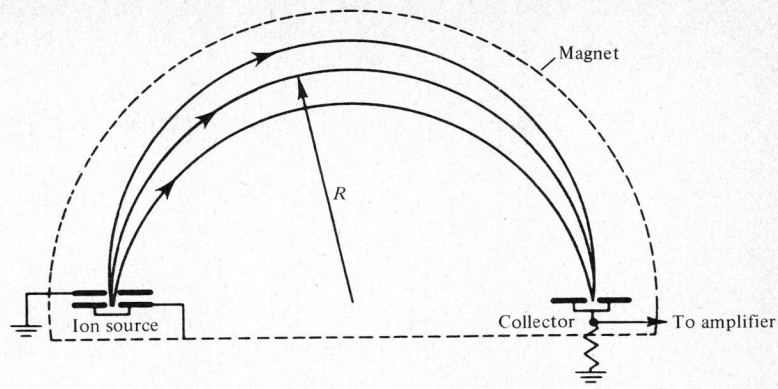

FIGURE 22-1 / Ion paths in a 180° deflection mass spectrometer. Used by permission of the Picker Nuclear Corp.

of H and V satisfies equation (22-5). In instruments with permanent magnets, such as that shown in Figure 22-1, the relation becomes

(22-6) $$\frac{m}{e} = \frac{k}{V}$$

k = apparatus constant

Scanning through a range of voltages therefore focuses a series of particles of different m/e ratios at the detector, giving a mass spectrum.

In actual practice, each species of ion is focused at the detector over a range of voltages rather than at one definite voltage. The detector signals corresponding to particular species therefore overlap and in the limit cause difficulty in reading and analyzing the spectrum. This overlapping results from (1) the finite width of the collimating slits, (2) nonuniformity of the electric and magnetic fields, (3) space charge repulsions (i.e., the ions in the beam repel each other), and (4) variations in the kinetic energy of ions with the same mass/charge ratio. The ability of instruments to separate ions of different mass/charge ratios is called the *resolution*. For any instrument it varies with the mass of the ion, being smaller at higher mass values. It is expressed as $m/\Delta m$, where m is the mass of a particle and Δm is the difference in mass between this and the nearest resolvable particle. For the MS 10 in basic form, the resolving power is 100, but for high resolution research instruments, resolving power goes to 70,000.

Since the ion currents are of the order of 10^{-10}–10^{-13} A, the detector circuit must be very sensitive, with a very low noise level. The output of the detector circuit is amplified and fed into a meter and also into a recorder, to provide a permanent record.

Besides the Dempster direction focusing type of mass spectrometer discussed above, there are several other types commercially available. These include instruments with electromagnets, which permit scanning over greater ranges than instruments with permanent magnets; double focusing instruments, in which an electric field focuses ions with respect to velocity and a magnetic field focusses with respect to direction; cycloidal focusing spectrometers, in which the electric and magnetic fields are crossed; quadrupole mass spectrometers that do not use magnets, sorting the ions by radio frequencies; and time-of-flight spectrometers, in which a sample of gas is ionized by a pulse of energy and then accelerated toward a detector; the heavier ions move more slowly and arrive at the detector later. This instrument records the separate pulses arriving at the detector at various time intervals after the original burst of ionizing energy. Each pulse of electricity lasts only microseconds, so that repeated pulses are necessary, up to perhaps 50 kHz. Time-of-flight instruments can analyze samples within milliseconds. However, the resolution is limited by the physical dimensions of the apparatus. The longer the path, the sharper the separation between particles of different m/e ratios, but, obviously, path length cannot exceed a few hundred centimeters in most cases. The rapidity of the time-of-flight analysis makes this the instrument of choice for studies of shock waves, flash photolysis, and other rapid reactions.

Each make and model of mass spectrograph has its own peculiarities, and for best results the instructions supplied by the manufacturer should be followed carefully. Nevertheless, there are a few procedures common to most apparatus. The analyzer chamber should be periodically baked out—that is, heated to 100–400°C under vacuum for 12–24 hr to drive out all adsorbed gases, water vapor, pump oil vapors, and so on. The vacuum system should be kept operating constantly. The cold trap should be kept filled at all times and not allowed to warm up; otherwise, the trapped vapors and pump oil vapor will back up into the system. If the trap does warm up, the system should be baked out. The electronic system should be kept running constantly, for greater stability. Finally, test runs should be made frequently with known gases or volatile liquids to check the calibration of the various settings.

Applications of Mass Spectrometry

QUALITATIVE GAS ANALYSIS

Each gas has its own particular fragmentation pattern. For example, propane shows the fragmentation pattern of Figure 22-2. The peak at 44 is that of the molecular ion $C_3H_8^+$, containing three ^{12}C atoms and eight ^1H atoms. The small peak at 45 is that of the molecular ion containing one ^{13}C isotope. The peak at 43 is for the fragment $C_3H_7^+$. The peak at 29 is for the $C_2H_5^+$ fragment, the ethyl ion, and that at 15 is for the CH_3^+ methyl ion. The 20.5 peak is for the doubly ionized mass 41 particle, $C_3H_5^{2+}$, and the peak at 21 is

	m/e	Relative Intensity	m/e	Relative Intensity
	2	6.87	27	37.9
	12	0.35	28	59.1
	13	0.48	29	100.0 (base peak)
	14	2.48	30	2.09
	15	3.9	36	0.43
	16	0.20	37	3.06
C_3H_8	19	0.82	38	4.91
	19.5	0.48	39	16.2
	20	0.94	40	2.8
	20.5	0.25	41	12.4
	21	0.015	42	5.15
	24	0.11	43	22.3
	25	0.71	44	26.2
	26	7.58	45	0.81

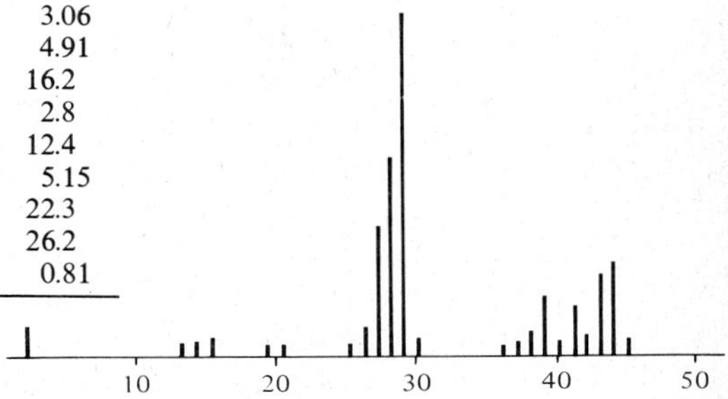

FIGURE 22–2 / Fragmentation pattern of propane ionized at 70 eV.

for the doubly ionized $C_3H_6^{2+}$. Peaks found at fractional mass values are usually doubly ionized fragments with odd mass numbers. The intensity of each peak is compared to that of the largest peak, called the base peak.

It is not always possible to find the peak corresponding to the parent ion, since in some gases it may be extremely short-lived, or the probability of its formation may be small. Furthermore, in a gas mixture the peak of the molecular ion of the lighter constituent may perhaps coincide with that of a fragment of the heavier constituent. For identification of an unknown gas or for qualitative analysis of a mixture of gases, it is necessary to know the cracking pattern for each possible constituent, so that each gas may be detected by a fragment characteristic of it alone. If it is not possible to select a characteristic peak for each possible constituent from the known cracking patterns, the heights of the various peaks can be substituted in a set of simultaneous linear equations that can be solved to give the composition of the mixture. This has been done by petroleum companies for years to analyze hydrocarbon mixtures. The method has now been supplemented by gas chromatography, which separates the components of mixtures before analysis by mass spectrometry.

In the qualitative analysis of gas mixtures or the identification of unknown gases mass spectrometry is especially valuable for the identification of positional isomers, such as 1,2- and 1,3-disubstituted propanes. The spectrum of the 1,2 compound will show a methyl fragment, while that of the 1,3 compound will show the $CH_2XCH_2^+$ fragment but no methyl fragment.

Leak detection by mass spectrometry should also be included under the heading of qualitative gas analysis. A small portable spectrometer with a fixed voltage set at mass value of 4 is used. The apparatus being searched for leaks is filled with helium and all possible escape paths are traced with the mass spectrometer.

QUANTITATIVE ANALYSIS

The detector current is proportional to the pressure of the gas producing the fragment, the proportionality constant depending on the energy and intensity of the ionizing beam as well as on the characteristics of the individual spectrometer. For absolute determinations the spectrometer must be calibrated for each gas. For relative abundances, a standard gas of known pressure may be used and the ratio of the peak height of a characteristic fragment of the unknown gas to that of a fragment of the standard gas may be easily obtained. This ratio must be multiplied by a correction factor, since different gases have different sensitivities toward ionization, the sensitivity being a function of the ionization potential. For example,

Gas:	H_2	O_2	He	Ar	CH_4	CO_2	Ne	Kr	N_2	CO
Sensitivity:	0.68	0.72	0.25	1.6	1.1	1.1	0.29	1.0	1.0	1.1

In analyzing mixtures of gases one must be careful to avoid fractionation of the mixture. If a viscous leak is used in the inlet system, equation (22-2) must be used to correct the observed ratios.

As mentioned previously, in the analysis of a gas mixture each isotopic species shows up at its characteristic m/e ratio, and from the magnitude of the detector currents at these ratios the relative abundances of the isotopes can be calculated.

As an example, the three different isotopic chlorine molecules have masses of 70, 72, and 74, corresponding, respectively, to $^{35}Cl^{35}Cl$, $^{35}Cl^{37}Cl$, and $^{37}Cl^{37}Cl$. These will all show up in the mass spectrum. (There will also be peaks at 35 and 37, corresponding to atomic chlorine and to doubly ionized particles of mass 70 and 74, and peaks at 17.5 and 18.5, due to doubly ionized atoms of chlorine.) The three different peaks at mass 70, 72, and 74 will have relative intensities that depend on the probabilities of forming the particular isotopic molecules.

Since the fraction of ^{35}Cl atoms in natural chlorine is 0.754, the fraction of $^{35}Cl^{35}Cl$ molecules is $0.754 \times 0.754 = 0.569$. In similar fashion the fraction of $^{37}Cl^{37}Cl$ molecules is $0.246 \times 0.246 = 0.0605$, and the fraction of $^{35}Cl^{37}Cl$ is $2 \times 0.756 \times 0.256 = 0.371$. The factor 2 is necessary because this species can be formed in two equivalent ways, $^{35}Cl^{37}Cl$ and $^{37}Cl^{35}Cl$. (Note that the sum of the fractions adds up to 1.)

The relative abundances of the isotopic particles may be obtained from the binomial theorem, using the total number of isotopic atoms in the fragment, the number of atoms of each isotopic species in the particle, and the relative abundance of each particle:

$$(22\text{-}7) \quad (a + b)^n = a^n + \frac{na^{n-1}b}{1!} + \frac{n(n-1)a^{n-2}b^2}{2!} + \frac{n(n-1)(n-2)a^{n-3}b^3}{3!} + \cdots + b^n$$

a, b = relative abundances of the two isotopic atoms
n = total number of isotopic atoms in the fragment being studied
$(a + b)^n$ = sum of relative abundances of the isotopic fragments = 1
a^n = relative abundance of fragment with n atoms of isotopic species a
$n(n-1)a^{n-2}b^2/2!$ = relative abundance of fragment with 2 atoms of b and $n - 2$ atoms of a

In the case of the CCl_4^+ molecule ion, the species $C^{35}Cl_2{}^{37}Cl_2{}^+$ has four atoms, the relative abundance of ^{35}Cl is 0.754, and that of ^{37}Cl is 0.246. The relative abundance of the species is

$$\frac{4 \times 3 \times (0.754)^2 \times (0.246)^2}{2 \times 1}$$

The formula may, of course, be generalized for fragments with more than two different isotopic species.

KINETICS

Mass spectrometry is especially valuable in determining the kinetics of gas phase reactions that do not result in any pressure change and/or are so rapid that they must be run at very low pressures.

The mass spectrum of each reactant and product is determined and a characteristic product peak is selected for the analysis. The detector current at this peak is calibrated against known pressures of the product gas. The spectrometer is then set at the m/e ratio for the characteristic product peak and the reaction mixture is connected to the instrument, with the detector output being fed into a recorder.

With isotopic tracers mass spectrometry can be used to determine reaction mechanisms. Such information as the position of the bond being cleaved, the migration of an atom, or the occurrence of an internal rearrangement can often be determined. Short-lived intermediate species can also be detected.

Mass spectrometry is used to detect compounds labeled with stable isotopes in studies of reaction mechanisms, kinetics, and metabolic pathways.

IONIZATION POTENTIALS AND DISSOCIATION ENERGIES

The formation of a *molecular ion* by electron impact may be written

$$M + e \rightarrow M^+ + 2e$$

and the energy balance summed up as

(22–8) $\qquad KE_r + V = V_i + KE_{pr} + EE$

$\qquad KE_r$ = kinetic energy of the reactant
$\qquad KE_{pr}$ = kinetic energy of the product
$\qquad V$ = electron impact energy
$\qquad V_i$ = ionization potential
$\qquad EE$ = excitation energy

The kinetic energy term of the reactant is usually of the order of room temperature thermal energy (i.e., about 1 kcal/mole), which is small compared to the other energies. The V term is also a kinetic energy term, that of the electron. The KE_{pr} term arises from the fact that the three particles on the right-hand side of the equation (the ion and the two electrons) may have considerable kinetic energy. The EE term refers to the excitation energy required if the molecule ion is formed in an excited state. At the potential at which the fragment first appears—the *appearance potential*—the kinetic energy and the excitation energy terms may be assumed to be zero. The appearance potential is therefore the ionization potential of the molecule; that is, the energy needed to remove an electron from the highest occupied molecular orbital:

(22–9) $\qquad V_a = V_i$

$\qquad V_a$ = appearance potential

To determine appearance potentials, ionization efficiency curves are plotted. The ionization efficiency curves for the sample and for a reference gas are determined separately by electron impact over a wide potential range, starting preferably at energies too low to cause ionization. Graphs are then plotted, such as that shown in Figure 22–3. The purpose of the reference gas, such as Ar, Ne, Kr, or CO_2, is to calibrate the electron energy scale of the particular mass spectrometer being used. Owing to contact potentials, and side effects, the actual electron energy may be considerably different from that read on the instrument dial. The appearance potential may be obtained (1) visually, as the potential at which the current vanishes or reaches an arbitrarily low value; (2) by extrapolating the straight line portion of the curve; or (3) by plotting the differences in potential between the two curves. At several values of current these differences in electron energy are obtained from the graph. Another graph is then drawn whose axes are the detector current and the electron energy difference at each current. This graph is then extrapolated to zero

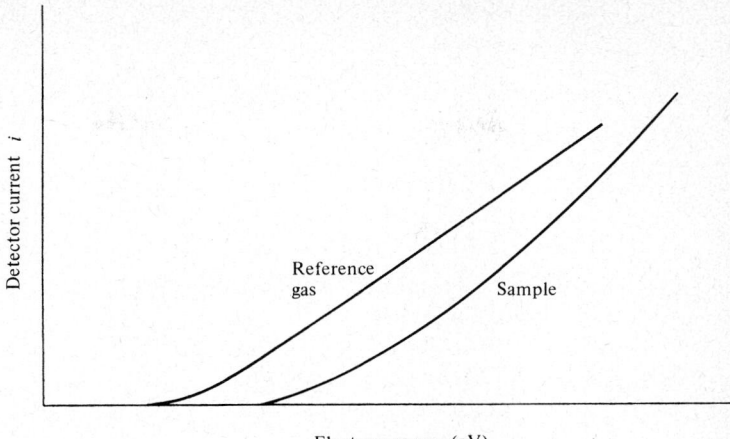

FIGURE 22-3 / Ionization efficiency curve for determining appearance potential.

current and the extrapolated difference value is added to or subtracted from the accepted appearance potential of the reference gas. Actually, none of these three methods is completely justified by theory. However, since they all involve extrapolations in which the gas being studied is subjected to the same procedures as a reference gas, they all give fairly consistent results.

Appearance potentials of fragments resulting from the dissociation of the parent molecular ion can furnish accurate values of dissociation energies. For example, for the process

$$XY + e \rightarrow X^+ + Y + 2e$$

the energy balance, neglecting the initial kinetic energy, is

(22-10a) $\quad V = V_{i,X} + D_{XY} + KE + EE$

V = observed potential
$V_{i,X}$ = ionization potential of X
D_{XY} = dissociation energy of XY

The appearance potential of the dissociation process is therefore

(22-10b) $\quad V_a = V_{i,X} + D_{XY}$

V_a = appearance potential

The appearance potential is therefore the sum of the ionization energy of the radical X and the dissociation energy of the molecule XY (alternatively, it might be considered as the sum of the ionization potential of the molecule and the dissociation energy of the ion). If the appearance potential of the X^+

fragment can be determined directly through studies of the free radical X, it may be subtracted from V_a to give D_{XY}. Extensive tables of free radical ionization energies are found in the literature under the heading of $\Delta H_f(X^+)$, the heat of formation of the ion [Field and Franklin].

In cases where there are no data on the ionization potentials of the free radicals, dissociation energies can be calculated from the appearance potentials of two or more processes. For example, the energy of the dissociation reaction

$$CH_4 \rightarrow CH_3 + H$$

can be calculated from the appearance potentials of the reactions

$$C_2H_6 \rightarrow C_2H_5^+ + H + e$$

and

$$C_3H_8 \rightarrow C_2H_5^+ + CH_3 + e$$

For the first reaction the energy balance is

$$V_{a1} = \Delta H_f^\circ(H) + \Delta H_f^\circ(C_2H_5^+) - \Delta H_f^\circ(C_2H_6)$$

ΔH_f° = standard heat of formation

For the second reaction,

$$V_{a2} = \Delta H_f^\circ(CH_3) + \Delta H_f^\circ(C_2H_5^+) - \Delta H_f^\circ(C_3H_8)$$

On subtracting,

$$V_{a2} - V_{a1} = \Delta H_f^\circ(CH_3) - \Delta H_f^\circ(H) - \Delta H_f^\circ(C_3H_8) + \Delta H_f^\circ(C_2H_6)$$

By definition, the dissociation energy $D(CH_4)$ is

$$D(CH_4) = \Delta H_f^\circ(CH_3) + \Delta H_f^\circ(H) - \Delta H_f^\circ(CH_4)$$

On combining this with the previous expression,

$$D(CH_4) = V_{a2} - V_{a1} - \Delta H_f^\circ(CH_4) + \Delta H_f^\circ(C_3H_8) - \Delta H_f^\circ(C_2H_6) + 2\Delta H_f^\circ(H)$$

Since accepted values of all the heats of formation are found in the literature, the determination requires only that the appearance potentials be measured. The accuracy also depends on the accuracy of the appearance potential measurement. To generalize the expression above, make the following definitions:

$$C_2H_5 = R_1 \quad H = R_2 \quad CH_3 = R_3$$

Therefore, $CH_4 = R_2R_3$, and so on. The final expression then becomes

(22-11) $$D(R_2R_3) = V_{a2} - V_{a1} + \Delta H_f^\circ(R_1R_3) - \Delta H_f^\circ(R_1R_2) - \Delta H_f^\circ(R_2R_3) + 2\Delta H_f^\circ(R_2)$$

Using this technique, extensive tables of heats of formation of various hydrocarbons have been obtained, and are available in the literature.

References

Beynon, J. H., *Mass Spectrometry and Its Application to Organic Chemistry*. Elsevier, Amsterdam, 1960.

Field, F. H., and J. L. Franklin, *Electron Impact Phenomena and Properties of Gaseous Ions*. Academic Press, New York, 1957.

Kiser, R. W., *Introduction to Mass Spectrometry and Its Applications*. Prentice-Hall, Englewood Cliffs, N.J., 1965.

Lambert, J. B., et al., *Organic Structural Analysis*, Part Four. Macmillan New York, 1976.

McDowell, C. (ed.), *Mass Spectrometry*. McGraw-Hill, New York, 1964.

McLafferty, F. W. (ed.), *Mass Spectrometry of Organic Ions*. Academic Press, New York, 1963.

Reed, R. I., *Ion Production by Electron Impact*. Academic Press, New York, 1962.

Weissberger, A., and B. W. Rossiter, *Physical Methods of Chemistry*, Part VI. ch. 3. Wiley–Interscience, New York, 1977.

Chapter 23

Thermodynamics of Irreversible Processes

All natural processes are irreversible on a macroscopic level. To be reversible, they would have to occur *infinitely* slowly. Classical thermodynamics, which was developed for equilibrium processes, says little about such irreversible processes except that $dS_{\text{total}} > 0$ and that for a process occurring at a finite rate $dS/dt > 0$. It cannot predict rates or relations between rates, and it cannot predict conditions of stability for nonequilibrium processes. A general *irreversible* or *nonequilibrium* thermodynamics has been developed to extend thermodynamic reasoning to irreversible processes. It is limited to systems sufficiently close to equilibrium for linear equations to be written relating dynamic processes to generalized thermodynamic forces. Fortunately, many important phenomena follow such linear relationships over a wide enough range of conditions for irreversible thermodynamics to be useful.

Two types of processes are treated by irreversible thermodynamics: *steady state* and *relaxation* processes. In steady state processes the properties of the system (e.g., temperature, pressure, concentration, chemical potentials) at each point do not change with time, even though the system is not at equilibrium. Familiar examples include diffusion of solute between two solutions maintained at different concentrations; the flow of heat along a metal rod with the two ends maintained at different temperatures; the flow of electricity along a conductor when a constant potential difference is maintained across it. Because steady state processes must be dynamic, systems such as diamond or white phosphorus, which are frozen in a metastable state, are not examples of steady state systems. Steady state processes take place in open systems. The concentration difference, temperature difference, or potential difference is maintained by a reaction or process in which equilibrium is approached elsewhere in the universe. Like systems at equilibrium, steady state systems are stable, provided the boundary conditions responsible for the steady state are maintained. They follow Le Châtelier's principle. If perturbed, they react in a manner to oppose the perturbation, and when the perturbation is removed, they return to the initial steady state.

Relaxation processes are those in which a system that is not initially at equilibrium gradually approaches equilibrium. Examples are equalization of levels in a U tube containing a viscous liquid that initially is at a different level in each arm, diffusion between two solutions of different concentrations when the concentration difference is not maintained, and the dissipation of an electrical charge on the plates of a capacitor connected through a resistor. (See also the discussion of fast reactions in Chapter 24.)

Dynamic nonequilibrium processes involve a "flow" that can be expressed by a "flux" equation, that is, one that describes the rate of flow of matter or energy across a boundary. Consider the one-dimensional cases of the three simple steady state processes listed above. (Although any consistent system of units may be used, for relevance to chemistry SI units are used here.) In mass transport, the flux equation is *Fick's first law of diffusion*,

$$(23\text{-}1\text{a}) \qquad J_m = -D\frac{dc}{dx}$$

J_m = mass flux, moles/m^2
D = diffusion coefficient, m^2/sec
c = concentration, moles/m^3
x = distance, m

In heat transport, the flux equation is *Fourier's law*,

$$(23\text{-}1\text{b}) \qquad J_\theta = -k\frac{dT}{dx}$$

J_θ = thermal flux, J/m^2
k = thermal conductivity, J/m^2-K
T = temperature, K

In charge transport, it is *Ohm's law*,

$$(23\text{-}1\text{c}) \qquad J_{el} = -\sigma\frac{d\phi}{dx}$$

J_{el} = electric flux A/m^2
σ = electrical conductance, Ω^{-1}m^{-1}
ϕ = electrical potential, V

In each of these cases the flux equation has the form

$$(23\text{-}2) \qquad J_i = L_i X_i$$

J_i = flux of i
X_i = corresponding generalized force
L_i = linear coefficient

(If ψ_i is the generalized potential, for the one-dimensional case $X_i = -d\psi_i/dx$, and for the more general case $X_i = -\nabla\psi_i$. In this case X_i is a vector.)

Several points should be noted. The "forces" are generalized forces, not mechanical forces, and in general are *not* measured in newtons. Unlike mechanical forces, but like pressure, they are intensive properties, not extensive properties. The coefficient L_i is the reciprocal of some "resistance" (i.e., $L_i = 1/r_i$). In the three examples the corresponding resistances are "diffusive resistivity" (there is no accepted name for this), thermal resistivity, and electrical resistivity. The flux may be flow of matter as in equation (23–1a), of energy (in the form of heat) as in (23–1b), of electrical charge as in (23–1c), of free energy, and so on. Finally, equation (23–2) says that *the flux is directly proportional to the "force."* This can be taken as the definition of an irreversible process "close to" equilibrium. Whether or not the flux is proportional to the force so that the equations of irreversible thermodynamics may be applied must in every case be determined by experiment. In some systems a linear relation exists over a wide range. In others, such as the flow of current through a solid state rectifier or other nonohmic resistor, there is no range that may be considered "close to" equilibrium.

Equation (23–2) can be extended to chemical systems. Here, however, the concentrations of reactants and products in a homogeneous system do not depend on position. Hence the chemical force cannot be defined as the spatial gradient of some potential. Instead, consider the generalized chemical reaction

(23–3) $\qquad r_1 R_1 + r_2 R_2 + \cdots + r_i R_i \rightleftharpoons p_1 P_1 + p_2 P_2 + \cdots + p_j P_j$

$R_i = i$th reactant
$P_j = j$th product

Assume that the process in equation (23–3) has a rate equation of the form

(23–4) $\qquad \text{rate} = \dfrac{d(\text{products})}{dt} = \prod [R_i]^{\rho_i} - \prod [P_j]^{\pi_j}$

$\prod$ = product over all terms
$[R_i]$ = concentration of R_i
ρ_i = order with respect to R_i
$[P_j]$ = concentration of P_j
π_j = order with respect to P_j

In general, ρ_i need not equal r_i and π_j need not equal p_j. Let $\overline{R}_i$ and $\overline{P}_j$ be equilibrium concentrations. Then, at equilibrium,

(23–4a) $\qquad \prod \overline{R}_i^{\rho_i} - \prod \overline{P}_j^{\pi_j} = 0$

Assume that the system is perturbed so that it is no longer at equilibrium but that it is still "close to" equilibrium. Let the new concentrations be

$$[R_i] = \overline{R}_i + \delta R_i \quad \text{and} \quad [P_j] = \overline{P}_j + \delta P_j$$

(Note that only one concentration need be perturbed.) The rate is now

(23–5) $\quad \text{rate} = \prod (\overline{R}_i + \delta R_i)^{\rho_i} - \prod (\overline{P}_j + \delta P_j)^{\pi_j}$

By the binomial theorem,

$$(\overline{R}_i + \delta R_i)^{\rho_i} = \overline{R}_i^{\rho_i} + \rho_i \overline{R}_i^{\rho_i - 1} \delta R_i + \text{higher order terms in } \delta R_i$$

If each such term in equation (23–5) is expanded and like terms collected, the result will be

(23–6) $\quad \text{rate} = \prod \overline{R}_i^{\rho_i} - \prod \overline{P}_j^{\pi_j} + \sum \alpha_i \, \delta R_i - \sum \beta_j \, \delta P_j$
$\quad\quad\quad\quad + \text{higher order terms in } \delta R_i \text{ and } \delta P_j$

α_i, β_i = complex functions of equilibrium concentrations

If equation (23–4a) is substituted into this, it becomes

(23–7) $\quad \text{rate} = \sum \alpha_i \, \delta R_i - \sum \beta_i \, \delta P_j$
$\quad\quad\quad\quad + \text{higher order terms in } \delta R_i \text{ and } \delta P_j$

Since only small displacements from equilibrium are assumed, as required for linearity, the higher order terms in δR_i and δP_j may be neglected and the equation becomes

(23–8) $\quad \text{rate} = \sum \alpha_i \, \delta R_i - \sum \beta_i \, \delta P_j$

Here the rate may be considered a chemical flux, the displacements from equilibrium, δR_i and δP_j, may be considered to be the forces, and the α_i's and β_j's the coefficients relating the flux to the displacement from equilibrium. This relation is analogous to equation (23–2) and becomes identical to (23–2) in the special case that the concentration of only one reagent or product has been displaced from its equilibrium value.

Equation (23–8) shows that regardless of the order of the reaction with respect to a reactant or a product, the rate of approach to equilibrium for a small displacement from equilibrium is proportional to the first power of the displacement. Since for any specific case, the coefficients α_i and β_j can be found from the rate equation, equation (23–8) is also the basis of relaxation methods in chemical kinetics.

Equation (23–8) can be reformulated in terms of the thermodynamic function, chemical potential, μ, instead of the concentration. From thermodynamics

(23–9) $\quad \mu_i = \mu_i^\circ + RT \ln \gamma_i c_i$

$\quad\quad\quad \mu_i^\circ$ = chemical potential in the standard state
$\quad\quad\quad \gamma_i$ = activity coefficient
$\quad\quad\quad c_i$ = concentration

PRINCIPLES

For a change that is so small that γ_i may be assumed to remain constant,

(23-10) $\qquad \delta\mu_i = \dfrac{RT\,\delta c_i}{c_i} \quad \text{or} \quad \delta c_i = \dfrac{c_i\,\delta\mu_i}{RT}$

We could substitute this directly into equation (23-8) and derive an equation analogous to equations (23-2). However, we can make the analogy clearer by defining two quantities, the *extent of reaction*, ξ, and the *affinity*, A.

(23-11) $\qquad d\xi = v_\gamma\, dn_\gamma$

$\qquad\qquad v_\gamma$ = the stoichiometric coefficient of component γ in the balanced equation
$\qquad\qquad\;\;\;$ = $-r_i$ or $+p_j$, depending on whether γ is a reactant or a product.
$\qquad\qquad n_\gamma$ = number of moles of component γ

The rate of a chemical reaction is therefore just $d\xi/dt$. A is essentially the net difference of the chemical potentials from their equilibrium values.

(23-12) $\qquad A = -\sum v_\gamma \mu_\gamma$

The rate of a chemical reaction in terms of affinities becomes

(23-13) $\qquad \dfrac{d\xi}{dt} = L_{\text{chem}}\dfrac{A}{T}$

$\qquad\qquad L_{\text{chem}}$ = chemical analog of linear coefficient in equation (23-2)
$\qquad\qquad A/T$ = driving force of the reaction

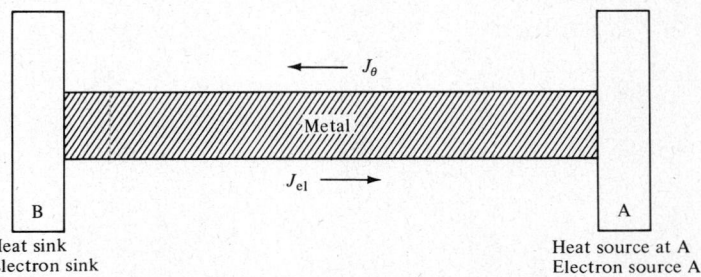

FIGURE 23-1 / Heat and electron conduction through a metal bar. (J_{el}, the flow of current, is opposite in direction to the flow of electrons.) Note that J_θ and J_{el} could be in the same or opposite directions, depending on the electrical circuit.

Chapter 23. Thermodynamics of Irreversible Processes

The three steady state irreversible processes listed above—the flow of an electric current, heat flow, and mass transport by diffusion—like all irreversible processes, produce entropy. In irreversible thermodynamics, the consequences of the production and flow of entropy are considered and developed. Consider the flow of heat along a metal rod with the two ends maintained at absolute temperatures T_A and T_B with $T_A > T_B$ (Figure 23–1). Assume that a quantity of heat δQ_A enters at A. The flow of entropy into the rod at A is $dS_A = \delta Q_A/T_A$. Similarly, the flow of entropy out of the rod at B is $dS_B = \delta Q_B/T_B$. Since there is no change in temperature, $\delta Q_A = \delta Q_B = \delta Q$. Therefore, the net increase in entropy due to heat exchange with the outside is

$$(23\text{–}14) \qquad d_e S = dS_B - dS_A = \delta Q \left(\frac{1}{T_B} - \frac{1}{T_A} \right)$$

Entropy is also produced within the bar by the heat flow. The internal entropy production can be written as equal to a force times a flux, analogous to equation (23–2).

$$(23\text{–}15) \qquad d_i S = \left(\frac{1}{T_B} - \frac{1}{T_A} \right) h$$

h = energy flow (in joules) due to transfer of heat from one end to the other in a unit time interval, dt.

The total entropy production for the open system is

$$(23\text{–}16) \qquad dS = d_e S + d_i S$$

The second law of thermodynamics can now be stated as

$$(23\text{–}17) \qquad d_i S \geq 0$$

with the equal sign applying to equilibrium conditions.

Consider a second example, the flow of electricity, again shown in Figure 23–1. The rate of heat production by the electric current as it flows along a wire with finite resistance is

$$(23\text{–}18a) \qquad \frac{dQ}{dt} = \frac{J_{el}^2}{\sigma}$$

and consequently the rate of entropy production is

$$(23\text{–}18b) \qquad \frac{dS}{dt} = \frac{J_{el}^2}{\sigma T}$$

By Ohm's law (equation 23–1c), $J_{el} = \sigma \Delta \phi$, and therefore the rate of entropy production along the wire is

$$(23\text{–}19) \qquad \frac{dS}{dt} = J_{el} \frac{\Delta \phi}{T}$$

In differential form this becomes

$$(23\text{-}20) \qquad \frac{dS}{dt} = J_{el} \frac{1}{T} \frac{d\phi}{dx}$$

Again, this equation has the form of equation (23-12); that is, the rate of entropy production can be written as the product of some flux and some generalized thermodynamic force. In this case the flux is the current and the force is the gradient of the electrical potential. In general, by a suitable choice of the flux and the force, it is possible to express the rate of production of entropy in this form. For any system the choice of the generalized force and the corresponding linear coefficient in the flux equation (23-2) is not unique. By a suitable transformation it is possible to choose a generalized force and a linear coefficient so that the product of that force and the flux is dS_i/dt, the rate of production of entropy (equation 23-15). In charge transport (equation 23-1c) the obvious choices of $L = T\sigma$ and $X = (1/T)(d\phi/dx)$ fulfill this condition. In heat transport (equation 23-1b) the choices are less apparent. One choice for a generalized thermodynamic force for heat conduction that meets the condition is $(1/T^2)(dT/dx) = -d(1/T)/dx$. With this choice, equation (23-1b) can be rewritten in the form

$$(23\text{-}21) \qquad J_\theta = kT^2 \frac{d(1/T)}{dx}$$

A suitable choice of L and X for the third example, mass transport by diffusion, is

$$L = \frac{Dc}{R} \quad \text{and} \quad X = \frac{1}{T}\frac{d\mu}{dx}$$

With this choice, equation (23-1a) becomes

$$(23\text{-}22) \qquad J_m = -\frac{Dc}{RT}\frac{d\mu}{dx}$$

for ideal solutions at constant temperature.

In all three of these systems, as presented here, there is only one flux and only one force. In such cases other methods are just as effective as irreversible thermodynamics. However, for systems where more than one flux and more than one generalized force must be taken into account, one must use the thermodynamics of irreversible processes, especially when a flux may arise from the action of more than one force and a force may contribute to more than one flux. If, for example, the rod of metal with the two ends maintained at different temperatures were replaced by a thermocouple, that is, by two rods of dissimilar metals joined at the ends, there would be a flow of current as well as a flow of heat. Or, if the rod were replaced by a tube containing a solution, there would be a flow of solute as well as a flow of heat, until a concentration gradient was produced. In systems where there is a coupling between forces and fluxes, equation (23-2) becomes

(23-23) $$J_i = \sum_j L_{ij} X_j$$

In the special case of two forces and two fluxes, this becomes

(23-24) $$J_1 = L_{11} X_1 + L_{12} X_2$$
$$J_2 = L_{21} X_1 + L_{22} X_2$$

If L and X are chosen properly,

(23-25) $$\frac{d_i S}{dt} = \sum J_i X_i$$

In the special case of two forces and two fluxes, this becomes

(23-26) $$\frac{d_i S}{dt} = J_1 X_1 + J_2 X_2$$

Although, as equation (23-23) shows, each flux may be a linear function of all the forces, one force may be singled out from the others. This force (thermal gradient for heat flow, emf for electrical current, etc.) may be considered to be the force directly related to the given flux. It is the product of this force and its corresponding flux that is used to calculate the rate of entropy production (equation 23-25) associated with that flux. The total rate of entropy production is the sum of these products. It may be more meaningful, physically, to consider the coupling expressed by equation (23-23) to result from interaction among the several fluxes; that is, a flow of heat tends to accompany a flow of matter, and so on. However, there is an important restriction, the Curie condition, on the coupling between forces and fluxes. The forces and fluxes must all have the same vector nature. Vector forces and fluxes in which a directional gradient is involved, such as charge flow, do not couple to scalar processes, such as chemical reactions, which are the same in all directions, i.e., isotropic.

For a system "close to" equilibrium, so that the fluxes fit equation (23-23) or one of its special cases, with the generalized forces and linear coefficients chosen so that $\sum J_i X_i$ gives the rate of production of entropy, there is a very important relation, the *Onsager reciprocal relation*, between the linear coefficients for the interaction terms,

(23-27) $$L_{ij} = L_{ji}$$

For the special case of two forces and two fluxes, this becomes

(23-28) $$L_{12} = L_{21}$$

Just as equilibrium thermodynamics allows general relations among the various static quantities to be deduced for systems at equilibrium (e.g., vapor pressure as a function of temperature), so the thermodynamics of irreversible processes allows general relations among various fluxes and forces to be deduced for systems near equilibrium. Just as equilibrium thermodynamics gives general conditions for stable dynamic equilibrium (e.g., $\Delta S = 0$ for adiabatic systems at constant volume, $\Delta G = 0$ for isothermal systems at constant pressure), so irreversible thermodynamics gives the general condition for a stable steady state—the rate of generation of entropy is at a minimum.

Finally, although irreversible thermodynamics provides relationships among fluxes, it says nothing about rates per se. Rates must be determined experimentally or found from fundamental kinetic theory. In this, too, irreversible thermodynamics resembles equilibrium thermodynamics, which does not give the magnitude of rate constants. Neither can say anything about individual rates because neither considers the kinetic details of processes.

To sum up—

1. In many irreversible processes there is a *flux* that is directly proportional to a generalized *force* (equation 23–2).
2. Whenever a flux is proportional to a generalized force, it is always possible to define the generalized force and the proportionality constant so that the product of the flux and the force is equal to the rate of production of entropy (equation 23–25). Forces and fluxes that are so related are termed *corresponding* forces and fluxes.
3. In addition to being a linear function of its corresponding force, a flux may be a linear function of other generalized forces (equation (23–23). In this case, if the forces are defined so that equation (23–25) is valid, the Onsager reciprocal relation (equation 23–27) holds for the coefficients of the interaction terms.
4. Processes far enough from equilibrium to be nonlinear must be treated by a more general theory.

Let us first discuss thermoelectricity. Consider a rod or wire of a substance that can conduct both heat and electricity. Assume that its ends are in contact with infinite sources or sinks of heat and electrons. Assume further that, except at its ends, it cannot exchange either electrons or heat with its surroundings (Figure 23–1). As the fluxes are coupled, a flow of electricity will produce a flow of heat, even if the two reservoirs are at the same temperature, and a flow of heat will be accompanied by a flow of electricity. The flux equations for this system are

(23–29) $$J_{el} = -L_{11}\left[\frac{1}{T}\left(\frac{d\phi}{dx}\right)\right] - L_{12}\left[\frac{1}{T^2}\left(\frac{dT}{dx}\right)\right]$$

and

(23–30) $$J_{\theta} = -L_{21}\left[\frac{1}{T}\left(\frac{d\phi}{dx}\right)\right] - L_{22}\left[\frac{1}{T^2}\left(\frac{dT}{dx}\right)\right]$$

The generalized forces in these equations, $(1/T)(d\phi/dx)$ and $(1/T^2)(dT/dx)$, have been chosen so that equation (23–26) holds. Therefore, the Onsager reciprocal relation (23–28) is valid. (These choices for the generalized forces and the relation of the generalized forces to the "natural" choices $d\phi/dx$ and dT/dx were considered in the preceding discussion of entropy production.) To develop the theory, two special situations will be considered. In the first there is no flow of electrons; that is, $J_{el} = 0$. Therefore,

$$\text{(23–31)} \qquad \frac{d\phi}{dx} = -\frac{L_{12}}{L_{11} T}\left(\frac{dT}{dx}\right)$$

and

$$\text{(23–32)} \qquad \frac{d\phi}{dT} = -\frac{L_{12}}{L_{11}}\left(\frac{1}{T}\right)$$

These equations relate the voltage gradient (electric field) to the temperature gradient or the voltage to the temperature, but they contain the unknown ratio L_{12}/L_{11}.

In the second special situation there is no temperature gradient; that is, $dT/dx = 0$. In this case π, the ratio of the heat flux to the electron flux, is given by

$$\text{(23–33)} \qquad \pi = \frac{J_\theta}{J_{el}} = \frac{L_{21}}{L_{11}}$$

Because L_{21} and L_{11} do not depend on the direction of flow, reversing the flux of electrons will reverse the flow of heat but will not change the rate at which heat is transported. Since from Onsager's relation $L_{21} = L_{12}$, equations (23–32) and (23–33) can be combined to give

$$\text{(23–34)} \qquad \frac{d\phi}{dT} = -\frac{\pi}{T}$$

This is the equation relating the transfer of heat by the flow of electricity to the voltage of a thermocouple, the Peltier effect.

Thermoelectricity cannot be studied in a single conductor because even approximately "infinite" electron sources or sinks are not available. The difference between the thermoelectric properties of two substances can be studied if the two rods or wires of different conducting materials are joined, as in Figure 23-2, to complete the circuit. If the two junctions are maintained at different temperatures, there will be a net potential difference, which can be measured with a potentiometer or similar instrument, because in general $d\phi/dT$ (or L_{12}/L_{11}) will differ for each of the two substances. Such a circuit, in which there is no flow of current, is a *thermocouple*. The quantity $\sigma_{AB} = (d\phi/dT)_A - (d\phi/dT)_B$ is defined as the *thermoelectric power* for the pair of materials comprising the thermocouple. If the two junctions are kept at the same temperature

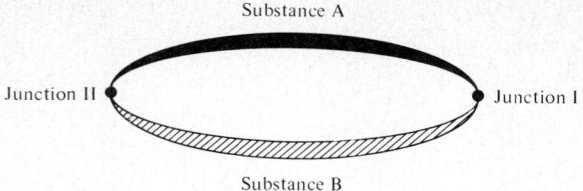

FIGURE 23-2 / Thermocouple junctions.

and a current flows through the circuit, there will be a flow of heat from one junction to the other, in addition to the heat produced by the resistance of the circuit. This is the *Peltier effect* for the pair of materials. These measurable thermoelectric effects obey the appropriate equations (23-32)–(23-34). In particular, the relation between the Peltier coefficient, π_{AB}, for the pair of materials and the thermoelectric power is given by equation (23-34).

A second application of the theory is to the development of relations among the electrokinetic properties of liquids. Consider a liquid that can conduct electricity, for example, mercury or an aqueous solution of an electrolyte. A number of relations can be defined among the flow of electricity, J_{el}, the flow of liquid, J_m, the electric field, $d\phi/dx$, and the pressure gradient, dP/dx (see Figure 23-3). Several of the most important are defined as:

1. The electroosmotic pressure

$$(23\text{-}35) \quad \text{EOP} = \left(\frac{dP}{d\phi}\right)_{J_m=0}$$

P = pressure

FIGURE 23-3 / Mass flow and electron flow through a barrier under pressure and potential gradients. Note that J_m and J_{el} could be in the same or in opposite directions.

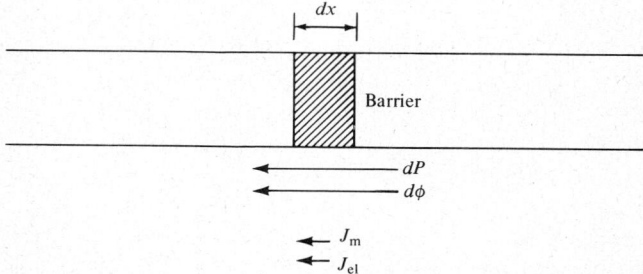

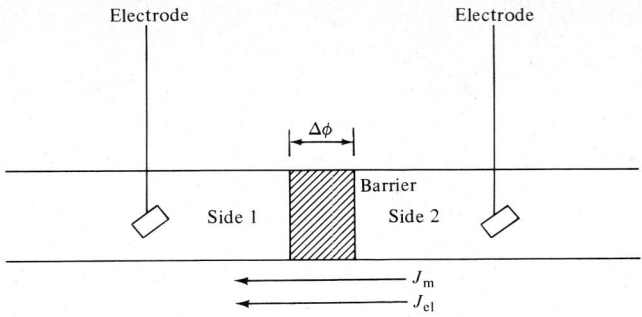

FIGURE 23-4 / Matter and current transport through a barrier.

2. The streaming current

(23-36) $$\text{SC} = \left(\frac{J_{el}}{J_m}\right)_{d\phi/dx=0}$$

$\Delta\phi$ = applied voltage

3. The streaming potential

(23-37) $$\text{SP} = \left(\frac{d\phi}{dP}\right)_{J_{el}=0}$$

4. Electroosmosis

(23-38) $$\text{EO} = \left(\frac{J_m}{J_{el}}\right)_{dP/dx=0}$$

5. and second electroosmosis

(23-39) $$\text{second EO} = \left(\frac{J_m}{d\phi/dx}\right)_{dP/dx=0}$$

The thermodynamics of irreversible processes can be used to derive relationships among these quantities. In terms of generalized thermodynamic forces, the flow of matter (actually volume flow) is given by

(23-40) $$J_m = -L_{11}\frac{1}{T}\left(\frac{dP}{dx}\right) - L_{12}\frac{1}{T}\left(\frac{d\phi}{dx}\right)$$

and the flow of electricity is given by

(23-41) $$J_{el} = -L_{21}\frac{1}{T}\left(\frac{dP}{dx}\right) - L_{22}\frac{1}{T}\left(\frac{d\phi}{dx}\right)$$

The generalized force corresponding to the flow of matter is $(1/T)(dP/dx)$, and, as before, the force corresponding to the flow of electricity is $(1/T)(d\phi/dx)$.

[Note that since PV has the dimensions of energy, PV/T has the dimensions of entropy and $J_m(1/T)(dP/dx)\,dx$ has the dimensions of entropy per unit of time if J_m is the volume flow.] From equations (23-35) and (23-40),

$$\text{(23-42)} \qquad \text{EOP} = \left(\frac{dP}{d\phi}\right)_{J_m=0} = -\frac{L_{12}}{L_{11}}$$

From equations (23-36), (23-40), and (23-41), SC is

$$\text{(23-43)} \qquad \text{SC} = \left(\frac{J_{el}}{J_m}\right)_{d\phi/dx=0} = \frac{L_{21}}{L_{11}}$$

Since from the Onsager reciprocal relation (23-28), $L_{21} = L_{12}$,

$$\text{(23-44)} \qquad \text{SC} = -\text{EOP}$$

In like fashion,

$$\text{(23-45)} \qquad \text{EO} = -\text{SP}$$

For the special case of passage of an electrolyte through a membrane or a porous plug, as in Figure 23-4, the voltage and pressure in each of the two reservoirs on either side of the barrier may be considered to be constant. The pressure gradient dP/dx and the electrical field $d\phi/dx$ within the barrier cannot be measured, but the pressure difference, ΔP, and the voltage drop, $\Delta\phi$, can. The theory given above may be applied to this case if $dP/d\phi$ is replaced by $\Delta P/\Delta\phi$, $d\phi/dP$ is replaced by $\Delta\phi/\Delta P$, $d\phi/dx = 0$ is replaced by $\Delta\phi = 0$, and $dP/dx = 0$ is replaced by $\Delta P = 0$. The coefficients L_{21} and L_{12} no longer refer to the properties of the liquid over the interval dx, but instead are functions of the "resistance" across the plug or barrier. Specifically, L_{11} is a function of the reciprocal of the viscosity of the fluid and is a reciprocal measure of the resistance of the barrier to fluid flow. (If the barrier is a capillary, L_{11} can be calculated from Poiseuille's equation.) The ohmic resistance r is related to L_{22} by

$$\text{(23-46)} \qquad r = \frac{T}{L_{22}}$$

and finally

$$\text{(23-47)} \qquad \text{SP} = r(\text{2nd EO})$$

References

Denbigh, K. G., *The Thermodynamics of the Steady State*. Methuen's Monographs on Chemical Subjects, London, and Wiley, New York, 1951.

deGroot, S. R., and P. Mazur, *Non-Equilibrium Thermodynamics*. North-Holland, Amsterdam, 1962.

Miller, D. G., Thermodynamics of irreversible processes. *Chem. Rev.* **60**, 15 (1960).

Prigogine, I., *Thermodynamics of Irreversible Processes*, 3rd ed. Wiley, New York, 1965.

Zemansky, M. W., *Heat and Thermodynamics*. McGraw-Hill, New York, 1957.

Chapter 24

Techniques of Chemical Kinetics

A knowledge of reaction rates is extremely important, both to the production chemist and chemical engineer, who want to control the rate at which reactions occur, and to the research chemist, who wants insight into the reaction mechanism. There are standard methods found in any physical chemistry or kinetics text for determining the rate equations and rate constants from experimental data. We are concerned here with how such data are obtained and will for the most part limit discussion to the more commonly used techniques for fast reactions in solution. Table 24–1 summarizes much of the discussion.

TABLE 24–1 / Methods for Study of Fast Reactions

Principle	Method	Minimum $t_{1/2}$ (sec)	Maximum k (liters/mole-sec)
Slow rate down to "normal" range	Low concentration Low temperature	— —	— —
Fast mixing, observation not fast	Quenching Continuous flow	0.05 10^{-3}	10^3 10^8
Fast mixing, fast observation	Stopped flow	10^{-3}	10^8
Relaxation: shift equilibrium, fast observation	Temperature jump Pressure jump	10^{-9} 10^{-4}	10^{11} 10^3
Initiation by irradiation	Flash photolysis (Fast obs.) rotating sector	10^{-5} 10^{-2}	10^{10} 10^{10}

Source: E. F. Caldin, *Fast Reactions in Solution*. Blackwell, Oxford, 1964.

Relatively simple procedures are adequate for reactions that take place over minutes or hours. The reagents are mixed and concentrations monitored, either continuously, by physical methods, or discontinuously, by chemical analysis. The continuous methods include those in which changes in spectral absorbance, in conductance, in emf, in volume, in pressure, or any other physical property can be measured. Measurements of these physical quantities can be made in situ without disturbing the reaction system. In discontinuous methods, samples are either withdrawn at given times and the reaction stopped (quenched) or the entire system is quenched. Common methods for quenching including chilling, diluting, adding a reagent that stops the reaction, or running the mixture into a container of such a reagent. The time taken for mixing the reactants and for quenching must be short in comparison to the time needed for them to react appreciably. For very slow reactions, samples can be withdrawn and analyzed without quenching.

Reactions that can be studied by these ordinary methods, have half-times (i.e., are 50% completed) ranging from 10 sec to several days. The rate constants for first order reactions in this range would be between 10^{-1} and 10^{-6} sec^{-1}.

Fast Reactions in Solutions

Some rapid reactions can be slowed sufficiently by dilution or cooling to permit use of conventional techniques. Others can be studied only with special techniques.

Fast Flow methods are those in which the reactants are both rapidly mixed and rapidly forced into the observation chamber, where the progress of the reaction is then monitored. In *relaxation methods*, the reactants are mixed and allowed to reach equilibrium. The equilibrium conditions are then perturbed rapidly and the reaction is monitored while the concentrations change to the new equilibrium values. *Rapid initiation methods* are those in which the reactants are mixed but do not react until some activating pulse is supplied. In *Steady state methods*, current or radiation or other outside controlling factor is supplied at a constant rate, so the reaction proceeds at a constant rate that need not be determined directly.

FAST FLOW METHODS

Reactions with half-times of at least 0.001 sec and up can be studied by fast flow methods if the reactants can be mixed rapidly and efficiently. It is, of course, necessary that the monitoring instruments have a sufficiently fast response time.

In the *rapid quenching method*, the reactants are mixed rapidly and allowed to react for only a short time before quenching. Figure 24–1 illustrates

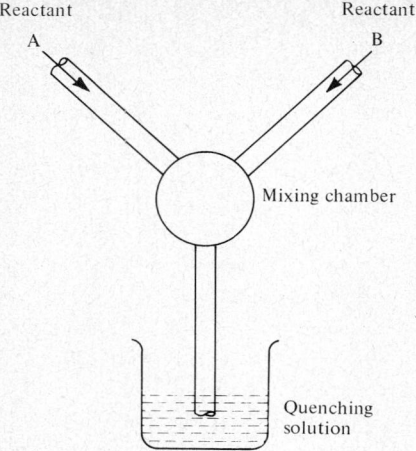

FIGURE 24–1 / Rapid quenching apparatus.

the principle. At t_0, pressure is applied to both storage chambers. The reactants are forced into and through the mixing chamber, then into a reaction tube, and finally into a quenching solution. They react only during the time in which they pass through the reaction tube. That time depends on the length and diameter of the reaction tube and the rate at which the contents of the storage chambers are expelled. After the sample has been quenched, the analysis can be done at leisure by any method desired. Half-times are in the range 5×10^{-2} to 10^3 sec. An advantage of this method is the fact that only a relatively few milliliters of each solution are needed. If there is any problem in measuring the length and diameter of the reaction vessel, it can be calibrated using a reaction of known rate constant. The method has been applied to isotope exchanges and to exchanges of water of hydration.

In the *continuous-flow method*, the apparatus used is very similar to that for rapid quenching. However, the solution is not quenched. Instead, the mixture is analyzed rapidly and continuously at several points along the reaction tube (which for this reason is called the *observation tube*), usually by spectrophotometry. The solution flows at the rate of meters per second. At each observation point the elapsed reaction time is

$$t = \frac{c}{u}$$

where c is the distance from the mixing point to the observation point and u is the linear flow rate. Mixing must be rapid and complete and the flow along the exit tube must be turbulent. Half-times down to 10^{-4} sec can be determined.

In the *stopped-flow method* (Figure 24–2), the solutions are mixed rapidly and forced into an observation chamber in about 10^{-3} sec. The flow is then stopped rapidly. The time for the reactants to go from the point of initial mixing to the point of observation, when the mixing must be 98% complete, is called the *dead time*. The reaction is then monitored as it proceeds toward completion by following some physical property, most frequently absorbance, as a function of time.

The cylinders are driven simultaneously by a driving block. The exit tube has a piston that is pushed by the solution, travels a short distance, and then slams against a stopping block, halting the flow. The time scale for the reaction varies from milliseconds to minutes. Only a few tenths of a milliliter of each reactant are required. The reaction is usually followed spectrophotometrically and the course of the reaction may be displayed on a storage oscilloscope.

RELAXATION METHODS

In these, the reactants are allowed to reach equilibrium and are then put into a reaction chamber. In the chamber, a sudden stress is applied, such as a change of pressure or temperature. When temperature is changed, the method is called *T jump*; when pressure is changed, the method is *P jump*. Under the stress, the equilibrium constant changes and the concentrations change to satisfy the

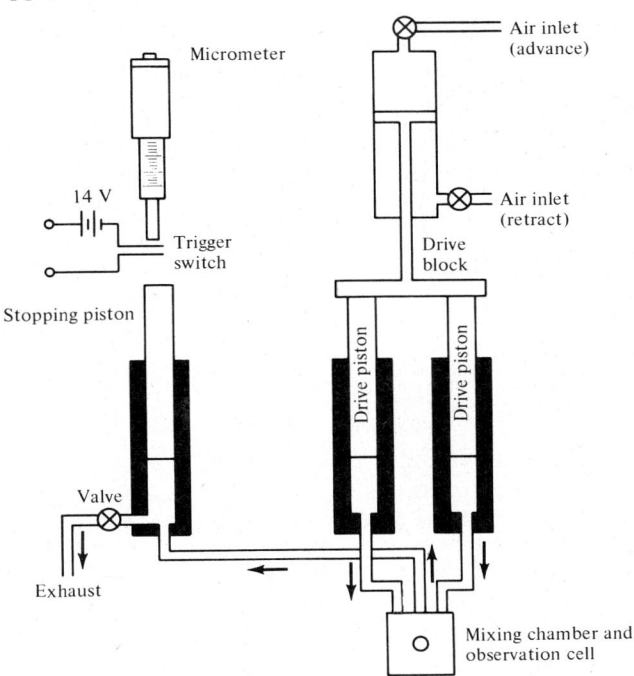

FIGURE 24–2 / Schematic Aminco-Morrow stopped-flow apparatus.

new equilibrium conditions. These concentration changes are followed by appropriate fast-monitoring devices. It must be emphasized that the displacement of the equilibrium must be small. In all relaxation methods, the return to equilibrium is first order, no matter what the order is with respect to each reactant (see pages 304–5).

In T jump, a sudden pulse of energy is supplied to the equilibrium mixture by discharging a high voltage capacitor through two electrodes placed across the solution. The temperature may be raised by several degrees in a microsecond. The return to equilibrium is usually followed spectrophotometrically. More rapid heating can be achieved with a Q-switched laser that can supply a sufficiently brief pulse of energy and extend the scope of the T-jump method to reactions with half-lives in the nanosecond range.

In P jump, the system is initially at high pressure in a cell equipped with a diaphragm. At t_0, the diaphragm is ruptured, releasing the pressure in about 20–100 μsec. Conductance is the most widely used physical property for following the progress of the reaction if ions are involved.

FLASH PHOTOLYSIS

In this technique, the reactants are premixed but do not react until an intense pulse of visible or ultraviolet light supplies the needed activation energy to trigger a reaction that would not normally take place in the dark. The flash is produced by discharging a capacitor, such as a 4–10 μF capacitor charged to 4–20 kV through a gas lamp. Typically, a large number of reactive species are formed by the flash, including molecules in excited states, radicals, and ions. One problem is to detect and, if possible, to identify all these substances. This is often done photographically. At a predetermined time, which can range from microseconds to seconds after the flash, a beam of light is passed through the reaction mixture and then through a prism or grating onto a photographic plate or paper. The beam can come from a continuous source or a photographer's flash bulb.

An alternative method of following flash photolysis is to pass light of a desired wavelength through the reaction mixture onto a photocell. The photocell output is measured by any convenient method. With each flash, one obtains a record of the production and decay of the particular intermediate or of the rate of production of the product. A combination of the photographic and spectrophotometric methods is very useful, the first giving a qualitative and the second a quantitative analysis of one component of the system at any given time.

A third method of monitoring the flash photolysis is by mass spectrometry. The entire range of reactants can be scanned rapidly for a qualitative analysis, or the instrument can be focused on a particular mass, for a quantitative analysis.

Photochemical Methods

The reactants are mixed but do not react appreciably unless radiation is supplied. Here, the radiation is not a high energy pulse but a low intensity beam, which may be either constant or intermittent. When the radiation is constant, the concentrations of reactant or product can be monitored by standard techniques.

Intermittent radiation can be produced by a rotating sector (a disk with a fraction of its area cut out) that is placed to interrupt a beam periodically as it rotates. Intermediates build up while the mixture is being irradiated and decay during the dark period. If the cycle is short compared to the decay time of the intermediates, the rate of the reaction is the same as if the system were being illuminated steadily by a source with the average intensity over the cycle; if the cycle is long compared with the decay time of the intermediates, the average rate of reaction over a cycle is the fraction of time the system is illuminated multiplied by the rate at the intensity of the uninterrupted illumination. By varying the frequency, the relaxation time and rate constants for the formation and decay of intermediates can be determined. Half-times of 10^{-2} sec or more can be studied by this technique.

Gas Phase Kinetics

Many of the methods surveyed above can be applied with suitable modifications to reactions involving gases. Most of the usual straightforward procedures of solution kinetics are readily adapted to reactions between gases and condensed phases and to reactions in the gas phase. In addition, if the total number of moles of gas is changed by the reaction, the course of the reaction can be followed by measuring the change in volume at constant pressure, or in some cases by measuring the change in pressure at constant volume. The classical Warburg methods for measuring the rate of respiration of tissue are examples of this. The *rapid quench* and *continuous flow* methods have been applied to rapid gas phase reactions. The reaction can be stopped by having the stream of reacting gases impinge on a cold surface, where, depending on their properties, unreacted materials, intermediates, or reaction products may be deposited or adsorbed. Relaxation methods are also suitable. One problem in the T jump is to avoid production of ions, radicals, and excited species by the electric discharge. This difficulty can be turned to advantage if one wishes to study the chemistry of these species. A special form of the P-jump method for gases is the shock tube. A gas or a reaction mixture is confined at high pressure behind a diaphragm at one end of a long tube. The rest of the tube contains gas, often an inert gas such as argon, at a lower pressure. When the diaphragm is ruptured, a shock wave analogous to a sonic boom passes down the tube. Because pressure energy is being dissipated, the gas at the front of the shock wave is at a high temperature. The reaction at this shock front can be followed by various techniques. Flash photolysis and most photochemical methods are equally suitable for gas phase reactions. There are several methods that are only used in the gas phase, notably ion beam and molecular beam techniques. A well collimated beam of molecules or ions, all with the same velocity, is produced. If the reactant is an ion, a beam is produced by basically the methods used to produce and select an ionic species in a mass spectrometer. If the reactant is uncharged, the gas or vapor from an oven is passed through a series of orifices to collimate it and through several rotating sectors which are so designed that only molecules within a narrow range of velocities can pass through. Two such beams, for example a beam of chlorine molecules and a beam of sodium atoms, will be allowed to interact. Detailed information about the nature of the elementary reaction process, in this case the reaction $Na + Cl_2 \rightarrow NaCl + Cl$, can be obtained by this method. The decay of unstable ions and radicals with lifetimes less than the time of flight of the ion beam has been studied in mass spectrometers. Further consideration of these two methods is beyond the scope of this text.

Very slow reactions present difficulties that are the opposite of those for fast reactions. There is no problem with mixing or sampling; the chief difficulty is measuring the very small amount of products formed. These reactions can sometimes be speeded up sufficiently by heating to allow conventional methods to be used. However, when this is done, or when fast reactions are slowed down by cooling or by dilution, a question arises as to whether the reaction that is observed is the reaction that one wishes to study. Very slow reactions are usually studied, when they can be studied, by using extremely sensitive methods to detect the products. There are four principal methods: gas chromatography, mass spectrometry, fluorimetry, and the use of radioactive isotopes. With some exceptions these methods can only be applied when the product, which preferably is gaseous or volatile, can be removed from the reactants. If, for example, the product can be continuously removed by a stream of inert gas, the stream with the product can be introduced into a gas chromatograph, a mass spectrometer, or a gas chromatograph–mass spectrometer combination, where the product can be identified, separated from volatile reactants if necessary, and measured. If the product reacts in trace amounts to yield a fluorescent derivative, this derivative can be prepared and measured. Finally, if the reactant can be synthesized with enough of a radioactive isotope incorporated, the product can be separated and the amount determined by counting. The most common and suitable isotopes for organic compounds are the weak β emitters, ^3H and ^{14}C. Radiochemical methods must be used with some reservations because the energy of radioactive decay may increase the reaction rate.

References

Caldin, E. F., *Fast Reactions in Solution*. Blackwell, Oxford, 1964.

Czerlinski, G. H., *Chemical Relaxation*. Marcel Dekker, New York, 1966.

Hammes, G. G. (ed.), *Investigation of Rates and Mechanisms of Reactions* (Vol. VI of *Techniques of Chemistry*, edited by A. Weissberger), 3rd ed., Part II, chs. 2–6. Wiley-Interscience, New York, 1974.

Vetter, K. J., *Electrochemical Kinetics, Theoretical and Experimental Aspects*. Academic Press, New York, 1967.

Chapter 25

Introduction to Computers

The rapidity and ease with which computers can measure, receive, store, process and report data, and send signals to control instruments have revolutionized science and technology in the last decade. Not only has the time spent in computation and data processing been greatly reduced but a number of new techniques and instruments have become feasible because of computer control. Approaches to experimental problems and methods of computation are now used that had previously been impractical. As a result, some knowledge of computers is now a necessity for most laboratory scientists.

The experimenter need not be an expert, even in programming, since more and more computers and computer languages are being developed for the untrained occasional user. However, one should know enough to be able to discuss computer problems with computer experts, and this requires some knowledge of the basic types of computers, their properties and capabilities, advantages and drawbacks, the major elements of computer systems, the types of components currently available, computer languages, and programming.

Analog Versus Digital Computers

There are two fundamentally different types of computers, *analog* and *digital*. In analog methods, the problem is translated into a physical system that can be easily built and manipulated and that is analogous mathematically to the system being studied. The most familiar analog computer is the slide rule. The mathematical operation, the addition or subtraction of logarithms, is translated into measuring off lengths on properly ruled scales. More sophisticated examples include flow tanks in which electrical potential problems are solved by using dyes to mark stream lines, wind tunnels, hydrodynamic tanks to study hull shapes, and dynamic scale models; electronic analog computers are also used.

In *digital methods*, numbers are manipulated arithmetically. The most primitive method is counting on one's fingers. The most common digital computer (apart from fingers) is the abacus. Electronic digital computers now have almost completely supplanted analog computers because they are more powerful and have inherently greater precision. The precision of an analog method is limited by the accuracy with which the problem can be modeled and the accuracy with which the measurements can be made, while the accuracy of a digital computer is limited by the number of significant figures it is designed to carry. Digital computers are also faster and more flexible because they can be programmed by general methods.

For some purposes analog methods may be preferable. Consider the following set of reactions:

$$2A \rightarrow B$$
$$A + B \rightarrow C$$
$$B + C \rightarrow D$$

One could write the simultaneous differential rate equations, program a digital computer to solve them, and compute the amount of A, B, C, and D as a function of time for each combination of rate constants and each set of starting conditions. Alternatively, one could build an analog system that followed the same set of differential equations, where the rate constants were presented by adjustable resistors, capacitors, or other electric circuit elements and the amounts of A, B, C, and D by measurable voltage or currents. One could display these voltage or currents on a cathode-ray tube, and vary the circuit elements to observe directly how the starting conditions and the rate constants affect the progress of the reactions. The digital computer would give more precise values, while the analog computer would likely provide greater insight into the behavior of this system. Note that all simulated experiments, whether analog or digital, are limited by the fidelity of the model to the actual physical situation.

Logic

There is no longer a sharp demarcation between the more sophisticated desktop calculators and the simplest minicomputers, at least from the point of view of the user. Some calculators can be programmed at the keyboard to accept several pieces of data and perform a series of computations. Within their limitations such calculators are often easier to program than computers. Functions are built in and are brought into use by a single command. Today, some calculators, like computers, may be programmed to make decisions that provide a flexible control over the program while it is being executed. Different parts of the program may be used, depending on the result of certain tests; that is, the program has different branches that can be executed according to a preprogrammed decision. Consider the quadratic equation $ax^2 + bx + c = 0$ with the well-known solution

$$x = \frac{-b \pm \sqrt{b^2 - 4ac}}{2a}$$

It will have two separate real roots, two equal real roots, or two complex or imaginary roots, depending on whether the discriminant $b^2 - 4ac > 0, = 0$, or < 0. We are so familiar with this that we substitute into the formula and write the roots in proper form without realizing that we have made a decision. A computer, however, must be programmed to calculate the discriminant, compare it to zero, and select the proper form for the roots depending on this comparison. If the capacity for branching were not included in its decision, it could not be programmed to solve the general quadratic equation automatically. A quadratic equation can be solved with a programmable calculator lacking branching capability, only if provision is made for the operator to intervene and make a decision, or if the sign of the discriminant is determined in advance by the problem (the physical situation may dictate one and only one positive real root) and the program written to comply with this restriction. Consider a simpler example. A machine is to be programmed to calculate the arithmetic mean of a set of values, $a_1, a_2, \ldots, a_n$, where n may vary but is known in advance. The program may be:

> Add each value as it is received to the sum of the values. In addition, keep a count of the number of values entered, and compare this count to n after each value is added. If the count is less than n, accept another value; if the count is equal to n, divide the sum by n, report the average, and end the computation.

If the machine cannot compare the count to n, the operator must intervene. Note that the averaging process described above is fundamentally different from the following program for averaging three measurements:

> Take the first value of the measurement, add the second, add the third, and divide by 3.

This second program, which is a typical one for a programmable calculator, is specific for the average of three measurements. It must be rewritten for two or four. The first program is general for n measurements, provided that n can be specified in advance.

Many computational methods depend on this ability to make decisions. In iteration techniques, an approximate estimate of some quantity is used to calculate a more accurate value, and so on. Some iteration procedures give results that converge to the desired value, regardless of the initial estimate. Others only converge in restricted circumstances or when the initial estimate is sufficiently close to the true value. Where convergence may be a problem, the computer may be programmed to stop the process if successive estimates drift more and more widely from the initial value or oscillate with progressively larger swings. As with other infinite methods, such as infinite series, the computer may also be programmed to stop when by some predetermined criterion the estimated value is expected to be sufficiently close to the true value.

The logic designed into computers may also be used to control experiments. Consider the problem of improving the precision of a measurement that contains random noise, such as an nmr spectrum, an esr spectrum, an x-ray diffraction pattern, or a measurement of radioactivity. Assume that this measurement is one that can be automatically repeated by an instrument and that the data can be obtained in a form that can be stored, read and processed by a computer. From elementary statistical theory, the signal/noise ratio is proportional to the square root of the number of determinations, and so the precision can be increased by averaging replicate measurements. The most straightforward procedure is to program the computer to recognize the end of a run and initiate another run a predetermined number of times, say 100, while automatically accepting, processing, and averaging the data. When the runs have been completed, the computer stops the experiment and reports the averaged measurement in some suitable format. In this application, the computer's capacity for making decisions may be essential for processing the data and for controlling the instrument during each run, but it does not use the results of the experiment to control the experiment. A simple counting circuit, or even a mechanical counter that signaled when a predetermined number was reached would be sufficient to control the number of replicate runs. An alternative approach, which makes use of the data to control the experiment, is to program the computer to make some test of precision and to initiate replicate runs until that precision has been achieved. Both procedures can be combined. The computer can be programmed to continue measurements until a certain number of replicates have been made or a certain precision has been achieved, whichever occurs first, or until both have occurred.

Uses

A digital computer, as used by most scientists, may be considered to be an idiot with a phenomenal capacity to do arithmetic rapidly and accurately if it is told explicitly what it is to do; with the ability to store and recall large quantities of information provided the information can be expressed in some numerical code; and with an ability to make simple decisions of the type continue/stop, yes/no, select this/that, and so on, provided that the criteria for the decision can be expressed explicitly in some numerical code. It cannot think, infer meaning from context, make intuitive judgments, supply obvious instructions that were omitted, or mimic other imprecisely formulated mental processes. It can think only when the rules for the type of thinking to be done, the meanings to be inferred, the judgments to be made, or the instructions to be supplied can be formulated explicitly, completely, and unambiguously in terms of the arithmetic operations, the storage and recall functions, and the decisions of which the machine is capable. Furthermore, all instructions, programs, and operations must be within the capabilities designed into the machine. If a minicomputer is designed to execute a program of 1000 steps or less, it cannot execute a 1001-step program; the program must be rewritten; if its logic is designed to distinguish between the two possibilities $x > 0$ and $x \leq 0$, it cannot distinguish between the two possibilities $x \geq 0$ and $x \leq 0$ or the three possibilities $x > 0$, $x = 0$, and $x < 0$; the choice must be reformulated. We cannot emphasize too strongly that *every step*, no matter how trivial, *must be explicitly included in the program*. Before a computer is programmed, the problem must be broken down to every single necessary step. Even such an obvious instruction as the computer equivalent of "Take a clean piece of paper" must be given.

Many applications of digital computers depend on their ability to process large quantities of numbers and to do arithmetic rapidly. Major uses by scientists include data processing, control of experiments, calculations, simulation of experiments, storage and recall of information, and literature searches. Data processing (i.e., computing results, averaging, smoothing observations, curve fitting, Fourier analyses of spectra, etc.) and calculations may be done either batchwise or in real time. In batch processing the data or the problem are "stored" in a format that could range from magnetic tape to a laboratory notebook. At a convenient time the data or problem plus the necessary programs are entered into the computer and the calculations performed. In real time data processing, measurements are automatically fed into a programmed computer and the desired computations made as the data are obtained. Consider, for example, the operation of an automatic amino acid analyzer. The output from the instrument is a graph with a number of peaks, one for each amino acid in the sample. The area under each peak is proportional to the amount of the particular amino acid in the mixture; the proportionality factor depends on the amino acid. Peaks must be identified by their position, integrated and multiplied by the proper factor, overlapping peaks must be resolved, and the data must be corrected for shifts in the base line between peaks and for contamination by the ammonia that is usually present. Although an elaborate program is necessary to perform these tasks, it is immaterial whether they be done as the run progresses or after it has been completed. By contrast, when the precision of a measurement is to be improved by averaging a large number of runs, there may be a distinct advantage to processing the data in real time. Obviously, this avoids the problem of storing a large amount of data. More important, the program can be designed to allow the worker to examine the data and to obtain the averaged results before all the replicates have been completed so that, if desired, the experiment can be modified or ended.

There is a similar flexibility in making calculations in real time. The process remains under the continuous control of the operator. If the program permits, he or she could obtain the results of each step of the calculation and continue, stop, or alter the calculations accordingly. Consider the problem of fitting an empirical power series,

$$y = a_0 + a_1 x + a_2 x^2 + a_3 x^3 + \cdots$$

by some method to a set of data where a few points may be seriously in error. The worker could fit a trial function, say a quadratic, and have the computed curve plus the data points shown on an oscilloscope or an X-Y recorder. Then, depending on the fit, he could accept the results, recompute with one or more outlying points discarded, or recompute with fewer terms. If the same problem were to be done in batch, the worker would have to analyze the problem in advance to determine all likely possibilities and fit the data for each case, or would have to define explicit criteria for the decisions he or she might make in the real time calculation and program the computer to make them. The ability to do real time calculations with a computer that has a repertoire of commonly used procedures, such as statistical tests, or that is easily programmed by personnel with limited training, means that the worker can make a variety of calculations that might not otherwise be done.

Components, Terminology, and Specifications

A digital computer consists in general of a *central processing unit* (CPU), which is its "brain," a *main memory*, which contains both the basic program instructions for the task and at least some of the data needed in the program, a *buffer memory* or memories in which information is stored temporarily, *input devices* to receive information or instructions, and *output devices* to report results or instructions. In addition, there may be a *peripheral memory* or memories in which are stored programs and information not immediately needed. The relations among these components are shown in Figure 25-1.

The signals used by the computer represent *binary numbers*, meaning that the number is expressed in terms of powers of 2 rather than of 10, as in the decimal system. In the decimal system, the number 8192 means the sum of $8 \times 10^3 + 1 \times 10^2 + 9 \times 10^1 + 2 \times 10^0$. The digit farthest on the right is multiplied by 10 to the power 0, the next digit on the left is multiplied by 10 to the power 1, the digit after that is multiplied by 10 to the power 2, and so on. In the binary system, the digits are multiplied by the number 2 raised to a power, rather than by 10 raised to a power. In binary notation, as in decimal notation, the power is determined by the position of the digit, the first on the right being

PRINCIPLES

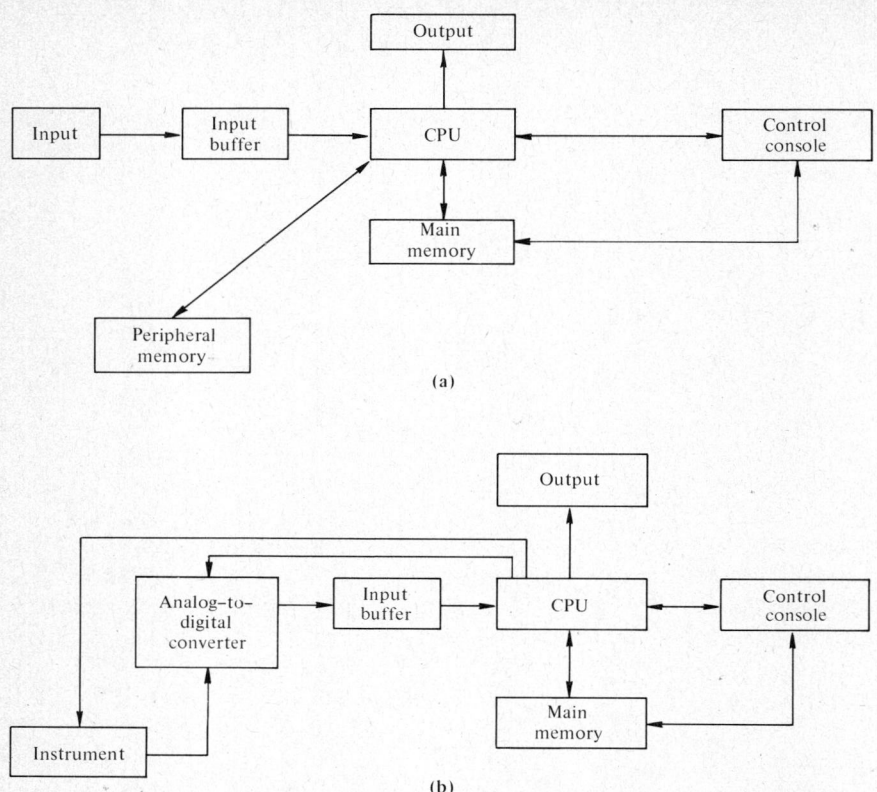

FIGURE 25-1 / (a) Typical computer configuration. (b) Typical configuration for computer control of instrument.

multiplied by 2^0, the second by 2^1, and so on. In binary notation, the digits are 0 and 1. These symbols have the same meaning as in decimal notation. The expressions for all other numbers differ in the two systems. Decimal 2 is binary 10, or $1 \times 2^1 + 0 \times 2^0$. Decimal 3 is binary 11, or $1 \times 2^1 + 1 \times 2^0$. Table 25-1 shows some decimals and their binary equivalents. Decimal 15 is binary $1111 = 1 \times 2^3 + 1 \times 2^2 + 1 \times 2^1 + 1 \times 2^0$.

Binary notation is used because within the computer the two symbols, or digits, can be represented by any two-valued function, such as voltage/no voltage, current/no current, and open circuit/closed circuit, and because they can be used for the basic operations of computer logic, which are normally limited to a choice between *two* alternatives, such as start/stop, yes/no, and this/that. Each binary digit is called a *bit* (binary digit). A string of n bits has 2^n possible combinations, each of which represents a binary number and is called an *n*-bit *word*. For example, there are 2^5 (32) 5-bit words, which can be used to

TABLE 25-1 / Decimal and Binary Numbers from 0 to 15

Decimal	Binary	Decimal	Binary
0	0	8	1000
1	1	9	1001
2	10	10	1010
3	11	11	1011
4	100	12	1100
5	101	13	1101
6	110	14	1110
7	111	15	1111

represent the 26 letters of the alphabet and 6 punctuation signs. An 8-bit segment of information is called a *byte*. A 4-bit segment of information is called a *nibble*. Byte oriented microcomputers are now available, even for hobbyists. Small laboratory computers usually use 12- or 16-bit words, medium-sized computers usually use 32 or 36 bits per word, and large "number crunchers" use up to 60 bits per word. The longer the word, the greater the number of different instructions and the greater the precision of any number. In *double precision operations*, two words are combined to hold a single datum. Since there are only 65,536 16-bit words, double precision is virtually a necessity with small computers.

The CPU has several temporary storage units called *registers* in which the actual operations take place. The instructions in the program are read in order. Depending on the sequence of instructions, a word (n-bit binary number) in a specific location in the main memory is brought to a register, where it is operated upon in a specified way, such as addition, subtraction, or comparison with another number. The result is either retained in a register for further operations, returned to a specific location, or both. Operations such as addition, subtraction, and usually multiplication and division are *hardwired*, that is, built into the computer with permanent circuits. Other processes, ranging from exponentiation to taking Fourier transforms, may be hardwired into *read only memories*, or ROM's, thereby speeding up programming and avoiding mistakes by eliminating steps in the written program (*software*). Note that because of the speed of operation, tables of functions such as sin, cos, and log are not stored. Values of these are calculated anew when needed. Inexpensive microprocessors may be combined with ROM's or PROM's (programmable ROM's) to control a particular instrument. Such *dedicated* computer systems are relatively inexpensive and easy to assemble and provide flexible control of a particular instrument and of preliminary processing of data from the particular instrument. Because they are not easily reprogrammed, they cannot be readily used as general purpose minicomputers or transferred easily to other instruments.

The capacity of a computer may be determined as much by the capacity of the main memory as by any other factor. In a minicomputer, the main memory may be limited to 4096–65,536 words, while in a large computer it may contain over 1 million words. (The number of words is usually expressed in terms of k's, k being defined as a unit of 1024 words; 4096 words is therefore 4k, and a 64k memory can contain 65,536 words.) Whatever the size, the main memory permits access that is both random and rapid, a microsecond or less being sufficient to transfer a word to the CPU. Information at each address in the memory can be stored or recalled independently of the other addresses. The main memory may be supplemented by one or more peripheral memories. There are four common types.

Magnetic disks take milliseconds to start transferring words to or from the CPU. However, since each transfer can be a large block of words, operations are still relatively fast at one machine cycle per word. Each address can be specified by two coordinates, the angle and the radius, providing random access because each coordinate can be varied independently. Disks are expensive but the price is gradually dropping.

Magnetic tapes can be almost as fast as the disks. Unlike magnetic disks, they do not provide random access, because the single coordinate, the distance along the tape, must be traversed continuously from one address to another. Slow tapes for minicomputers are reasonably priced.

Punched paper tapes are much slower than other types, taking about 5 minutes for 1000 words and they are therefore not used for peripheral memory except with small minicomputers without other types of peripheral memory. They are read through a teletype or a tape reader. Unlike more sophisticated devices, they are not usually called by the CPU but require an operator for transfers to or from the main memory. They cannot be erased and reused. Both paper tapes and the equipment for punching and reading them are much cheaper than other storage devices. In addition, paper tapes are usually edited by hand. They are widely used for storing programs for minicomputers and as output devices from instruments, if the data are to be processed further, by a minicomputer.

Punched cards serve the same purposes as paper tape but are much faster and much more expensive.

The slowest part of most computer operations is getting information into and out of the computer. For minicomputers, the most common input/output (I/O) device is the *teletype*. This is slow compared with most other devices but is dependable and reasonably inexpensive. If it is equipped with a tape reader and punch, it is the cheapest device for handling punched tape. Other automatic typewriters are now beginning to appear on the market. *Card readers* and punches are rapid but are too expensive for many laboratory uses. They are generally found in medium-sized and large installations. Like paper tape, they require some auxiliary device to translate the information into readable form. High speed *printers* are two orders of magnitude faster than the teletype but are too expensive for small installations and individual laboratories. *Graphical terminals* that function like television screens are also used for output. Relatively simple devices can only display alphanumeric characters, but more elaborate forms are available that can display graphs, vectors, and complex figures such as three-dimensional views or sections through solid objects. A molecular structure, such as part of a protein molecule, can be displayed and rotated on command to show the spatial relations of the different parts. The limits to the visual display are determined primarily by the computer and the program, not by the device. In some models, hard (permanent) copy can be provided. Graphical output can also be produced with an *X–Y plotter*. Teletypes and automatic typewriters are also used to print X–Y plots and bar graphs. This is the most inexpensive way of producing a graphical display, but the resulting figures are extremely coarse-grained.

For taking data and controlling experiments by computer, there must usually be *analog-to-digital conversion* and vice versa. Laboratory apparatus puts out continuous voltages, which are analogs of pressure, absorbance, temperature, and so on. These analog voltages are converted to binary digital signals by an analog-to-digital converter (ADC). The speed and accuracy of the conversion can be selected according to the requirements of the experiment. The digital voltages produced by the operations of the computer can be read out with any of the devices mentioned above or they can be converted back to analog voltage by a digital-to-analog converter (DAC). The analog voltages put out by the DAC can then be read out with an oscilloscope or a recorder. (Extremely accurate plotters that work directly with digital information are also available.)

Programming Digital Computers

Before a digital computer can be used for any purpose, it must be programmed; that is, a set of instructions must be stored in the central processing unit to specify what operations are to be done and in what order. If the problem is routine, it is often possible to look up an appropriate program in a *program library*, a collection of programs that may either be purchased separately, or is supplied by the computer manufacturer.

With nonroutine problems, the programmer must analyze the problem carefully to decide what is to be computed (averages, least squares, matrix inversion, etc.), where decisions must be made, where information is to be entered and results displayed, and the location of any control signals, if the computer is to be used to control an instrument. Unless the problem is trivial,

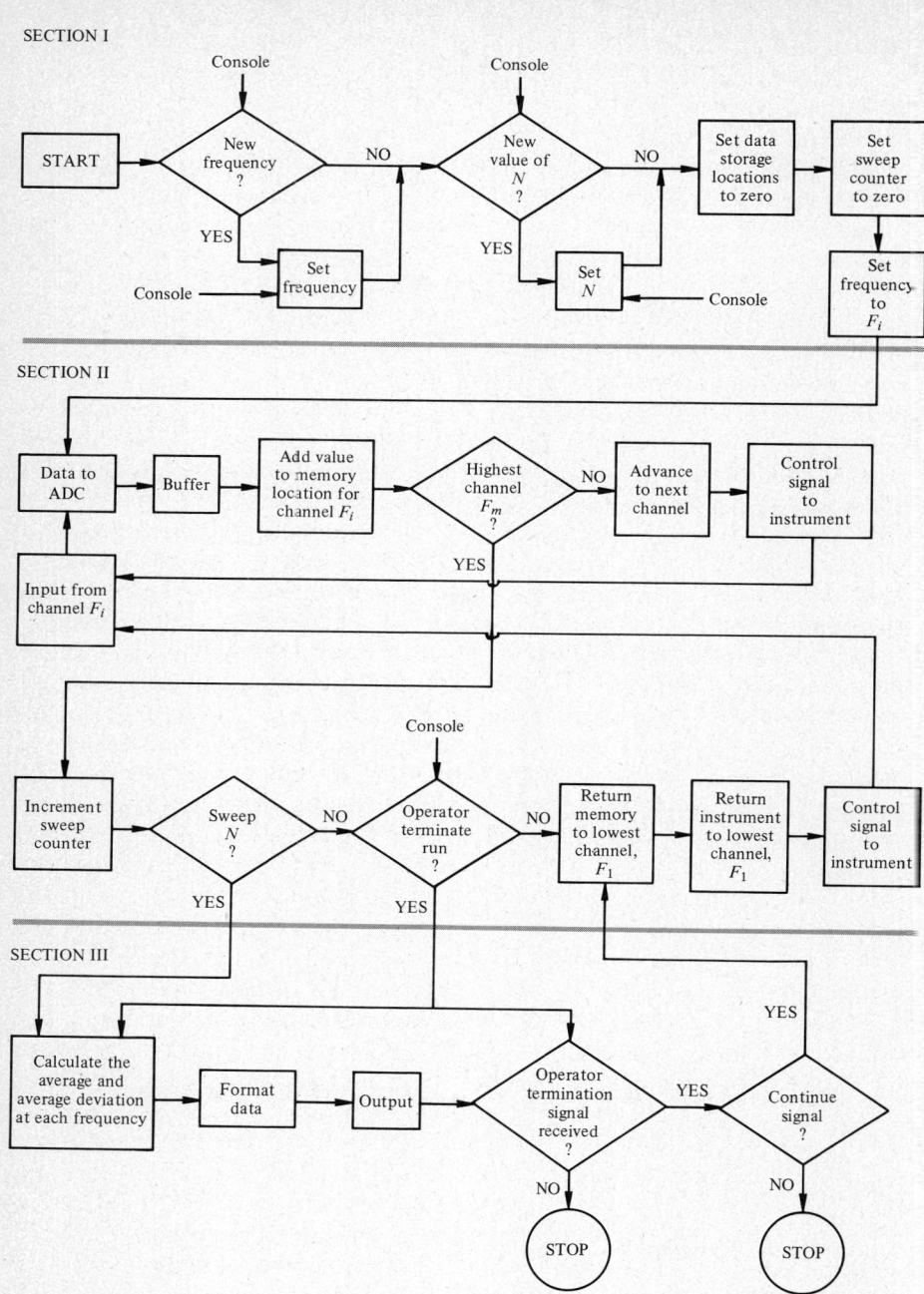

FIGURE 25-2 / Flow chart for program to time-average an nmr spectrum.

this analysis results in a flow chart. After the programmer has decided what is to be done, the next step is to decide on the order of the logical and arithmetical operations that are required to do what is to be done. This order is also shown in a flow chart. Finally, after the analysis has been done and the decisions made and the flow charts prepared, the program is written in the form of a series of instructions in some code or *language* that can be interpreted by the computer.

Following are examples of problem analysis and preparation of flow charts.

1. *The Operations to Be Done* / Consider the problem of improving the signal/noise ratio of an automatic nmr spectrometer by the method of time averaging. The instrument is to sweep through a range of frequencies, making measurements at a number of discrete frequencies, $F_1, F_2, \ldots, F_i, \ldots, F_m$, the sweep is to be repeated until n sweeps have been made, and the measurements at each of the frequencies are to be averaged. For flexibility, the number of sweeps and the number and value of the frequencies can be changed and provision is to be made for the operator to halt the experiment at any point, if desired. A flow chart devised for this process is shown in Figure 25-2.

The flow chart consists of three sections. Section I shows the initial program to set the instrument for a specific experiment. It has two decision points, one where the machine asks whether the frequencies are to be reset and one where it asks whether the number of sweeps is to be changed. Input from the experimenter is indicated by CONSOLE. Note the instructions SET DATA STORAGE LOCATIONS TO ZERO, SET SWEEP COUNTER TO ZERO, and SET FREQUENCY TO F_1. These are the machine equivalents of "Take a clean sheet of paper." If they are omitted, the machine will combine the data with the results of the previous runs. Section II shows the control of the individual sweeps. It contains three control points, one to determine whether F_m, the highest frequency, has been reached, one to determine whether N, the desired number of sweeps, has been made, and one to determine whether the operator has decided to stop the experiment before N has been reached. It also contains two program *loops*. The first repeats the process of measuring the signal at one frequency, but with the frequency increased from F_i to F_{i+1}. The second resets the frequency from F_m back to F_1, thereby initiating a complete new sweep. In this part, input comes from the spectrometer and signals are sent to the spectrometer to set the frequency and initiate a measurement at the frequency. Because the operation is in real time, there must be signals (not shown) to keep the computer from advancing to the next operation before the spectrometer has made the measurement and moved to the next frequency. Section III is the final section, where the average and standard deviation of the signals are to be calculated and where results are converted to the proper format for them to be printed in tabular form by a teletype or fed through a digital-to-analog converter to a recorder. A decision point is included so that if the run is stopped by the operator before N sweeps, it may be continued without starting over from the beginning.

PRINCIPLES

This flow chart illustrates a number of important ideas in programming. Although a program usually has a start and an end, it need not continue in a straight line. It may loop back on itself or even have loops within loops. The loops in this program illustrate the value of indirect addressing, that is, the ability of the computer to calculate a memory storage address used in the program, instead of requiring that the address be specified in advance. In this program, each time the measurement loop is repeated, the frequency is increased to the next higher value and the address where the data are to be stored is increased from i to $i + 1$. If the computer could not change the

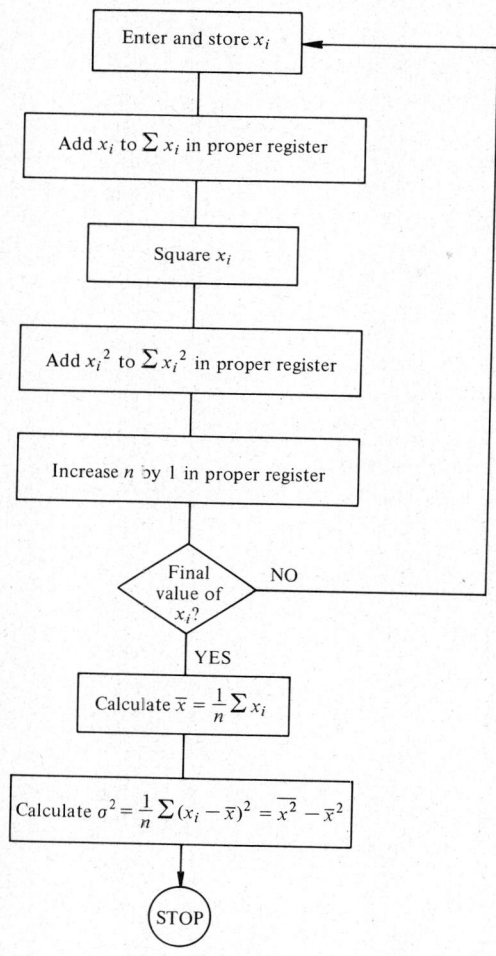

FIGURE 25–3 / Flow chart for computation of mean and variance for a set of n observations.

address, this part of the program would have had to be written m times, to specify addresses $1, 2, 3, \ldots, m$.

Besides loops, a program may have *branches* or *jumps*, which differ from each other in that the computer returns to the main program after a branch but not after a jump. Assume, for example, that step 57 in some program is "branch to step 200." The machine would go to step 57, then to step 200, and, after executing the program subroutine that starts at step 200, would then return to step 58. If the instruction at step 57 were to be "jump to step 200," the machine would go to step 200 and continue on, but would not return to step 58 unless specifically instructed to do so elsewhere in the program. Both jumps and steps can be conditional (i.e., depend on some logic) or unconditional. The instruction in Section III, CALCULATE THE AVERAGE AND STANDARD DEVIATION FOR EACH FREQUENCY, is an example of a *subroutine*. Rather than write a program for these calculations each time they are needed, which would be a total of m times in this program, this subroutine is written once and stored, with the main program branching to it whenever necessary.

2. *The Order of Operations* / Consider the computation of $\bar{x}$, the mean, and σ^2, the variance, for a set of n observations. The mean is defined by $\bar{x} = \sum (x_i/n)$ and the variance by $\sigma^2 = \sum [(x_i - \bar{x})^2/n]$. With pencil and paper or a desk calculator, we might compute $\bar{x}$, then calculate $x_i - \bar{x}$ for each value of x, and finally calculate σ^2. This is an awkward procedure for a computer because each value of x_i must be stored. Therefore, a more suitable algorithm for σ^2, (i.e., a recipe for finding σ^2) should be used, if possible.

By simple algebra it can be shown that

$$\sigma^2 = \sum \frac{x_i^2}{n} - \bar{x}^2 = \overline{x^2} - \bar{x}^2 = \sum \frac{x_i^2}{n} - \left(\frac{\sum x_i}{n}\right)^2$$

This suggests the series of steps in the flow chart of Figure 25–3 as a suitable procedure for finding $\bar{x}$ and σ^2. Note that although this flow chart shows how the machine could calculate the desired quantities, it does not show how the machine can be instructed to do the indicated operations.

Computer Language

To enter programs into a machine, we must use some computer *language*. The earliest programs were written in *machine language*. Each sequence of steps that the machine would have to execute in order to add, subtract, store a number in main memory, and so on, was specified. Even the simplest programs required a skilled programmer who was familiar with how the particular machine did arithmetic. Machine language, although the most primitive method of programming, remains the most powerful, because it specifies anything and everything that the machine is theoretically capable of doing. For most purposes, however, it has been replaced by higher level languages.

In *interpretive languages*, also called *assembly languages*, instructions are represented by numerical codes or by 3-letter mnemonics. "Add," for example, might be 21; "subtract," 22; "store in memory," 1307; "print," PNT; and so on. Routines to translate these codes into machine steps must be programmed into the main memory or hardwired into the machine. The syntax of the particular interpretive language, that is, the rules of the game, must be followed strictly in order for the computer to distinguish between numbers and instructions. Some interpretative languages were developed for a particular computer, others were developed for general use. For all but the smallest machines and certain real time applications, they have been replaced by still higher level languages, such as *Fortran* and *Basic*, in which the instructions consist of command words such as DO, JUMP, PRINT, and READ and mathematical expressions such as $z = x^2 + y^2$. These languages are designed to resemble, at least superficially, normal English and normal algebra, to be learned easily, to be usable after some fashion by persons with a very limited knowledge of programming and to be adaptable to various computers. As with all computer languages, the grammar and vocabulary must be adhered to strictly. Although the mathematical expressions appear to be algebra, they are not. In algebra, the expressions $z = x + y$ and $x + y = z$ are equivalent. In Fortran the first has meaning, the second has not. In particular, the sign " $=$ " does not mean " equals," it means " replace by." The instruction $z = x + y$ means "replace the number stored in memory location z by the sum of the numbers stored in locations x and y" (where x, y, and z have been previously defined). In Fortran, the expression $n = n + 1$ means "replace the value in register n by a number 1 greater" and is an acceptable routine for counting the number of times a particular set of instructions have been executed. In algebra this expression is nonsense. Appreciable space in the main memory is needed to store the routines to convert instructions in these languages to machine steps.

Desktop minicomputers and programmable calculators do not have sufficient memory for general high-level languages. Instead, easy programming is ensured by *keystroke programming*. Arithmetical operations " $+$," " $-$," " $\times$," etc., instructions such as HALT (to enter a value from the keyboard), PRINT (to print a value), and sometimes functions such as " e^x," " log x," " sin x," " cos x," etc., are programmed by depressing the appropriate key. There may be numerical codes for operations not included in the keyboard, such as BRANCH or RUN. There may also be keys for arithmetic mean, standard deviation, least squares, *Student's t*, and other standard statistical and algebraic routines that are hardwired in the machine. Programming these machines will be considered in some detail because the student is almost certain to use them at some time and because instruction manuals are often deficient in explaining the grammar of a particular language or in pointing out possible difficulties.

There are two principal formats for arithmetic, *algebraic* (actually pseudo-algebraic) and *adding machine*, which is commonly called *reverse Polish notation* (RPN). In algebraic notation, the normal sequence is "number, operation, number, operation," and so on. Machines differ, but in most cases each operation is performed on the result of all operations to that point. In this respect, the normal routines of algebra are not followed. Thus "$2 \times 3 + 4$" = 10 because this sequence is interpreted as $2 \times 3 = 6$ followed by $6 + 4 = 10$. On the other hand, "$4 + 2 \times 3$" = 18 because this is interpreted as $4 + 2 = 6$ followed by $6 \times 3 = 18$. In most machines with algebraic format, this difficulty can be circumvented by using parentheses, as in "$4 + (2 \times 3)$" = 10. With algebraic format, parentheses or storage of an intermediate result is necessary for such calculations as $(a + b)/(c + d)$. If parentheses are available, the machine may still not be able to accept programs with nested parentheses or may be able to accept certain arrangements of nested parentheses but not others. Thus the expression $((a + b)/c + d)e$ may be forbidden while the expression $(d + (a + b)/c)e$ may be allowed, or vice versa. Depending on the design, there are other hidden possibilities for error. In one widely used series of desktop computers, the sequence "$2 + 3$, store" results in 3 being stored, while the sequence "2×3, store" results in 6 being stored because the machine automatically supplies " = " after multiplication and division but not after addition or subtraction. To store the sum 5, the sequence "$2 + 3 =$, store" or the sequence "$(2 + 3)$, store" must be programmed. A few calculators use true algebraic format, which obeys the hierarchy of operations. Exponentiation is performed before multiplication and division, which are performed before addition and subtraction. The sequence of keystrokes is stored in a buffer memory and the calculation is made when the " = " key is depressed. With such calculators the value of both "$2 \times 3 + 4$" and "$4 + 2 \times 3$" is 10. Even in these machines, however, it is not always apparent in advance how a string of successive divisions such as "$5/4/3$" will be interpreted.

In reverse Polish notation, the normal sequence is "number, enter in arithmetic register, second number, operation to combine the two, next number, operation to combine it with the previous result," and so on. Also, there is *stacking*: when the second number is entered, the number in the arithmetic register is automatically placed in a temporary storage register so that it can be retained until the operation to combine the two numbers is entered. If a third number is now entered instead of an arithmetic operation, the number in the arithmetic register is placed in the temporary storage register and the number in the temporary storage register is automatically transferred to a second temporary storage register. This process can be continued until a stack of numbers (four, in one popular model) is built up in temporary storage in addition to the number in the arithmetic register. This permits the calculation of expressions like $(a + b)/(c + d)$. The sequence in steps is: "a, enter, b, + (giving $a + b$ in the arithmetic register), c, enter, d, + (giving $a + b$ in the first level of the stack and $c + d$ in the arithmetic register), ÷ (giving the desired quotient in the arithmetic register)." There is an exchange operation to interchange the number in the arithmetic register and the number in the lowest level of the stack. If, in the example above, the operation "exchange" is inserted immediately before " $\div$," the quotient would become $(c + d)/(a + b)$. Because it does not resemble algebra, RPN leads to fewer programming mistakes by the novice or occasional user. It also eliminates such questions as: When is " $=$ " implied and when is it not implied? When and with what restrictions can parentheses be nested? What algebraic expressions are acceptable? The most common calculators using RPN are those made by Hewlett–Packard. Regardless of the format, a worker should experiment with programs and calculations until he or she understands the language and the limitations of the machine.

With all computers, but especially with minicomputers and pocket calculators, one must guard against forbidden or meaningless operations such as requesting the square root of a negative number or the logarithm of a negative number or division by 0. Depending on the design of the machine, it may indicate an error, or take the square root or the logarithm of the corresponding positive number or compute a meaningless number. Most machines will not continue calculating indefinitely if asked to divide by zero. Not all will indicate that they have been asked to perform a nonsensical task. If there is a possibility that a forbidden or meaningless operation may sometimes occur when the program is being run, the program should contain tests for possible trouble spots.

Computerized Literature Searches

As the annual output of scientific papers grows, monotonically and without discernible bound, machine searches of the literature have become increasingly important. The chief problems connected with machine searches, involve, first, the practical restrictions on the size of the *data base* and, second, the difficulty of indexing the literature for novel ideas and theories.

Ideally, the literature should be stored in some form of immediately accessible memory. The coverage should be comprehensive, detailed and completely cross-indexed. Obviously, for a given amount of memory, the greater the coverage, the less detail that can be included about each item and the fewer terms that can be used to index the item. Less obviously, the larger the memory, the slower the search and the greater the cost in machine time to recover any one item. Most data bases currently available are restricted to one broad field, frequently one of practical importance, such as medicine, psychiatry, pharmaceuticals, or organic compounds. As a minimum, each item includes authors, title, journal or book, and the words, terms, formulas, and so on, used to index it. The item may also include a more or less detailed abstract. In many data bases, the literature for the last two or three years in the more important journals (of which there may be several hundred) is kept in an immediately accessible memory. The earlier literature and the literature in obscure journals is kept in peripheral memory. As new items are added to the immediately accessible memory, the older items are transferred to the periphery. When an item is placed in the data base, it should be indexed by author and cross-indexed for all matters (facts, theories, speculations, formulas, compounds, procedures, new instruments, etc.) that might conceivably be of interest to any worker at a future time.

The indexing terms must be restricted to a list, albeit a very long list, of standard terms that the computer has been programmed to recognize. These terms should be sufficiently broad to retrieve all pertinent references in any one search, yet sufficiently detailed and restrictive to exclude irrelevant items. Furthermore, they should be terms that the user who is not a trained indexer would naturally use to describe the desired information. A search is most likely to be fruitful when the desired item can be referred to by a specific term, name, or formula. A search is most difficult and has the poorest prospects if the information cannot be described by a few obvious specific terms. It should be a straightforward matter to find the nmr spectrum of acetone, the phase diagram for the system $CaO-SiO_2-TiO_2$, or the electrolyte content of crab blood, if it is in the data base.

New ideas and theories are extremely difficult to index and to find. If the searcher knows some specific workers or publications associated with them, it becomes easier. Even then, he may miss a substantial portion of the relevant information. A search might not uncover a theoretical treatment of linear free energy diagrams or earlier theories or observations that were applicable to a new theory of electrode kinetics in fused salts. The problem is that new concepts must be placed in old categories and described by specific indexing terms from a restricted list of terms that are often not applicable. For example, how could one have indexed Dalton's atomic theory, Planck's quantum theory, or the concept of the genetic code when they were first proposed? Furthermore, if these had been indexed, under what categories would one have looked for supporting evidence when searching through studies undertaken for other purposes? In new and startling theories, the important information may not even be recognized by the indexer and never get into the data base. Even if it is indexed, it may be under any of a number of standard terms, none of which may be really applicable. If the search is restricted to a few terms, the information may well be missed. If it is extended to all possible terms, the information may be buried in a flood of irrelevant references.

A machine search of the literature is best done in real time with the continuing assistance of a librarian trained in the techniques. The searcher selects a few terms and asks the computer how many items are in those categories. If it says 30 or so, the searcher can then ask for the complete list, or even for a few items to see if the list is relevant. If the machine says 20,000 items, the searcher must further restrict the search. One good way to find suitable index terms is to ask how a paper, or a few papers, that are known to be relevant are indexed. If the search cannot be made in real time, as, for example, when the peripheral memory must be searched, the searcher must select suitable index terms and hope for the best.

References

Arnold, J. T., Microprocessor applications, a less sophisticated approach. *Science* **192**, 519–523 (1976).

Dessy, R. E., Microprocessors—an end user's view. *Science* **192**, 511–518 (1976).

Part Two

EXPERIMENTS:

Applications of Laboratory Techniques

Experiment 1

Viscosity of a Gas

Pressure–Time Measurement

OBJECTIVES / To determine gas pressure as a function of time in a system being evacuated and to use the results to calculate the viscosity of the gas.

READING ASSIGNMENT / Chapter 12, pages 121–123, Chapter 9, pages 79–80, and 83.

Viscosity in a gas is caused by transfer of momentum from gas molecules to one another. It depends on mean free path, molecular velocity, collision diameter, and density, as shown in equation (12–13). In this experiment the velocity of a gas, relative to that of air, is computed from the rate at which the gas in a container is evacuated through a capillary. From the gas viscosity, the mean free path, collision diameter, and collision number of the gas at standard temperature and pressure are calculated. If time permits, the temperature coefficient of the viscosity of the gas may also be determined.

EQUIPMENT AND MATERIALS

The pressure is determined with either a manometer or a transducer whose output is fed into a strip chart recorder. The gas is supplied from cylinders of argon, ethane, butane, or other commercially available gas.

DIRECTIONS

Calibrate the apparatus with dry air at 0°C. Then determine the viscosity of the gas being studied relative to that of air at 0°C.

If time permits, determine the viscosity of the gas at 100 and 150°C.

CALCULATIONS AND GRAPHS

For each run, plot $1/p$ versus time.

Calculate the viscosity of each gas, relative to that of air, from the slope of the graph.

If measurements were made at 100°C, calculate the ratio of viscosity at 0°C to that at 100°C and compare the result with the ratio of the square roots of the absolute temperatures. Then plot the log of the viscosity against the log of the Kelvin temperature and compare the slope with the theoretical value of $\frac{5\pi}{2}$ expected for ideal gases.

Assume that the gas is ideal and follows elementary kinetic theory. Calculate the mean free path, collision diameter, and collision number at STP using the experimentally determined viscosity. Compare the results with accepted data wherever possible.

NOTES / Carbon dioxide is not suitable for this experiment (note this in the answer to Exercise 2).

The equations relating pressure and time with viscosity are derived for laminar flow and are not valid at very low pressures, when the mean free path is of the same order of magnitude as the radius of the capillary.

If the pressure is measured with a U-tube manometer, it must be remembered that the pressure change is twice the height change in either limb. If the pressure is measured with a transducer, the transducer must be calibrated with a manometer. Runs should be repeated until the precision in the calculated viscosity is 5% or better.

EXERCISES

1. Explain the temperature coefficient of viscosity in terms of molecular phenomena. Why is the temperature dependence of viscosity so different for gases from that for liquids?

2. On the basis of your procedure, is the viscosity of an ideal gas independent of pressure? Explain your answer.

3. Explain why the Poiseuille equation is not valid at very low pressures.

Experiment 2

Velocity of Gas Molecules

Speed-of-Sound Method

OBJECTIVES / To determine the frequencies of standing waves in a gas-filled tube by means of an oscilloscope and a sine wave generator and to calculate the speed of sound in the gas and the average velocity of the gas molecules.

In this experiment the speed of sound in a gas is determined and the average velocity of the gas molecules is calculated. Sound is carried by transfer of energy from one molecule to another. Obviously, therefore, the speed of sound is related to the rate at which molecules collide and, therefore, to the molecular velocity.

In an ideal gas, the speed of sound is a function of temperature, molecular weight, and heat capacity. The relationship is

(1) $$V_s = \left(\frac{C_p}{C_v}\right)^{1/2} \left(\frac{RT}{M}\right)^{1/2} = \left(\frac{\pi}{8}\right)^{1/2} \left(\frac{C_p}{C_v}\right)^{1/2} \bar{c}$$

V_s = velocity of sound
C_p = heat capacity at constant pressure
C_v = heat capacity at constant volume
$\bar{c}$ = average molecular speed

For real gases, the departures from ideality are large enough to make equation (1) only an approximation. For real gases at ordinary temperatures, a simple approximate relationship is

(2) $$V_s = 0.7\bar{c}$$

where the constant 0.7 has been obtained from the averaged values of heat capacities of a number of polyatomic gases at room temperature.

The method here is a modification of that originally developed by Kundt, with the position of nodes determined electrically, rather than visually (see Diagram A).

The vibrating loudspeaker produces sound waves that are reflected back and forth from the microphone. Unless the speaker and the microphone are each at a node, a standing wave is not formed; that is, unless the distance between the microphone and the speaker is an integral number of half-wavelengths there is interference resulting in partial or complete cancellation of the sound waves. When the frequency is such that a standing wave is formed, the output from the microphone is at a maximum, as observed with an oscilloscope or a high input impedance voltmeter.

When the observed signal is at a maximum,

(3) $$L = \frac{n\lambda_m}{2}$$

L = distance from speaker to microphone
n = number of half-wavelengths
λ_m = wavelength of sound at the observed maximum

Since $V_s = \lambda v$,

(4) $$v_m = nV_s/2L$$

v_m = frequency at the observed maximum

Frequencies can be measured more accurately than distances. Therefore, instead of moving the microphone back and forth to locate a node and then measuring the distance between microphone and speaker, the experimenter fixes the speaker and microphone in place and varies the frequency of the sound until standing waves are observed. Calculations are often made, using the accurately known frequency.

Equation (4) cannot be used directly to calculate V_s because n is too difficult to determine. Instead, the frequency is varied, until a series of successive maxima are observed, on the oscilloscope or meter. The frequency at each of these successive maxima is recorded. If the first of these frequencies is assigned an arbitrary value of n, the number of half-wavelengths, the successive frequencies then correspond to $n + 1$, $n + 2$, $n + 3$, and so on. A graph of v_m against successive integers will therefore be a straight line whose slope is

$$\frac{V_s}{2L}$$

Rather than measure L, calibrate the apparatus with a gas for which the speed of sound is known, such as laboratory air.

MATERIALS

The apparatus shown in Diagram A.
 Cylinder of ethane, propane, or other commercially available gas.

Experiment 2. Velocity of Gas Molecules

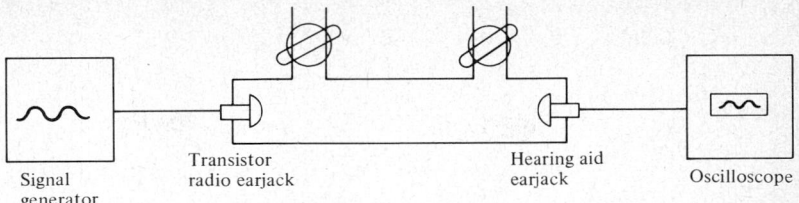

DIAGRAM A / Apparatus for measuring the speed of sound.

DIRECTIONS

Fill the tube with dry air at 1 atm pressure. Vary input frequency in the audio range until a sound is heard. Then adjust the controls of the oscilloscope or meter for proper observation of the output signal.

Decrease the frequency until it is no longer possible to distinguish maxima, adjusting the meter or oscilloscope controls as necessary. Then, slowly and carefully increase the frequency, recording each value at which a maximum in output signal is observed, until it is again no longer possible to distinguish maxima.

Replace the calibrating gas with methane, ethane, or another gas, as desired, and again determine the frequencies at which maxima are observed.

CALCULATIONS AND GRAPHS

For each gas used, plot a graph of v_m versus arbitrary successive integers. The slope of the straight line is $V_s/2L$.

Determine L from the speed of sound in the calibrating gas and use it to determine the speed of sound in the test gases.

For each gas, use equation (2) to calculate the average molecular velocity.

NOTES / Do not evacuate the cell to clear out the gas. The diaphragms of the speaker and microphone might break if subjected to atmospheric pressure on one side and a partial vacuum on the other. Instead, flush out the air or gas with the new gas. The best way to tell if you have completely removed the old gas is to take your measurements, flush the cell out again with the same gas, and repeat the measurements. If the first sample has not been mixed with some air, the results of the two consecutive samples will be the same within experimental precision.

You can establish the experimental precision with repeated runs on one sample of gas.

If you use a monatomic gas, the constant in equation (2) becomes 0.8.

EXPERIMENTS

An oscilloscope is preferred to a voltmeter because the shape of the sound wave can be seen, in addition to the amplitude.

For an inexpensive speaker, use the earjack from a transistor radio, and for an inexpensive microphone, use a hearing aid earjack.

EXERCISES

1. Compare the ratios of your observed molecular velocities with those expected from Graham's law. Can Graham's law of effusion apply to the results of this type of measurement?
2. Why should the sound wave, on the average, go slower than the molecules?
3. The accuracy of the slope of the graph may be determined to one significant figure more than any of the measurements. Explain.

Experiment 3

Viscosity of a Liquid

Ostwald Viscometry

OBJECTIVES / To determine liquid viscosities, using an Ostwald viscometer, over a series of temperatures and to calculate the activation energy for viscous flow.

READING ASSIGNMENT / Chapter 12, pages 116–118, Chapter 8, pages 75–78.

The viscosity of liquids is of great practical importance in the chemical industry, medicine, and biochemistry. Energy is needed to pump blood through the vascular system, fuel through a fuel line, and coolant through a reactor. In these cases, high viscosity is undesirable. On the other hand, for lubricants and sealers, high viscosity may be essential.

Clearly, a knowledge of how the viscosity varies over a range of temperature is often necessary. If equation (12–8) is put in semilogarithmic form,

(1) $$\ln \eta = \ln a + \frac{b}{T} = \ln a + \frac{E}{RT}$$

a, b = constants
T = absolute temperature
E = apparent activation energy flow

Measurement of viscosities at a few temperatures can furnish data for a linear graph that can be extrapolated over a range of temperatures.

In this experiment, the viscosity of a liquid is measured at several temperatures, its temperature dependence is determined, and the activation energy for viscous flow is calculated.

EQUIPMENT AND MATERIALS

The Ostwald viscometer, a pycnometer or Westphal balance, a thermostat, and a timer.

Distilled water and an organic liquid such as methanol or toluene.

DIRECTIONS

Determine the time for liquid to flow through the capillary of the viscometer. Use distilled water at 20°C as the standard. With the organic liquid, make measurements at several convenient temperatures in the range 0–50°C.

At two temperatures about 30°C apart, determine the specific gravity of the liquid.

CALCULATIONS AND GRAPHS

Calculate the viscosity of the organic liquid at each temperature relative to that of water at 20°C. Assume that the specific gravity is linear with temperature over the experimental range of temperatures.

Calculate the absolute viscosity of the liquid at each temperature and plot a graph of viscosity versus the Celsius temperature.

Plot the graph of the logarithm of viscosity versus the reciprocal of the Kelvin temperature. From the slope estimate the activation energy for viscous flow.

NOTES / Select combinations of a liquid and a temperature that will avoid the problem of rapid evaporation of the liquid.

The flow time of distilled water at 20°C must be carefully determined, since it is the reference point. The measurement should be repeated at the end of the experiment as a check.

Specific gravities may be obtained directly without calculation using the Westphal balance. However, since the temperature of the liquid must be carefully controlled during the measurements, it may prove easier to use a pycnometer.

Additional measurements at different temperatures may be necessary.

Repeat each time measurement until a precision of 0.5% is obtained. If there is marked irregularity in the flow rates, check the temperature control. Clean the viscometer again if necessary.

Repeat the specific gravity readings until the experimental precision is adequate for an overall 1% precision in the computed viscosity.

If the organic liquid is immiscible with water, rinse the viscometer with acetone and then with the organic liquid when changing from water to organic. Rinse with acetone to remove the organic liquid when changing back.

EXERCISES

1. Explain the temperature dependence of the viscosity.
2. Explain why the Ostwald method is not applicable for very fluid liquids.
3. Explain the term "activation energy for viscous flow."
4. Why does the viscosity of a gas increase with temperature while the viscosity of a liquid decreases?

Experiment 4

Vapor Pressure Versus Temperature

Ramsay–Young, Isoteniscope, or Internal Manometer Method

OBJECTIVES / To measure the vapor pressure of a liquid over a range of temperatures and to calculate the heat of vaporization.

READING ASSIGNMENT / Chapter 10, pages 100–103; Chapter 9, pages 80–82.

The variation of vapor pressure with temperature can be described by the various forms of the Clausius–Clapeyron equation. The differential form really gives only dp/dT, the slope of the p/T curve at each point. Over short temperature ranges $\Delta p/\Delta T$ can be taken as approximately equal to dp/dT. However, for any appreciable change in temperature and pressure, the integral form should be used. This gives a much better approximation to actual data and permits some extrapolation. Nevertheless, it must be remembered that even the equation for the integral form (10–1b) is obtained from the differential form (10–1a) by a reasonable assumption, that ΔH_v is constant. However, this assumption does not hold good over more than a few score degrees because

$$\Delta H_v = \Delta H_{v,\,298K} + \int_{298K}^{T} (C_{p,\,vapor} - C_{p,\,liquid})\,dT$$

and the heat capacity of the liquid and the vapor are not the same.

In this experiment, the temperature range is not too large for the integral form (10–1b) to be used.

EQUIPMENT AND MATERIALS

The Ramsay–Young apparatus, the isoteniscope apparatus, or the internal manometer apparatus.

Distilled water or a volatile liquid such as toluene, or 1-propanol, or 2-propanol.

Experiment 4. Vapor Pressure

DIRECTIONS

Obtain vapor pressure and temperature data for a sample of liquid over as wide a range of pressures and temperatures as the apparatus permits. Repeat with several other samples of the same liquid.

CALCULATIONS AND GRAPHS

Plot a graph of vapor pressure vs. Celsius temperature. Determine the slope at several temperatures and use the differential form of the Clausius–Clapeyron equation to calculate the heat of vaporization.

Plot the graph of log p vs. $1/T$, where T is the Kelvin temperature. Use the integrated form of the Clausius–Clapeyron equation to calculate the heat of vaporization.

Compare your calculated heats of vaporization with accepted values.

NOTES / Correct the manometer and barometer readings for thermal expansion of mercury only if absolute pressures are desired in order to compare the experimental results with literature values.

Volatile solutes may raise the observed pressure. Nonvolatile solutes lower the vapor pressure.

EXERCISE

1. Assume that the vapor pressure of a given sample is measured first with increasing temperatures and then with decreasing temperatures. Assume further that the results with rising temperatures differ from those with falling temperatures. Discuss possible causes of this phenomenon. Which values would you expect to be lower? Explain.

Experiment 5

X-Ray Diffraction Photographs of a Cubic Solid

Debye–Hull–Scherrer Method

OBJECTIVES / To make X-ray powder diffraction photographs of cubic solids and to use the results to calculate lattice constants.

READING ASSIGNMENT / Chapter 20, pages 258–259 and 264–268.

From X-ray data, atomic and ionic diameters can be calculated. In the unit cell of a simple cubic crystal of an element, the edge of the cube is

$$a = 2r$$

a = edge of cube
r = atomic radius

In the unit cell of a face-centered cube of an element, the diagonal of a face is

$$d_f = \sqrt{2}\,a = 4r$$

d_f = diagonal through face

In the unit cell of a body-centered cubic crystal of an element, the diagonal through the center of the body is

$$d_b = \sqrt{3}\,a = 4r$$

d_b = diagonal through body

If the cell has atoms of A at the corners and atoms of B in the center of the body, the diagonal distance through the center is

$$d_b = 2r_A + 2r_B$$

If the cell has atoms of A at the corners and an atom of B in the center of the face, the diagonal across the face is

$$d_f = 2r_A + 2r_B$$

The density of the solid can also be calculated from X-ray data, using the volume of the cell and the mass of the atoms in the cell. Since the solid consists only of a large number of unit cells, the observed solid density is the same as that of the cell. Therefore,

$$\rho = \frac{\sum n_i AW_i}{Na^3}$$

ρ = density of the solid
n_i = number of atoms of species i in the unit cell
N = Avogadro's number
AW_i = atomic weight of species i
a = edge of the cube

The lattice type of the cubical crystal is indicated by the Miller indices of the lines that appear in the X-ray powder photograph. Table 20–3 shows the values of the sum of $h^2 + k^2 + l^2$ that show up in each type of pattern. Since λ and a are constants in the relation,

$$\sin^2 \theta = \frac{\lambda^2}{4a^2} (h^2 + k^2 + l^2)$$

to index the photograph, it is convenient to calculate $\sin^2 \theta$ for each line and then from the ratios of the $\sin^2 \theta$ values to decide which lines are present and which are absent. This, of course, shows the lattice type. Once the lattice type is known, a value of $h^2 + k^2 + l^2$ can be assigned to each line and the edge of the cube can be calculated.

EQUIPMENT AND MATERIALS

An X-ray generator, powder camera, film cutter, and viewer. Film and developing supplies. Al, Cu, KCl, and NaCl powders.

DIRECTIONS

Prepare and mount a sample of powder on a fiber and photograph its X-ray diffraction pattern. After the film is developed and dry, mount the film on the viewer and locate the positions of each line to the limit of precision of the viewer.

EXPERIMENTS

CALCULATIONS AND GRAPHS

Determine the positions of the inlet and outlet beams by averaging the positions of several pairs of lines symmetrically placed around the holes for the inlet and exit tubes. (The beams are not always in the centers of the holes.)

Calculate the film shrinkage factor from the distance in millimeters between the inlet and outlet beams. For 57.3 mm cameras, the shrinkage factor is

$$\frac{90.00}{\theta_2 - \theta_1}$$

$\theta_1 =$ position of inlet beam
$\theta_2 =$ position of outlet beam

For 114.6 mm, the shrinkage factor is

$$\frac{180.0}{\theta_2 - \theta_1}$$

The distance, in millimeters, of each line from θ is multiplied by the shrinkage factor to give the corrected distance. In 57.3 mm cameras, this corrected distance, in millimeters, is θ, the diffraction angle, in degrees. In 114.6 mm cameras, it is $2 \times \theta$, in degrees.

Calculate $\sin^2 \theta$ for each line. Use Table 20-3 and the ratios of $\sin^2 \theta$ for the four or five strongest lines to determine, by trial and error, the lattice type of the powder. Then index all the observed lines.

Calculate the edge of the unit cube, using the accepted value of the wavelength.

Calculate the atomic radii or the sum of the ionic radii, depending on whether the substance studied is an element or a salt.

Calculate the density of the solid.

NOTES / *X-rays are extremely dangerous.* **Follow the safety precautions in the manufacturer's manual. Do not start the generator until the instructor is satisfied that you are familiar with the safety precautions.**

Be careful to notch one edge of the film before loading the camera so as to be able to identify the forward and backward reflections. If you forget, you may be able to locate the inlet beam by the pattern since the inlet usually has fewer lines around it than the exit beam.

Expose the film for about ½–1 hour at 35 kV and 15 mA if the small camera is used. Heavy atoms such as copper and bromine reflect better than lighter atoms like aluminum and chlorine and so need shorter exposures. With the 114.6 mm camera, the exposure time is four times that of the smaller camera. (Why?)

Experiment 5. X-Ray Diffraction Photographs of a Cubic Solid

A toothbrush bristle makes a satisfactory fiber.

Not all lines fall into the pattern. Some weak lines might be due to the fiber or to impurities.

Medical no-screen X-ray film gives good results.

The accepted value of λ for copper target tubes is 1.539 Å.

Experiment 6

Repeat Distance in a Fiber

Rotating-Crystal X-Ray Method

OBJECTIVES / To use the rotating-crystal X-ray method to take diffraction photographs of a fiber and to determine the repeat distance.

READING ASSIGNMENT / Chapter 20, pages 260–263.

In this experiment a fiber is exposed to X rays, with the beam perpendicular to the long axis of the fiber. The photograph of the diffraction pattern is similar to that in Figure 20–4. From the radius of the camera and the distance between layer lines, the repeat distance along the fiber axis may be calculated.

Since the fiber used here is actually a bundle of parallel crystals, repeat distances may be determined without rotating the fiber.

EQUIPMENT AND MATERIALS

An X-ray generator, a rotation camera, or a Weissenberg camera. X-ray film and developing supplies. A stretch-pulled nylon, polypropylene, or other polymer fiber.

DIRECTIONS

Mount the fiber on the goniometer with the long axis perpendicular to the beam. Expose the sample to radiation for about 3 hr if no heavy atoms are present, less if the fiber contains chlorine or other heavy atoms.

CALCULATIONS

The long axis of the fiber corresponds to the c axis of the unit cell. Calculate the repeat distance from the radius of the camera and the distance between the layer lines on the X-ray photograph.

Calculate the pitch of the chains from the calculated repeat distance and the chain length expected from normal bond angles and distances.

NOTES / *X-rays are extremely dangerous.* **Do not attempt to perform the experiment without the instructor present or until the instructor is satisfied that you know all the safety precautions.**

Normally powder cameras are not used for rotation or oscillation photographs since repeat distances must be at least 3 Å. However, they can be used for polymer fibers, since these do have long repeat distances.

Experiment 7

Repeat Distance in a Single Crystal

Rotating-Crystal X-Ray Method

OBJECTIVES / To use the rotating-crystal X-ray method to determine the repeat distance along the long axis of a single crystal.

READING ASSIGNMENT / Chapter 20, pages 260–263.

In this experiment a single crystal is exposed to X-rays while being rotated around an axis perpendicular to the beam. The diffraction pattern is similar to that in Figure 20–4. From the radius of the camera and the distance between layer lines, the repeat distance along the crystal axis is calculated.

If time permits, the procedure can be repeated with the crystal oriented along the other two axes. If this is done, the volume of the unit cell may be calculated. The number of formula units per cell can then be calculated from the cell volume and the density.

EQUIPMENT AND MATERIALS

An X-ray generator, 57.3 mm powder camera, film cutter, and viewer. One or more single crystals of ammonium oxalate hydrate, 35 mm X-ray film, and developing supplies.

DIRECTIONS

Mount a single crystal directly in the sample holder of the powder camera without attaching it to a glass or nylon fiber. Expose the film to X radiation for about 10–15 min while rotating or oscillating the sample.

For exposures along the short axis, glass, copper, or nylon fiber supports must be used.

Experiment 7. Repeat Distances in a Single Crystal

CALCULATIONS

The long axis of the macroscopic crystal is parallel to the c axis of the unit cell. Calculate the length of the lattice parameter c from the radius of the camera and the distance between layer lines on the X-ray photograph.

NOTES / *X-rays are extremely dangerous.* **Do not attempt to perform the experiment until the instructor is satisfied that you are familiar with the safety precautions.**

The crystal used should be inspected under polarized light with a microscope to make sure that it is perfect and has not grown around a smaller crystal with a different orientation. To prepare ammonium oxalate hydrate crystals, simply let a water solution of ammonium oxalate evaporate slowly over a period of several days.

If the axis of the crystal is not parallel to a unit cell axis, the calculated repeat distance will not be an edge of the unit cell.

Normally, powder cameras are not used for rotation or oscillation photographs since repeat distances must be at least 3 Å for the first layer line to appear on a 35 mm film. However, for crystals with long repeat distances, powder cameras may be used to obtain layer lines.

The 114.6 mm camera cannot be used because the distance between the zero and the first layer lines would be too great to appear on the 35 mm film.

Experiment 8

Molecular Weight of a Solid

Cryoscopy

OBJECTIVES / To determine the depression of the freezing point by a solute and to calculate its molecular weight.

READING ASSIGNMENT / Chapter 11, pages 105–108; Chapter 3, pages 23–24.

Cryoscopy is a rapid and inexpensive method of determining the molecular weight of materials below 500 g/mole to within 3–5%. In this experiment, the molecular weight of an organic solid is determined from the extent to which a measured weight of solid lowers the freezing point of a weighed amount of benzene.

EQUIPMENT AND MATERIALS

The Beckmann cryoscopic apparatus. Dry benzene, to be used as solvent, and a solid or liquid whose molecular weight is unknown to the student.

DIRECTIONS

Determine the freezing point of a sample of about 20–40 g of dry benzene, measured to ± 0.1 g. Use an ice–water mixture as the coolant.
 Dissolve about 0.2–0.3 g of solute, measured to ± 0.001 g, in the benzene and determine the freezing point of the solution.
 Repeat with fresh samples of solvent and solute until 2% precision is obtained for the calculated molecular weight.

CALCULATIONS AND GRAPHS

Plot cooling curves for the solvent and the solution on the same graph. Determine the freezing point by extrapolation.

Calculate the molality of the solution from the molal freezing point depression constant and the observed freezing point depression.

Calculate the molecular weight of the solute.

NOTES / The extrapolation locates the temperature at which freezing would have occurred had there been no supercooling.

If time is short, it is permissible to add a second portion of solute to the solution instead of starting over with fresh solvent.

Always determine the freezing point of each fresh sample of solvent. The solvent in the stock bottle may have been contaminated between runs, or a new bottle may have been substituted.

The Beckmann thermometer is calibrated for temperature differences, not for absolute values of temperature.

The solvent may be measured out with a pipet and the weight calculated from the volume and density. A liquid solute may also be measured volumetrically.

Benzene is very toxic. The experiment must be performed in a hood. Do not pipet by mouth.

EXERCISES

1. Comment on the purity of the solvent. Base your comments on the shape of your experimental cooling curve.

2. Is it necessary that the solvent be completely free of impurities? Explain your answers.

3. What errors are caused by the presence of too much solute?

4. If the solute is partially associated or dissociated in the solvent, how would you go about determining the monomeric molecular weight?

5. If a solid solution separated out, rather than pure solid solvent, would your experimental results be different? Explain.

Experiment 9

Cryoscopic Constant

Beckmann Method

OBJECTIVES / To determine the depression of the freezing point by a solute and to calculate the molal freezing point depression constant of the solvent.

READING ASSIGNMENT / Chapter 11, pages 105–108; Chapter 3, pages 23–24.

For very dilute solutions, the freezing point depression is

$$\Delta T_f = K_f m_2$$

K_f = molal freezing point depression constant
m_2 = molality of solute

The constant in this equation is a compound constant, whose value is

$$K_f = \frac{RT_0^2}{1000 L_f}$$

T_0 = freezing point of pure solvent, K
L_f = heat of fusion per gram of solvent

Obviously, K_f will be higher for solvents with higher freezing temperatures and lower heats of fusion.

In this experiment, measurements are made of freezing points of solutions of known concentration. Since the observed K_f varies with the solute concentration, the observed values of K_f must be plotted against m_2 and extrapolated to zero concentration.

In an optional experiment, the apparent molecular weight of a carboxylic acid such as benzoic acid is determined over a series of concentrations and the molecular weight of the monomer calculated. Carboxylic acids are chosen since they dimerize reversibly in nonpolar solvents.

Experiment 9. Cryoscopic Constant

EQUIPMENT AND MATERIALS

The Beckmann cryoscopy apparatus and a timer.

Dry benzene, reagent grade naphthalene, and benzoic acid or other solid carboxylic acid.

DIRECTIONS

Use the Beckmann apparatus and timer to determine the freezing temperature of a sample of dry benzene. Use an ice–water mixture as the coolant. Depending on the size of the apparatus, the sample of benzene is 20–40 g, measured to ± 0.1 g.

Dissolve about 0.5 g of naphthalene, weighed to ± 0.0001 g, in the benzene and determine the freezing point of the solution. Repeat with two additional portions of naphthalene, adding them successively to the solution. If there is sufficient time, repeat the procedure with a fresh batch of solvent.

If there is enough time to run the optional experiment, repeat the procedure, using a solid carboxylic acid as the solute.

CALCULATIONS AND GRAPHS

Plot cooling curves for the freezing of the solvent and the solutions, in each case determining the freezing point by a suitable extrapolation.

From the data for the naphthalene solutions, compute the cryoscopic constant for benzene at each concentration. Compare these values with the accepted value and with the value computed from the accepted freezing point and the heat of fusion of benzene.

Use your *measured* cryoscopic constants to compute the apparent molecular weight of the organic acid in the benzene solution. If your measured cryoscopic constants change consistently with concentration, decide whether it is better to use an average value or the value at a concentration corresponding to that of the organic acid. Use whichever value you think is preferable, but state your choice explicitly.

Compare the observed molecular weight to the accepted value and explain any significant discrepancy. If there is any *consistent* change with concentration, plot the apparent molecular weight against concentration, extrapolate to zero concentration, and discuss the reason for this change.

NOTES / Several cooling curves may be drawn on the same graph, provided they are readable and do not interfere with each other.

The temperature scale may be shifted for each graph if necessary.

If time permits, it is preferable to prepare fresh solutions rather than to add successive portions of solute.

An error in weighing one portion of the solute will not completely invalidate all data for solutions with additional portions, because the equation for the freezing-point depression may be applied to the change in the freezing point and the change in the concentration. However, such an error will make it impossible to extrapolate to zero concentration if desired. It is always better practice to avoid errors rather than to compensate for them.

Be cautious in drawing conclusions about the changes in the cryoscopic constant or in the observed molecular weight. Analyze your data carefully before deciding that any trend is or is not the result of experimental error.

Always determine the freezing point of each sample of solvent. The solvent in the reagent bottle may have been contaminated between runs, or a new bottle may have been substituted.

The Beckmann thermometer is calibrated for temperature differences, not for absolute values of the temperature.

Benzene is very toxic. Work in a hood. Do not pipet by mouth.

EXERCISES

1. Comment on the purity of the solvent, basing your observations on the shape of your cooling curve.
2. Discuss whether or not it is necessary to use completely pure solvent.
3. Explain why it is preferable to use your observed cryoscopic constant rather than the accepted value even if these two differ significantly.
4. If your observed cryoscopic constant varied significantly with the concentration, justify the value you chose to compute the molecular weight of the organic acid.
5. List and briefly discuss the errors, if any, caused by the presence of too much solute.
6. Tell how you would determine the monomeric molecular weight of a solute that partially dissociated or associated in the solvent.
7. Explain how, if at all, the separation of a solid solution rather than pure solid solvent would affect your results.
8. Predict the molecular weight of the organic acid in ethyl alcohol solution and in aqueous solution.
9. Predict what might happen if this experiment were tried using terephthalic acid (1,4-benzenedicarboxylic acid).

Experiment 10

Activity of a Solvent

Vapor Pressure by Ramsay–Young Method

OBJECTIVES / To determine the vapor pressure of a series of solutions of a nonvolatile solute and to calculate the boiling point elevation constant and activity coefficient of the solvent.

READING ASSIGNMENT / Chapter 10, pages 100–104.

The addition of a solute lowers the vapor pressure of a solvent as shown in Diagram B. For ideal solutions of nonvolatile solutes, *Raoult's law* is obeyed in the form

$$p_i = p_0 X_i$$

$p_0 =$ vapor pressure of pure solvent
$X_i =$ mole fraction of solvent
$p_i =$ vapor pressure of solvent in solution of mole fraction X_i

For a nonideal solution,

$$p_i = p_0 a_i = p_0 X_i \gamma_i$$

$a_i =$ activity of solvent of mole fraction X_i
$\gamma_i =$ activity coefficient of solvent of mole fraction X_i

In this experiment a family of vapor pressure–temperature curves is obtained for a solvent and a series of solutions of known mole fraction. For the pressure vs. mole fraction data along line A–A, a pressure vs. mole fraction curve is obtained from which activity and activity coefficients are calculated.

From the data along line B–B the boiling point is obtained as a function of solute mole fraction and the molal boiling point elevation constant calculated.

EXPERIMENTS

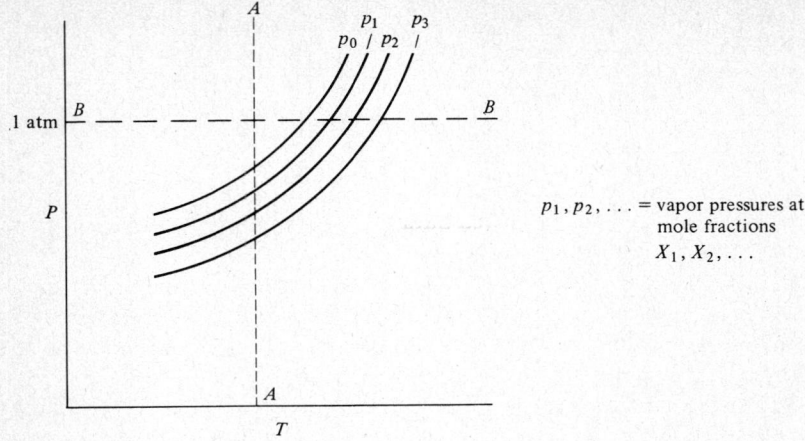

DIAGRAM / Family of vapor pressure vs. temperature curves for solvent and solutions. p_0 is the pressure of pure solvent. p_1, p_2, etc., are the pressures of the *solvent* in solutions of concentration 1, 2, etc.

EQUIPMENT AND MATERIALS

The Ramsay–Young apparatus equipped with a thermometer calibrated to 0.2°C or better.

Methanol as the solvent and either glycerol or ethylene glycol as the solute.

DIRECTIONS

Use the Ramsay–Young method to obtain vapor pressure–temperature data for the pure solvent.

Repeat the vapor pressure–temperature measurements with solutions of known mole fractions in the range 0.01–0.3 mole fraction of solute.

CALCULATIONS AND GRAPHS

On one graph plot the family of vapor pressure–temperature curves and on another graph plot the family of $\log p$ vs. $1/T$ curves.

From these graphs, obtain vapor pressure–mole fraction data for any convenient temperature and plot the graph of vapor pressure versus mole fraction of *solvent*. Compare this graph with that expected from Raoult's law and with the graph obtained from the accepted data found in the literature. Note the range over which the solution is ideal.

Experiment 10. Activity of a Solvent

From the vapor pressure and mole fraction data calculate the activity of the solvent at each mole fraction, using the relationship

$$a_i = \frac{p_i}{p_0}$$

p_0 = vapor pressure of pure solvent

Compare your results with accepted data, if any are available. Plot the graph of activity versus mole fraction of solvent.

Use the family of vapor pressure versus temperature graphs for the data needed to plot a graph of the boiling point versus the mole fraction of the *solute*. Use this graph to calculate the molal boiling point elevation constant of the *solvent*. Compare it with the accepted value.

For each solution, estimate from the graph of log p vs. $1/T$ the heat of vaporization of the solvent from the solution. If there is any systematic change in the heat of vaporization as the composition of the solution changes, suggest an explanation.

Calculate the activity coefficients of the methanol and plot the graph of activity coefficient as a function of concentration.

NOTES / Weigh out enough of the solvent and the solute to prepare about 100 ml of each solution.

It is necessary to work with mole fractions because of the temperature changes, which make molarities meaningless. Explain.

Check the boiling point of the solvent to make sure it is pure. It may be necessary to distill it.

Be careful. Both methanol and ethylene glycol are toxic.

EXERCISE

1. Explain any observed deviations from Raoult's law in terms of molecular phenomena.

Experiment 11

Mean Ionic Activity Coefficient

EMF Measurement

OBJECTIVES / To measure the potential of an electrochemical cell and to calculate the mean ionic activity coefficient of the electrolyte.

READING ASSIGNMENT / Chapter 16, pages 189–192, 194–195; Chapter 14, pages 147–150.

In this experiment, the standard potential of the $Zn|Zn^{2+}$ electrode is calculated and the mean ionic activity coefficient of $ZnCl_2$ is determined at a series of concentrations.

The emf of the cell

$$Zn|ZnCl_2||AgCl(s)|Ag$$

is

(1) $$E = E^\circ_{Ag|AgCl(s)} - E^\circ_{Zn|Zn^{2+}} - \frac{RT}{2F} \ln(a_{Zn^{2+}})(a_{Cl^-})^2$$

$$= E^\circ - \frac{RT}{2F} \ln a_{ZnCl_2}$$

The activity product, $(a_{Zn^{2+}})(a_{Cl^-})^2$, is *defined* as the activity of the electrolyte, $ZnCl_2$. In terms of molalities and activity coefficients

$$a_{ZnCl_2} \equiv (a_{Zn^{2+}})(a_{Cl^-})^2 = (m_{Zn^{2+}})(m_{Cl^-})^2(\gamma_{Zn^{2+}})(\gamma_{Cl^-})^2$$

γ = activity coefficient

The term $(\gamma_{Zn^{2+}})(\gamma_{Cl^-})^2$ is defined as the activity coefficient of the electrolyte $ZnCl_2$. Two other relevant quantities are the mean ionic activity $a_\pm$ and the mean ionic activity coefficient, $\gamma_\pm$

$$a_\pm \equiv (a_{ZnCl_2})^{1/3} = [(a_{Zn^{2+}})(a_{Cl^-})^2]^{1/3}$$
$$\gamma_\pm \equiv (\gamma_{ZnCl_2})^{1/3} = [(\gamma_{Zn^{2+}})(\gamma_{Cl^-})^2]^{1/3}$$

Experiment 11. Mean Ionic Activity Coefficient

In terms of molalities and activity coefficients,

(2) $$E = E° - \frac{RT}{2F} \ln(m_{Zn^{2+}})(m_{Cl^-})^2 - \frac{RT}{2F} \ln \lambda_\pm^3$$

If $E°$ for the cell is known, mean activity coefficients can be calculated but not, of course, individual ion activities or coefficients. $E°$ is obtained from a series of measurements of E at different molalities. If (2) is rewritten as

(3) $$E + \frac{RT}{2F} \ln(m_{Zn^{2+}})(m_{Cl^-})^2 = E° - \frac{RT}{2F} \ln \gamma_\pm^3$$

the right-hand side represents a quantity that approaches $E°$ as the concentration gets smaller and smaller, because at infinite dilution, activity coefficients become unity. Therefore, a series of measurements of E are made at decreasing molalities. A graph of $E + (RT/2F)\ln(m_{Zn^{2+}})(m_{Cl^-})^2$ vs. molality is plotted and extrapolated to zero molality. The extrapolated value of the ordinate is $E°$. Once $E°$ is known, from the measured emf at each molality, the mean ionic activity coefficient can be calculated.

EQUIPMENT AND MATERIALS

A potentiometer, a thermostat, and a 150 ml beaker with Teflon or rubber cap or a 150 ml three-neck flask.

A zinc chloride electrode, a silver wire, and a saturated calomel or a standard silver/silver chloride reference electrode.

Saturated potassium chloride solution, 1.00 m $ZnCl_2$, concentrated HCl, and 1% mercuric chloride solution.

DIRECTIONS

Prepare a silver|silver chloride wire electrode by anodizing a silver wire in concentrated HCl solution. Connect the silver wire to the positive terminal of a battery or DC power supply and connect a copper wire to the negative terminal. Electrolyze in concentrated HCl until a visible layer of silver chloride has formed on the wire. Calibrate the electrode in saturated KCl solution against either a saturated calomel electrode or a standard Ag|AgCl(s) electrode.

Amalgamate the zinc electrode by dipping it into dilute $HgCl_2$ solution for a few minutes. Remove and rub it with filter paper until all oxides have been removed and the surface is shiny. (Be careful! *Mercuric chloride is extremely poisonous if taken internally.*)

Set up the cell, making provision for removing dissolved air.

Measure the emf of the cell Zn|$ZnCl_2$ ‖ AgCl(s)|Ag at several different accurately known molalities between 1.00 and 0.00100 m.

CALCULATIONS AND GRAPHS

Add to or subtract from each measured emf the difference between the potential of the standard $Ag|AgCl(s)$ electrode and the measured potential of your electrode; for example, if your electrode is 10 mV negative to the accepted value, add +10 mV to each measured emf. In calculating the correction use the $E°$ for a reference electrode with Cl^- at unit activity.

Plot a suitable graph from which $E°$ for the cell can be found by extrapolation to infinite dilution.

From the observed $E°$ for the cell find $E°$ for the $Zn|Zn^{2+}$ half-cell. Compare this to the accepted value.

Using the observed emf's and the extrapolated $E°$ for the cell, compute the mean ionic activity coefficient and the mean ionic activity for each concentration of $ZnCl_2$ studied. Compare the observed mean ionic activity coefficients with accepted values and with values calculated from the Debye–Hückel limiting law. Comment on the range of applicability of the limiting law to solutions of $ZnCl_2$.

NOTES / The emf of the $Ag|AgCl_{(s)}$ wire electrode varies slightly with the conditions of preparation and with aging. The electrode should be calibrated when first used and the calibration should be checked several times during the laboratory period. The calibration is done against a standard in a saturated KCl solution, to eliminate the liquid junction potential.

The potential of the amalgamated electrode is the same as that of pure zinc.

A commercial silver chloride electrode cannot be used in the actual measurement because the commercial electrode has a salt bridge and the potential measured therefore includes a liquid junction potential, which is an unknown function of concentration of the solution.

Control the cell temperature to within 1°.

Do not let the electrodes touch each other.

Twisting the silver wires produces strains that affect the potential.

Do not record the emf until the potential has stopped drifting steadily and has begun to fluctuate randomly.

Prepare the solution by diluting the stock 1.00 m $ZnCl_2$ gravimetrically, weighing out the samples.

Add 1 or 2 drops of concentrated HCl to the $ZnCl_2$ stock solution to prevent hydrolysis.

The smaller the mass of the cell and contents, the shorter the time needed for it to reach the thermostat temperature.

Experiment 11. Mean Ionic Activity Coefficient

EXERCISES

1. Explain why the $E°$ of the reference electrode with Cl^- ion at unit activity is used to calculate the corrections for the $Ag\,|\,AgCl(s)$ electrode, even though the measurement is actually made in saturated KCl.
2. What is the activity of the zinc in the saturated zinc amalgam on the surface of the zinc electrode?

Experiment 12

Transference, Ionic Conductance, and Ionic Mobility

Conductance, Moving-Boundary Method

OBJECTIVES / To determine the conductance of a solution, by a resistance measurement, using an impedance bridge, and to determine the cation transference number by the moving-boundary method. From the data, the cationic mobility is calculated.

READING ASSIGNMENT / Chapter 16, pages 200–206.

In this experiment, the mobility and conductance of the ions of HCl are determined as a function of concentration.

The mobility of an ion is defined as its velocity in the direction of an electric field of unit strength.

(1) $$u_i \equiv \frac{\text{velocity}}{\text{field strength}}$$

Rather than attempt to measure velocity and field strength directly, u_i is calculated from the relationship

(2) $$u_i = \frac{\Lambda_i}{F}$$

Λ_i = equivalent ionic conductance of species i
F = the faraday

The equivalent ionic conductance, in turn, is obtained from the equivalent conductance of the electrolyte and the transference number of the ion.

(3) $$\Lambda_i = \Lambda t_i$$

t_i = transference number of species i

Experiment 12. Transference, Ionic Conductance, and Ionic Mobility

Λ, the equivalent conductance of the electrolyte, is obtained from the specific conductance by equation (16–18)

$$(4) \qquad \Lambda = \frac{1000\kappa}{c}$$

κ = specific conductance
c = equivalents per liter

The transference number is defined as the fraction of the current carried by a particular ion.

In this experiment, both the specific conductance of the electrolyte and the transference number of the cation are measured, the latter by the moving boundary method. From the results, Λ, Λ_i, and u_i are calculated.

EQUIPMENT AND MATERIALS

The moving-boundary apparatus with a 200 V, 5 mA DC power supply, a cadmium anode, and a silver/silver chloride wire cathode. A potentiometer, an electric timer, and a 100-ohm precision resistor. An impedance bridge, a conductance cell, and a thermostat.

Concentrated HCl stock solution, methyl violet indicator, and a 0.0200 M KCl reference solution.

DIRECTIONS

Measure the resistance of the 0.0200 M KCl reference solution.

Prepare and measure the resistances of several different HCl solutions in the 0.005–0.5 M range, with each concentration known to three significant figures.

On separate portions of each of the HCl solutions, measure the velocity at which the interface moves at constant current in the moving-boundary apparatus.

CALCULATIONS AND GRAPHS

Calculate the cell constant from the resistance of the reference solution.

From the cell constant and the observed resistance calculate the specific conductance and the equivalent conductance of each HCl solution.

From the velocity of the interface at constant current in the moving boundary apparatus, calculate the ionic transference numbers at each concentration.

Calculate the equivalent ionic conductances at each concentration. If the data are sufficient, plot the equivalent ionic conductance against the square root of concentration and extrapolate to zero concentration to obtain the limiting equivalent ionic conductance.

Calculate the mobilities of the ions at each concentration and plot the graph of this function.

Plot the graph of cation transference number versus concentration. Compare this with the graph of mobility versus concentration.

NOTES / To measure the transference numbers, use currents of 1–5 mA, as determined by the ohmic drop across the resistor. Control the current if necessary with the aid of an additional rheostat in series with the power supply.

Use enough methyl violet to impart a barely visible color to the liquid in the capillary of the moving boundary apparatus. Add the indicator before or during dilution, not afterward.

Make all measurements at 25 °C.

EXERCISES

1. Compare the cation and anion mobilities. Explain the difference in terms of ionic phenomena.

2. Does the transference number of the hydronium ion increase or decrease on dilution? Does the velocity change markedly? Compare the trend for the hydronium ion with that for the chloride ion and explain any differences.

Experiment 13

Energies and Heats of Combusion and Formation

Constant Volume Calorimetry

OBJECTIVES / To determine the energy of combustion of an organic substance, using an oxygen bomb calorimeter, and to calculate the enthalpy of formation.

READING ASSIGNMENT / Chapter 7, pages 63–67 and 72–74.

In this experiment a known weight of a solid is burned at constant volume (in a bomb) in an atmosphere of about 20 atm of oxygen. The molar energy of combustion, ΔE_c, is calculated from the heat capacity of the system and the temperature rise of the water in the calorimeter (and of course, the molecular weight of the substance burned). From ΔE_c, ΔH_c, the enthalpy of combustion, is calculated; and from that, ΔH_f, the enthalpy of formation of the solid, is obtained.

EQUIPMENT AND MATERIALS

The Parr oxygen bomb calorimeter. If a different instrument is used, follow the manufacturer's instructions for calibrating and using the calorimeter.

Combustion standard crystalline benzoic acid and sucrose, diphenyl, or naphthalene.

DIRECTIONS

Measure the temperature rise of the calorimeter on burning a 1 g sample of benzoic acid, either adiabatically or with an isothermal jacket.

Determine the temperature rise on burning 1 g samples of the solid being studied.

CALCULATIONS AND GRAPHS

From the temperature rise and the known ΔE_c of the benzoic acid, calculate C_v, the heat capacity of the system.

From C_v and the temperature rise on combustion of the solid being studied calculate ΔE_c at the calorimeter temperature.

Calculate ΔE_c at 25°C from Kirchhoff's law using the accepted heat capacities of reactants and products as found in the literature. If the accepted values are not available, use Kopp's rule, based on the law of Dulong and Petit.

Calculate ΔH_c, the enthalpy of combustion, at 25°C.

From the enthalpies of formation of CO_2 and H_2O, calculate ΔH_f for solids being investigated at 25°C. Compare this with the accepted value.

NOTES / Weigh the samples to ±0.5 mg.

In making your calculations, note that the water formed by the combustion is in the liquid phase.

If there are unburned carbon particles on the walls of the bomb or in the sample pan, discard the run.

EXERCISES

1. Why is 1 ml of water added to the bomb?
2. Assume that impure naphthalene containing 2% of anthracene by weight was burned in the calorimeter. Compute the resulting error in the heat of combustion per mole.

Experiment 14

Resonance Stabilization Energy of o-Phthalic Anhydride

Constant Volume Calorimetry

OBJECTIVES / To determine the energies of combustion of phthalic anhydride and tetra- and hexahydrophthalic anhydrides by combustion in an oxygen bomb calorimeter. From the results, the enthalpies of formation and the resonance stabilization are calculated.

READING ASSIGNMENT / Chapter 7, pages 63–67 and 72–74.

The difference between the heats of formation of tetrahydrophthalic anhydride, $C_8H_8O_3$, and of hexahydrophthalic anhydride, $C_8H_{10}O_3$, is the heat of formation of a double bond. If the aromatic nucleus contained three double bonds, the difference between the heats of formation of o-phthalic anhydride and of hexahydrophthalic anhydride would be three times the heat of formation of a double bond. Actually, the energy of one o-phthalic anhydride is less than that expected from three double bonds, the difference being the resonance stabilization energy.

In this experiment, weighed samples of o-phthalic anhydride and of the hexa- and tetrahydrophthalic anhydrides are burned with oxygen in a bomb calorimeter. From the heat capacity of the system and the temperature rise due to the combustion, the molar energy of combustion is calculated for each material, and the enthalpies of combustion and of formation calculated. These values, in turn, are used to estimate the resonance stabilization energy.

The heat capacity of the system is obtained from the combustion of a standard substance, in this case benzoic acid.

EQUIPMENT AND MATERIALS

The Parr oxygen bomb calorimeter. If a different instrument is used, follow the manufacturer's instructions for calibrating and using the calorimeter.

Combustion standard crystalline benzoic acid and crystalline *o*-phthalic, tetrahydrophthalic and hexahydrophthalic anhydrides.

DIRECTIONS

Use a 1 g sample of crystalline benzoic acid for the calibration. Then measure the temperature rise of the calorimeter on combustion of 1 g samples of each of the other solids.

CALCULATIONS AND GRAPHS

From the temperature rise and the known ΔE_c of the benzoic acid calculate C_v, the heat capacity of the system.

From C_v, and the temperature rise on combustion of the solid anhydrides, calculate ΔE_c for each anhydride at the calorimeter temperature.

Calculate ΔE_c at 25°C from Kirchhoff's law, using the accepted heat capacities of reactants and products as found in the literature. If these values are not available, follow Kopp's rule and assume that each atom in the formula of the solid contributes 25 J/mole-deg to the heat capacity.

Calculate the enthalpies of combustion and of formation of each anhydride at 25°C.

From the enthalpies of formation, calculate the resonance stabilization energy of the aromatic nucleus.

NOTES / Weigh the samples to ±0.5 mg.

In making your calculations, note that the water formed by the combustion is in the liquid phase.

If there are unburned carbon particles on the walls of the bomb or in the sample plan, discard the run.

EXERCISES

1. Why is 1 ml of water added to the bomb?

2. Compare your calculated value of the resonance stabilization energy with the 205 kJ/mole accepted for benzene. Suggest an explanation for any difference.

Experiment 15

Heat of Neutralization and Dilution

Constant Pressure Calorimetry

OBJECTIVES / To determine the heat evolved in a neutralization reaction in an atmospheric pressure reaction calorimeter and to calculate the molar enthalpy of neutralization.

READING ASSIGNMENT / Chapter 7, pages 65–72.

In this experiment, the heats evolved by an acid–base reaction and subsequent dilution are determined and the heat of neutralization calculated.

EQUIPMENT AND MATERIALS

A styrofoam calorimeter equipped with a Beckmann thermometer in the range 22–28°C, a 50° thermometer, graduated to ± 0.1°C, a timer, and several Pasteur pipets.
    Stock solutions of 1 M HCl, 8 M NH$_3$, standardized 1 M NaOH, and 0.1 M HCl. Phenolphthalein and methyl red indicators.

DIRECTIONS

Determine the heat capacity of the calorimeter.
    Measure the temperature change when a sample of 8 M NH$_3$ is added to about 100 ml of 1 M HCl.
    After the reaction is complete, weigh the container and contents, withdraw an aliquot of about 20 ml of the reaction mixture, weigh it, and titrate it with standard 1 M NaOH to the methyl orange end point. Obtain the weight of the aliquot, either by the difference in the weight of the calorimeter and contents or by running it into a previously weighed vessel and determining the weight increase.

Titrate a weighed sample of the original 1 M HCl.

Use the same overall procedure to determine the heat of dilution of the same amount of 8 M NH_3 in the same amount of water as in the neutralization. In this case, titrate the weighed aliquot of calorimeter contents with standard 0.1 M HCl, to the methyl red endpoint.

Repeat each part of the procedure until a precision of about 2–3% is attained.

CALCULATIONS AND GRAPHS

The heat observed in the neutralization reaction is the sum of the heat of dilution of the ammonia and the heat of neutralization of the diluted ammonia. To obtain the heat of neutralization, subtract the heat of dilution of ammonia in water, obtained from the second part of the experiment.

For each procedure—heat capacity measurement, heat of reaction, and heat of dilution—construct temperature–time graphs and use the extrapolation method to calculate the temperature change, corrected for heat transfer.

The heat capacity of the system is the sum of the heat capacity of the calorimeter and that of the solution it contains. For the precision of this experiment, the specific heat of the solutions can be assumed to equal that of the water.

$$Q = \Delta T C_s$$

Q = heat evolved
C_s = heat capacity of the system (i.e., heat capacity of calorimeter and solution)

To calculate the molar heat of reaction and of solution, divide the heat evolved by the number of moles of ammonia used.

In the neutralization reaction, the number of moles of ammonia that has been neutralized is the same as the number of moles of HCl consumed. The original quantity of HCl is obtained from the original weight of the HCl solution and the results of the titration of a weighed sample. The final amount of HCl is obtained from the final weight of solution and the results of the titration of the aliquot.

In the dilution experiment, the number of moles of ammonia diluted is calculated from the results of the titration of the aliquot with 0.1 M HCl.

NOTES / To obtain 2–3% precision, a certain amount of digital dexterity is needed. It is a good idea to practice the injection procedure beforehand, using water, rather than reagents.

If you want to know the concentration of the 8 M NH_3 to better than 1 significant figure, pipet 10 ml into a volumetric flask and dilute to 250 ml. Then, titrate an aliquot with standard 0.1 M HCl.

EXERCISES

1. Would you expect much difference between the heat of neutralization of 8 M NH_3 and that of 0.1 M? Explain or justify your answer.

Reference

Condon, F. E., R. T. Reece, D. G. Shapiro, D. C. Thakkar, and T. B. Goldstein, Influence of hydration on base strength. Pt. V. Hydrazines and oxamines, *J. Chem. Soc. Perkin Trans.* **II**, 1112–21 (1974).

Experiment 16

Standard Free Energy, Enthalpy, and Entropy of Reaction

EMF Measurement

OBJECTIVES / To measure the emf of an electrochemical cell over a range of temperatures and to calculate the entropy, enthalpy, and free energy of the reaction.

READING ASSIGNMENT / Chapter 14, pages 147–150; Chapter 16, pages 189–193 and 195.

In this experiment, the emf of the electrochemical cell

$$Pb(s)|PbCl_2(s)|HCl|AgCl(s)|Ag(s)$$

is measured over a range of temperatures. The cell reaction is

$$Pb(s) + 2\ AgCl(s) \rightleftharpoons PbCl_2(s) + 2\ Ag(s)$$

Since reactants and products are all in the standard state (i.e., solids) the observed emf is simply $E°$, the standard potential.

From the Gibbs–Helmholtz equation,

$$\left(\frac{\partial \Delta G}{\partial T}\right)_p = -\Delta S$$

and the fundamental relationship

$$\Delta G = -nFE$$

one obtains the relationship

$$\left(\frac{\partial E}{\partial T}\right)_p = \frac{\Delta S}{nF}$$

Therefore, from the emf and its temperature coefficient, one can easily calculate $\Delta G°$, $\Delta S°$, and $\Delta H°$ for the reaction.

Experiment 16. Standard Free Energy, Enthalpy, and Entropy of Reaction

EQUIPMENT AND MATERIALS

A potentiometer, a 150 ml three-neck flask, or a 20 ml vial with a plastic cap. An adjustable thermostat.

A saturated calomel electrode, a silver wire, and a lead rod or strip.

Concentrated HCl, 1% $HgCl_2$ solution, crystalline $PbCl_2$ or $Pb(NO_3)_2$, and crystalline KCl.

DIRECTIONS

Prepare the Ag|AgCl(s) electrode by anodizing the silver wire in concentrated HCl solution. Connect the wire to the positive terminal of a battery or DC power supply. A copper wire connected to the negative terminal makes a good cathode. Electrolyze the HCl solution until a deposit of silver chloride is formed on the silver wire. Calibrate the electrode in saturated KCl against the saturated calomel electrode.

Amalgamate the lead electrode by dipping it into a 1% $HgCl_2$ solution for about 5 min. Rub the amalgamated electrode with filter paper until all the oxide has been removed and the electrode is shiny. (Be careful!. *Mercury chloride is very poisonous if taken internally.*)

Set up the cell with the electrodes mounted in fixed position, dipping into dilute HCl, with a small excess of solid $PbCl_2$ present.

Determine the emf at several convenient temperatures over the range from 0°C to about 30°C. Obtain sufficient data for a temperature–emf graph.

CALCULATIONS AND GRAPHS

Calculate the empirical correction to be added to or subtracted from the measured emf's to compensate for the fact that, owing to the method of preparation, the Ag|AgCl(s) electrode will probably not have the exact standard potential. In calculating this correction use the $E°$ of the Hg|Hg_2Cl_2(s) electrode, that is, the standard potential for the electrode with Cl^- at unit activity. Do *not* use the $E°$ for the saturated calomel electrode, although this is the electrode used in the actual measurement.

The potential of the amalgamated lead electrode is 6 mV more negative than that of an electrode that has not been amalgamated.

Plot the $E°$ of the cell reaction against the temperature and from the graph evaluate $(\partial E°/\partial T)_p$ and calculate $(\partial \Delta G°/\partial T)_p$.

Use the Gibbs–Helmholtz equation to calculate $\Delta S°$ and $\Delta H°$ at 25°C.

NOTES / Check the standardization of the Ag|AgCl$_{(s)}$ electrode from time to time. The emf of a freshly prepared electrode may fluctuate somewhat.

Record the emf only after it has stopped drifting.

EXPERIMENTS

Make sure the solution is saturated with $PbCl_2$; otherwise there will be a different reaction, with a different temperature coefficient. This may be done most easily by starting with the higher temperature measurements and working down. It is better to saturate the solution with $PbCl_2$ by adding a small amount of concentrated $Pb(NO_3)_2$ solution to the HCl than to try to dissolve the $PbCl_2$ directly.

Control the temperature to within $0.2°C$.

EXERCISES

1. Why should the emf be independent of HCl concentration?

2. If the solution were not saturated, would $(\partial E/\partial T)_p$ still be equal to $(\partial E°/\partial T)_p$? Explain.

3. $PbCl_2$ dissolves slowly. The presence of solid $PbCl_2$ may indicate either saturation or an unsaturated solution with the solid not yet dissolved. Explain how to distinguish between these two cases.

4. In this experiment why are the calculated values of the $\Delta H°$ more accurate than those of the $\Delta S°$? Explain.

5. Explain from molecular considerations why $\Delta S°$ could be expected to be small for this type of reaction.

6. Explain why the standard potential of the $Hg\,|\,Hg_2Cl_2(s)$ electrode at unit Cl^- activity is used for the calibration of the $Ag\,|\,AgCl(s)$ electrode, although the measurement is made with a saturated calomel electrode in saturated KCl. Refer back to the equations for the electrode reactions.

Experiment 17

Partial Molar Volume

Pycnometry

OBJECTIVES / To determine, accurately, the densities of a series of solutions and to calculate the molar volumes of the components.

READING ASSIGNMENT / Chapter 8, pages 75–77.

Since the environment of a substance in a solution is not the same as its environment in the pure state, properties such as its volume, entropy, and enthalpy are functions of the composition of the solution. In a solution, the "true volume" of a mole of component cannot be measured and hence has no meaning. By definition, this quantity is taken to be its *partial molar volume*, given by the expression

(1) $$\tilde{V}_1 = \left(\frac{\partial V}{\partial n_1}\right)_{T,p}$$

V = volume of solution
n = number of moles
1, 2 refer to species 1 and 2

In this experiment, mixtures of methanol and water are prepared and the densities measured. From this, the molar volume of the solutions are calculated and by a graphical method, the molar volumes of the components are calculated.

EQUIPMENT AND MATERIALS

Six or more 50 ml glass-stoppered flasks and several 10 ml Weld pycnometers.
 Distilled water and anhydrous methyl alcohol.

DIRECTIONS

Prepare mixtures of methyl alcohol and water of accurately known composition by weight. Suggested mole fractions of methyl alcohol are 0.01, 0.02, 0.03, 0.05, 0.07, 0.10, 0.15, 0.2, 0.3, 0.4, 0.6, and 0.8.

Using the Weld pycnometer, determine the density at 30°C of pure water, of pure methyl alcohol, and of each mixture.

CALCULATIONS AND GRAPHS

Compute the mole fraction of water and of methyl alcohol in each mixture.

Compute the number average molecular weight of each mixture using the definition.

(2) $$\overline{M} = M_1 X_1 + M_2 X_2$$

$\overline{M}$ = number average molecular weight
M_1, M_2 = molecular weights of the two components
X_1, X_2 = mole fractions of the two components

Calculate the molar volume of each liquid, including the pure components from the density and the molecular weight or the number average molecular weight.

Plot a graph of the molar volume as a function of the mole fraction of water, covering the entire range from 0 to 1 mole fraction.

Draw tangents to the graph at the two ends and at intervals of 0.1 mole fraction. The intercept of the tangent at 0 mole fraction of water (pure methyl alcohol) is the partial molar volume of the alcohol. The intercept at 1 mole fraction of water is the partial molar volume of the water. (This method of calculation is the "method of intercepts.")

For each liquid plot the partial molar volume vs. the mole fraction.

Check your results at several mole fractions, such as 0.2, 0.6, and 0.8, by comparing your measured molar volume with the molar volume computed from the equation

(3) $$\tilde{V} = \tilde{V}_1 X_1 + \tilde{V}_2 X_2$$

$\tilde{V}$ = molar volume of the mixture
$\tilde{V}_1, \tilde{V}_2$ = partial molar volumes of the two components

Draw a graph of the molar volume as a function of composition for the range 0.9–1 mole fraction of water. Find the partial molar volumes of water and methyl alcohol at intervals of 0.02 mole fraction, using some suitable algebraic method to extrapolate the tangent to 0 mole fraction water. Plot these partial molar volumes as a function of composition.

The Gibbs–Duhem equation applied to partial molar volumes of two-component solutions is

(4) $$-X_1 \left(\frac{\partial \tilde{V}_1}{\partial X_1} \right)_{T,p} = X_2 \left(\frac{\partial \tilde{V}_2}{\partial X_2} \right)_{T,p}$$

At some typical composition, such as 0.4 or 0.6 mole fraction of water, test your results to see whether they are consistent with this relationship.

NOTES / The mole fractions suggested in the directions can be approximated by, respectively, 45 ml of water + 1 ml of alcohol, 45 ml of water + 2 ml of alcohol, 30 ml of water + 2 ml of alcohol, 25 ml of water + 3 ml of alcohol, 30 ml of water + 5 ml of alcohol, 20 ml of water + 5 ml of alcohol, 25 ml of water + 10 ml of alcohol, 18 ml of water + 10 ml of alcohol, 15 ml of water + 15 ml of alcohol, 10 ml of water + 15 ml of alcohol, 6 ml of water + 20 ml of alcohol, and 3 ml of water + 25 ml of alcohol.

To prepare the mixtures, pipet the desired volume of the liquid of the smaller mole fraction into a clean, dry, weighed flask and weigh the flask and liquid. Then pipet the required volume of the other liquid into the flask. It is not necessary to weigh it unless the mole fractions are about equal. (Why not?) Try to achieve a precision of three significant figures (0.3%) in the mole fraction of the minor constituent.

Weigh the pycnometer and contents to ±1 mg.

In drawing the graphs, connect the points with a smooth curve, but be careful not to smooth out any significant changes of slope or inflection points, especially in the region 0.8–1 mole fraction of water.

There will be many weighings. Budget your time wisely, weighing out some liquids while others are coming to temperature. If you work in pairs, one person can weigh while the other fills the flasks and pycnometers, or each person can work with his own solution, handling the entire procedure.

Use pure water as the reference to calibrate the pycnometer. From Smithsonian tables the specific volume of water at 30°C is 1.00534 ml/gram.

The Gibbs–Duhem equation is a severe test of your experimental work, because not only must the partial molar volumes, which depend on the first derivative of the molar volume with respect to mole fraction, be precise, but the derivatives of the partial molar volume, which depend on the second derivative of the molar volume, must also be precise.

EXERCISES

1. Explain why partial molar volumes vary with solution composition.
2. Predict the dependence of the molar volume and the partial molar volumes on the concentration for a solution whose volume is equal to the sum of the volumes of the components.
3. Explain the sharp change in the partial molar volume of water at low alcohol content.

Experiment 18

Liquid–Solid Equilibrium

Thermal Analysis

OBJECTIVES / To obtain temperature–time data for a mixture of two molten metals and to construct the phase diagram.

READING ASSIGNMENT / Chapter 6, pages 58–60; Chapter 11, pages 108–109.

In this experiment mixtures of tin and lead are melted and cooled and temperature–time data are obtained. From these data, cooling curves are constructed. From the cooling curves, a phase diagram of the system is constructed.

EQUIPMENT AND MATERIALS

A thermocouple, a thermocouple potentiometer or a potentiometric recorder, a furnace, a refractory chamber, and a Dewar flask.
   Graphite crucibles containing bismuth, cadmium, lead, tin, and lead–tin mixtures of known composition.

DIRECTIONS

Calibrate the thermocouple, using the boiling point of water and the melting points of bismuth and cadmium as the calibration temperatures. Use an ice–water mixture in the Dewar flask for the reference temperature, if the thermocouple potential is being measured with a potentiometer.
   Melt each solid sample in the furnace and let it cool in the refractory chamber.
   Measure the thermocouple potential at regular intervals until the emf has dropped to a value corresponding to about 150°C. If a recording potentiometer is used, obtain a continuous emf–time curve.

GRAPHS AND CALCULATIONS

Plot an emf–temperature calibration curve. The formula is $E = at + bt^2$ (where t is in °C). Obtain a and b by plotting the graph of E/t vs. t for the calibration data.

Plot emf–time cooling curves for each mixture.

From the emf's at each break point and halt on each cooling curve determine the temperatures corresponding to phase changes, using the emf–temperature curve.

From the temperatures of the phase changes, construct the temperature–composition graph for the condensed system. Use mole fraction as the abscissa and temperature as the ordinate.

From the slopes of the phase diagram at each experimental mole fraction calculate the differential heat of solution of the solids using the equation

$$\frac{d \ln X_A}{dT} = -\frac{\Delta \overline{H}_{soln}}{RT^2}$$

X_A = mole fraction of component A
$\Delta \overline{H}_{soln}$ = differential molar heat of solution for a saturated solution

NOTES / Use Vycor or pyrex guard tubes on the thermocouple junctions. Never insert a junction into the molten metal or the ice–water without a guard tube.

Keep the metal surface covered with carbon. Why?

Use wet shaved ice and a small quantity of distilled water in the ice–water bath.

Because tin and lead form a limited series of solid solutions in each other, the differential heat of solution observed here is not the heat of solution of the pure metal. However, the difference between the two values is within the range of experimental errors in student experiments.

The constants in the expression $E = at + bt^2$ are overspecified in this experiment, since you determine two constants with three temperatures.

Metal vapors, especially lead, are highly toxic.

EXERCISES

1. Explain the significance of each portion of a typical cooling curve.
2. Interpret the experimental phase diagram in terms of solubilities and the presence or absence of compounds and solid solutions.
3. For solutions containing roughly equal amounts of both components, the first break in the curve will be difficult to determine accurately, but the

eutectic temperature will be accurately determined. For solutions that have very little of one component, the situation is reversed. The first break will be accurately measured, but the second will not. Explain.

4. Compare your observed heats of solution with the heats of solution that would be expected if the tin and lead formed an ideal system with no solid solutions.

Experiment 19

Phase Diagram of a Binary Liquid-Vapor System

Choppin–Cottrell Method

OBJECTIVES / To obtain small samples of liquid and vapor from a boiling two-component mixture and to analyze these by refractive index. The results are applied to constructing the boiling point–composition curve.

READING ASSIGNMENT / Chapter 18, pages 221–224.

In this experiment, the phase diagram of a two-component liquid–vapor system is obtained experimentally and compared with the theoretical diagram based on Raoult's law.

The theoretical phase diagram is calculated from vapor pressure vs. temperature data for the two pure liquids. The data are obtained, using the Ramsay–Young apparatus, and the results plotted in the form $\log p$ vs. $1/T$. These graphs are linear and can be extrapolated to pressures above atmospheric. At each of several arbitrary temperatures between the boiling temperatures of the pure liquids, p_0 is estimated for each pure liquid. Then, at each of these arbitrary temperatures, the p_0's are used with *Raoult's law*, in the form

$$p = p_{0,A} X_A + p_{0,B}(1 - X_A)$$

X_A = mole fraction of component A in the mixture that would boil

to calculate the composition of the mixture boiling at the arbitrary temperature. The boiling temperature and X_A are then used in a graph of the liquidus portion of the ideal phase diagram. *Dalton's law*, in the form

$$X_{A,\text{vapor}} = \frac{p_A}{p} = \frac{p_{0,A} X_{A,\text{liquid}}}{p}$$

is used to calculate the vaporus portion of the ideal phase diagram.

The data for the experimental phase diagram are obtained by analyzing samples of liquid and of vapor in equilibrium at the boiling temperature of

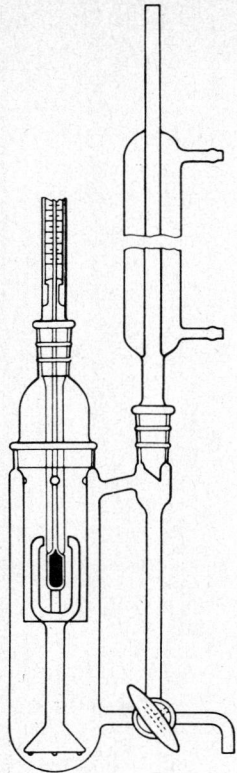

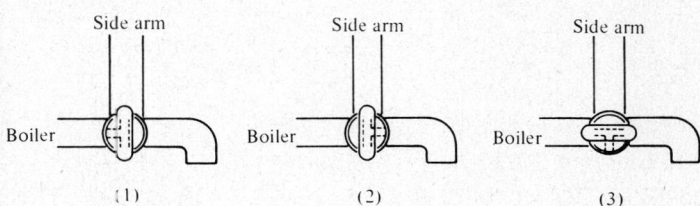

DIAGRAM C / Choppin–Cottrell still.

arbitrary mixtures of the components. The samples are analyzed by refractometry, comparing the observed refractive indexes of the mixtures with a calibration curve of refractive index versus composition. The samples analyzed are obtained by boiling the mixtures in a Choppin–Cottrell equilibrium still and sampling the liquid and vapor at the steady state.

Experiment 19. Phase Diagram of a Binary Liquid–Vapor System

In the Choppin–Cottrell equilibrium still (Diagram C) the vapor and boiling liquid are not initially in equilibrium, since the condensed vapor flows into the side arm and not directly back into the boiler. Consequently, the composition of boiler liquid and of condensate gradually changes. However, as more and more liquid is condensed and flows back through the side arm and stopcock into the boiler, a steady state is soon reached in which the escaping vapor has the same composition as the reentering liquid. Under these conditions, liquid and vapor are in equilibrium, and the side arm liquid has the same composition as the condensing vapor.

The three-way stopcock is manipulated to deliver first a sample of condensate and then a sample of boiler liquid. With the stopcock in position 1 (Diagram C) liquid circulates from the condenser to the boiler, with the stopcock in position 2 the side arm drains, and with the stopcock in position 3 the boiler drains.

OPERATING INSTRUCTIONS FOR THE CHOPPIN–COTTRELL STILL

Boil the liquid slowly, with the stopcock in position 1, until the vapor temperature remains constant for at least 5 min or fluctuates randomly over a narrow range.

Drain about 0.5 ml of side arm liquid. Discard the first few drops and collect the next 5–10 drops in a sample tube. Cork the tube immediately to prevent evaporation.

Drain about 0.5 ml of boiler liquid. Again discard the first few drops, catch the next 5–10 drops in a tube, and cork the tube.

Set the stopcock back in position 1 and continue boiling.

NOTES / A gas flame gives better control, but an electric heater must be used for flammable liquids. Always use an asbestos pad under the boiler for uniform heating.

Boil gently. To minimize superheating use a Cottrell pump and boiling chips.

Do not grease the stopcock and joints unless you have checked and found the grease to be insoluble in the boiler liquid. It is preferable to use a still with a Teflon stopcock, or, if none is available, to lubricate the joints with the boiler liquid itself.

EQUIPMENT AND MATERIALS

The Choppin–Cottrell equilibrium still, the Ramsay–Young vapor pressure apparatus, and the Abbe refractometer.

A pair of miscible liquids selected from methanol, 1-propanol, 2-propanol, carbon tetrachloride, chloroform, benzene, and water.

DIRECTIONS

Measure the refractive index of each component and of several mixtures of known composition.

Boil each component and some arbitrary mixtures of the two. Record the boiling temperature and use the Choppin–Cottrell still to obtain samples of liquid and vapor at equilibrium. Start with pure component A and add an arbitrary amount of B through the condenser and side arm after each pair of samples has been taken. Then boil and sample the new mixture and add another increment of B. Repeat until the boiler liquid is about 50% A, then empty the still and start the process over with pure B.

Determine the refractive index of each sample of condensed vapor and boiler liquid.

Use the Ramsay–Young apparatus to obtain vapor pressure–temperature data for two pure liquids.

CALCULATIONS AND GRAPHS

Construct a graph of refractive index vs. composition. Use this graph to analyze the samples taken from the boiling mixtures.

From the boiling temperatures and the experimentally determined compositions of the liquid and vapor samples, construct a graph of boiling temperature vs. composition for the mixture.

For purposes of comparison draw on the same graph the ideal boiling point vs. composition curve, calculated as discussed above. (If there is not enough time to obtain the vapor pressure data for the pure liquids, use the handbook values of boiling points and heats of vaporization to plot the $\log p$ vs. $1/T$ graphs.)

Construct a graph of vapor composition as a function of liquid composition. This is a McCabe–Thiele plot. On the same graph draw a 45° straight line through the origin. This line is the function for which vapor composition equals liquid composition. The azeotropic composition, if there is one, is that at which the experimental curve crosses the 45° line.

NOTES / Control the temperature of the samples during the measurement of the refractive index by circulating water through the refractometer.

Mixtures of known composition should be used if there are significant gaps in the data for the phase diagram.

The refractive indices of the two components must differ by at least 0.04 unit for the mixture to be analyzed by refractometry.

Some of the vapors are very toxic, so use adequate ventilation.

Experiment 19. Phase Diagram of a Binary Liquid—Vapor System

EXERCISES

1. Explain why the composition of the liquid and vapor should differ.
2. Briefly discuss the causes of minimum and maximum boiling points.
3. Explain the deviations, if any, of your mixture from Raoult's law. Base your explanation on molecular structure if possible.
4. Explain why a much more accurate value for the azeotrope composition can be obtained from the McCabe–Thiele plot than from the phase diagram, although the same data are used to construct both.

Experiment 20

Vapor and Sublimation Pressure

Internal Manometer Method

OBJECTIVES / To measure the vapor pressure of a sample of benzene over a range of temperatures below and above the triple point and to calculate the enthalpies of the various changes of state.

READING ASSIGNMENT / Chapter 10, pages 99–100; Chapter 9, pages 80–82.

Benzene, in the solid state, has an appreciable vapor pressure near the melting temperature. Both the solid–vapor and the liquid–vapor equilibria obey the Clausius–Clapeyron equation. In this experiment the vapor pressures of liquid and solid benzene are determined at several temperatures; the triple point and the heats of sublimation, vaporization, and fusion are calculated; and the molal freezing point depression constant is determined.

EQUIPMENT AND MATERIALS

The internal manometer apparatus is used for the vapor pressure–temperature measurements. The benzene should be C.P. or else redistilled.

DIRECTIONS

Make direct pressure measurements on a sample of about 10 ml of benzene at temperatures between -5 and $+5°C$ and from 10 to 30°C. Then repeat the procedure with a fresh sample. The pressures at any temperature should be reproducible to within 2 mm.

CALCULATIONS AND GRAPHS

Construct a graph of the logarithm of pressure vs. the reciprocal of the absolute temperature.

From the graph calculate the average heats of vaporization and of sublimation in the temperature range studied.

From the heats of vaporization and sublimation, calculate the heat of fusion. From this and the accepted value of the melting point, calculate the molal freezing point depression constant. Estimate the triple point of benzene and compare this to the accepted value.

NOTES / Vapor pressure equilibrium is attained more rapidly working from higher temperatures down to lower ones. Allow about 5-10 min at each temperature for thermal equilibrium to be attained.

If low temperature thermostatic units, such as the Formatemp unit, are not available, chopped ice-salt mixtures are convenient low temperature baths in the range of this experiment. As more salt is added to the ice, the bath temperature becomes lower (until the eutectic temperature is reached).

Benzene is very toxic. Work in the hood.

Experiment 21

pK_a of an Indicator

Spectrophotometry

OBJECTIVES / **To determine the visible absorption spectrum of an acid–base system and to measure the absorbance as a function of concentration and of pH.**
From the results, the pK_a of the acid is calculated.

An indicator is a weak acid or base, whose ionized form has one color and whose molecular form has another. (Alternatively, one form may be colorless.) The equilibrium

$$HA \rightleftharpoons H^+ + A^-$$

can therefore be studied spectrophotometrically. The pH is either measured or else fixed with a buffer and then the concentration of each form is determined from its absorbance at some appropriate wavelength. If only one form absorbs radiation, the concentration of the other form can be determined by difference, provided that the total indicator concentration is known.

To calculate concentration from absorbance, one uses the *Beer's law* equation,

$$A = abc$$

A = absorbance
a = absorptivity
b = length of light path
c = concentration of absorbing substance

The constant b is the thickness of the cell or cuvet containing the solution. Even if it is not known, it need not be measured, provided that the same cuvet or

Experiment 21. pK of an Indicator

identical cuvets are used in all runs. The constant a, however, varies from wavelength to wavelength and from substance to substance and so a or the product ab must be determined for each system. In other words, the spectrophotometer must be calibrated for use at the desired wavelength.

In order to decide on the wavelength to use, the spectrum of each form of the indicator must be determined and graphed. The optimum wavelength for analysis is one at which one form of the dye absorbs and the other does not, or else absorbs only very slightly.

The indicator in this experiment is bromophenol blue (BPB). The acid and the base spectra are obtained and the wavelength for analysis is selected. Then, at this wavelength, solutions of known concentration are used to obtain the Beer's law curve (i.e., the constant ab). Once the analytical curve is obtained, solutions of BPB are made up at pH's near the pK_a. From the absorbance of each solution, and the total concentration of indicator, the equilibrium concentration of each form is obtained and the pK_a is calculated.

One additional minor point: in determining the spectrum of each form, you do not care how much BPB you have in either tube because the amount remains constant over the range of wavelengths. For all other measurements, it is necessary to know the total concentration.

EQUIPMENT AND MATERIALS

A spectrophotometer and a pH meter.

Stock solutions of 0.05% (by weight) bromophenol blue (BPB), concentrated HCl, concentrated NH_3, and, if no pH meter is available, buffer solutions in the pH range 3.4–4.6.

DIRECTIONS

Determination of the Spectra of the Ionized and Nonionized Forms of the Acid. Prepare two samples of about 3 drops each of BPB in 10 ml of distilled water. Make one acidic and the other basic with 3 drops of concentrated HCl and concentrated NH_3, respectively.

Measure the absorbance of each sample over the region 375–650 nm at 25 nm intervals. Near the maximum, work at 5 nm increments.

Calibration of the Instrument for Analysis of the BPB Solutions. Select a wavelength at which one form of indicator absorbs light strongly and the other form absorbs slightly or not at all. At this wavelength obtain a set of absorbance–concentration values for the absorbing form. Start with stock solution diluted 1 : 10 and then make successive dilutions with water containing either acid or base, depending on the form of the indicator desired. Continue dilutions until at some concentration c the solution has an absorbance between

0.9 and 1.0. Then make up solutions of concentrations $0.8c$, $0.6c$, $0.4c$, and $0.2c$ and determine the absorbance of each.

Determination of the Ionization Constant. Select a concentration of indicator that would show an absorbance of 0.9–1.0 at the selected wavelength if it were all in the absorbing form. Prepare several solutions of this selected total indicator concentration at different pH's in the range 3.4–4.6.

Determine the absorbance and the pH of each BPB solution.

CALCULATIONS AND GRAPHS

On one graph draw the absorption spectra for both forms of the indicator.

At the wavelength selected for the analysis, plot the graph of absorbance versus concentration of the absorbing form of the indicator.

From the measured absorbances, calculate the concentration of the absorbing form in each solution of intermediate pH.

Calculate the pK_a of the acid (pK_a is $-\log K_a$) from the formula

$$pK_a = pH + \log \frac{[HA]}{[A^-]}$$

$[HA]$ = concentration of acid form
$[A^-]$ = concentration of base form

Alternatively, if pH is plotted against $\log([A^-]/[HA])$, the pH is the pK_a when $\log([HA]/[A^-])$ is zero.

There are two methods of calculating $[HA]/[A^-]$ at each pH. One way is to obtain the concentration of the absorbing form from the measured absorbance and the Beer's law curve. This concentration is then subtracted from the known total concentration of the dye to give the concentration of the nonabsorbing form. The disadvantage here is that any error in the concentration of the original stock solution has a large effect on the difference term, especially when the difference is small.

A much better method is to use the absorbances directly rather than concentrations. When the solutions are being made up at the various pH's, make up one that is strongly acid (or basic) so that all the dye is in the absorbing form. The absorbance of this solution is A_0. The absorbance of a solution of the same total concentration, but at an intermediate pH, is A. Then

$$\frac{A}{A_0 - A} = \frac{\text{concentration of absorbing form}}{\text{concentration of nonabsorbing form}}$$

and

$$\log \frac{A}{A_0 - A} = \log \frac{[HA]}{[A^-]} \quad \text{or} \quad \log \frac{A}{A_0 - A} = \log \frac{[A^-]}{[HA]}$$

depending on whether the acid form or the base form does the absorbing. In this second method, the absolute value of the original concentration need not be known at all. This method is preferred, if there is any doubt as to the accuracy of the concentration of the stock solution or of the dilutions.

NOTES / For the dilutions use volumetric apparatus. Do not use graduated cylinders for quantitative work.

Plot rough graphs of the spectra and of the calibration curve as you work in case more values are needed or some points are questionable.

Use three matched cuvets in determining the spectra. Use one filled with distilled water as a reference and place the samples in the other two. If the cuvets are not matched, use the same one for all samples and use another only for reference liquid.

In adjusting the pH it is most convenient to use buffer solutions, making all but the final dilution with distilled water. If buffer solutions are not available, simply add dilute acid and dilute base dropwise until the solution has a color intermediate between the acid and the base forms of the bromphenol blue indicator.

For easier calculation assume that the acid and the base form of the indicator have the same molecular weight. (What percentage error is introduced by this assumption?)

EXERCISES

1. Why is it preferable to calculate pK_a rather than K_a?
2. Derive an equation to be used if there is no wavelength at which only one form absorbs light. Outline the procedure to be followed in analyzing solutions, using this equation to calculate the concentration of each form of the indicator.

Experiment 22

Ionization Constant of a Weak Acid

Kohlrausch Conductance Method

OBJECTIVES / To make measurements of the conductance of a solution of a weak acid, HAc, and of the strong electrolytes NaAc, HCl and NaCl and to calculate the ionization constant of the acid.

READING ASSIGNMENT / Chapter 16, pages 200–204; Chapter 14, pages 150–152 and 165.

If a solution of a weak electrolyte is 10% ionized, the equivalent conductance will be only 10% of the conductance the solution would have if it were completely ionized (i.e., at infinite dilution). In terms of conductance, to a good first approximation we can define the fraction ionized as the conductance ratio

$$\alpha_c = \frac{\Lambda_c}{\Lambda_\infty}$$

α_c = degree of dissociation, or conductance ratio
Λ_c = equivalent conductance of electrolyte at concentration c
Λ_∞ = equivalent conductance of electrolyte at infinite dilution

From the degree of dissociation, the ionization constant, K, is easily calculated. For a binary electrolyte,

$$AB \rightleftharpoons A^+ + B^-$$

$$K = \frac{c_0 \alpha^2}{1 - \alpha}$$

c_0 = total electrolyte concentration

In this experiment, the ionization constant of acetic acid is calculated from the conductance ratio. At several different known concentrations, the

Experiment 22. Ionization Constant of a Weak Acid

specific conductance is measured and the equivalent conductance calculated, using the formula

$$\Lambda = \frac{1000\kappa}{c}$$

κ = specific conductance
c = equivalents of electrolyte per liter of solution

A problem arises in that for a weak electrolyte, Λ_∞ cannot be determined by the obvious method of plotting Λ vs. $(\text{conc})^{1/2}$ and extrapolating to zero concentration. Instead, *Kohlrausch's law* must be used:

$$\Lambda_{\infty,\,\text{acetic acid}} = \Lambda_{\infty,\,\text{HCl}} + \Lambda_{\infty,\,\text{NaAc}} - \Lambda_{\infty,\,\text{NaCl}}$$

The equivalent conductance of each of the three strong electrolytes HCl, NaAc, and NaCl is determined and the straight line graph of Λ vs. $\sqrt{c}$ is extrapolated to zero concentration. From the extrapolated values, Λ_∞ is calculated for acetic acid.

A better value of the fraction ionized is α', defined by

$$\alpha' = \frac{\Lambda_c}{\Lambda_c^\circ}$$

Λ_c° = hypothetical equivalent conductance the electrolyte would have if it were completely ionized at c', the actual concentration of ionized acetic acid at the formal concentration C

To determine $\Lambda_{\text{HAc}}^\circ$, use α to calculate c', the concentration of H_3O^+ and Ac^-. Then use Kohlrausch's law and Λ for HCl, NaAc, and NaCl at that concentration:

$$\Lambda_{\text{HAc}}^\circ = \Lambda_{c',\,\text{HCl}} + \Lambda_{c',\,\text{NaAc}} - \Lambda_{c',\,\text{NaCl}}$$

EQUIPMENT AND MATERIALS

An impedance bridge, a thermostat, and a conductance cell.
   A reference solution of 0.0200 M KCl, stock solutions of 0.100 M HCl, NaCl, and NaAc and 1.00 M HAc.

DIRECTIONS

Measure the resistance of the water to be used in the dilutions and the resistance of the 0.0200 M KCl.
   Measure the resistances of four or five dilutions of HCl, NaCl, and NaAc. Start the measurements with 0.100 M solutions and then continue with successive twofold dilutions.

Measure the resistances of several solutions of acetic acid in the concentration range between 0.3 and 1.0 M. The concentrations must be known to three significant figures.

CALCULATIONS AND GRAPHS

Subtract the conductance of the water from that of the electrolytic solutions whenever necessary.

Calculate the cell constant from the known specific resistance of the reference solution, and use the cell constant to calculate the specific conductance of each solution.

Calculate the equivalent conductance of each solution. Then for each strong electrolyte plot a graph of Λ, the equivalent conductance, against the square root of the concentration. If the data are scattered, make more measurements until a well defined curve is obtained. Extrapolate the graph to zero concentration to estimate Λ_∞, the equivalent conductance at infinite dilution.

Use Kohlrausch's law to calculate Λ_∞ for acetic acid from the values for the strong electrolytes.

Calculate the conductance ratio and ionization constant at each acetic acid concentration. Compare these with accepted values.

If the ionization constant differs appreciably from accepted data, recalculate it from α'.

NOTES / In making the dilutions, use volumetric apparatus, *not* graduated cylinders.

Rinse the cell each time with a portion of the solution to be used.

Recheck the cell constant if the cell is used by anyone else before your experiment is complete.

Use two different batches of each solution in order to calculate precision.

Control the temperature at $25.0 \pm 0.5°$ C and allow sufficient time for thermal equilibration. If necessary, keep the solutions in the thermostat while measurements are being made on other samples.

EXERCISES

1. Does the specific conductance increase or decrease as the concentration decreases? Explain your answer.

2. In using the conductance ratio as if it were equal to the degree of ionization, an implicit assumption is made concerning the ionic velocities. State this assumption and discuss its validity.

3. Explain why two different batches of each solution are used in calculating precision.

# Experiment 23

Formation Constant of a Complex Ion

Polarography

OBJECTIVES / To obtain polarograms for the reduction of Pb^{2+} at the dropping mercury electrode in acid and in basic solutions and to calculate the instability constant and the composition of the plumbite complex.

READING ASSIGNMENT / Chapter 16, pages 207–215.

The lead ion forms a very stable plumbite complex in alkaline solution, in accord with the reaction

$$Pb^{2+} + n\, OH^- \rightleftharpoons Pb(OH)_n^{(2-n)+}$$

In this experiment the equilibrium constant for the formation of the complex is determined polarographically and the number of OH^- ligands calculated, using the method described on pages 211–212.

DIRECTIONS

Use a dropping mercury electrode in a three electrode setup to obtain a polarogram of Pb^{2+} in acid solution at ionic strength 1.0.

Obtain polarograms for the reduction of Pb^{2+} in solutions whose OH^- concentrations range from 0.01 to 1.0 M, at ionic strength 1.0.

If time permits, obtain the polarograms of solutions with the same Pb^{2+} concentration and the same pH values but at different ionic strengths.

CALCULATIONS AND GRAPHS

Determine the half-wave potential of each polarogram. Plot the half-wave potentials in alkaline solution, at constant ionic strength, against the logarithm of the OH^- concentration. From the intercept, calculate the formation constant of the complex at each ionic strength. From the slope of the graphs calculate the number of OH^- ions bound to the Pb^{2+}.

If results have been obtained at several ionic strengths, plot the formation constant versus the square root of ionic strength and extrapolate to zero ionic strength to obtain the thermodynamic value of the stability constant.

NOTES / The mercury drop time should be between 2 and 6 sec. The solution should be purged with nitrogen. Temperatures should be kept at 25°C, but this is not critical.

The optimum lead concentration is 5×10^{-4} M. The ionic strength should be adjusted using KNO_3 as the neutral salt.

EXERCISES

1. How can the number of bound water molecules in the complex be determined?

2. Suggest an alternative way of determining the formation constant.

Reference

Lingane, J., Interpretation of the polarographic waves of complex metal cations, *Chem. Rev.* **29**, 1 (1941).

Experiment 24

Solubility of an Insoluble Salt

EMF Method

OBJECTIVES / To measure the potential difference between an Ag|Ag$^+$ electrode and an Ag|AgX(s) electrode and to calculate the solubility of the silver halide, AgX.

READING ASSIGNMENT / Chapter 14, pages 147–149; Chapter 16, pages 189–193 and 195.

When a metal electrode dips into a saturated solution of one of its insoluble salts, MX, it may be considered as either an electrode of the first kind, with potential depending on M$^+$, or as an electrode of the second kind, with potential depending on X$^-$. Depending on which kind of an electrode the metal is considered to be, the $E°$ is different, since in one case it would be based on unit activity of the cation and in the other case on unit activity of the anion. The difference in potential between the two $E°$'s is a measure of the activity product of the insoluble salt. The E's, of course, are the same, no matter what arbitrary assumptions we make about the standard state.

In this experiment, the cell

$$\text{Ag(s)}|\text{AgX(s)}|\text{KX}(0.00100\ M)\|\text{AgNO}_3(0.00100\ M)|\text{Ag(s)}$$

is set up and from its potential we calculate the difference between $E°_{\text{Ag(s)}|\text{AgX(s)}}$ and $E°_{\text{Ag(s)}|\text{Ag}^+}$, and from that the K_{sp} and the solubility.

EQUIPMENT AND MATERIALS

A potentiometer; two beakers; a salt bridge filled with saturated NH$_4$NO$_3$; solutions of 0.00100 M KCl, KBr, KI, and AgNO$_3$; and two silver wires.

DIRECTIONS

Prepare a silver / silver halide electrode by dipping a silver wire into $0.00100\ M$ KX solution and adding 1 or 2 drops of saturated $AgNO_3$.

Insert the other silver wire into some $0.00100\ M\ AgNO_3$ and connect the two half-cells by means of a salt bridge filled with saturated ammonium nitrate. Record the cell potential after the emf has stopped drifting.

Repeat the procedure with the other two halides.

CALCULATIONS

Write the equations for the cell reactions.

Assume that the activity coefficients of the ions are unity. Use the measurements of emf to calculate the $E°$ for the cell reaction. From the $E°$ for the cell, calculate the solubility product of the silver halide.

NOTES / Use small quantities of solution with 100 ml beakers as half-cells.

The liquid should be at the same level in both half-cells. Otherwise, solution will siphon over from one to the other, through the salt bridge.

Stopper the ends of the salt bridge with glass wool.

Try to keep oxygen away from the iodide solutions. Use a flask instead of a beaker and aspirate out the oxygen or flush it out with nitrogen.

The unknown liquid junction potentials are large enough to make the calculation of activity coefficient pointless.

EXERCISES

1. Why does the measured value of $E°$ agree with accepted values better than the calculated solubility product agrees with the accepted values?
2. Use the accepted value of the solubility of the silver halide and the activity coefficients, calculated from the Debye–Hückel limiting law, to estimate the liquid junction potential.
3. Explain how the presence of oxygen can interfere with the measurement in KI solution.

Experiment 25

Decomposition Pressure of Ammonium Carbamate

Mercury Isoteniscope Method

OBJECTIVES / To determine the decomposition pressure of ammonium carbamate over a range of temperatures and to calculate the thermodynamic quantities for the decomposition reaction.

READING ASSIGNMENT / Chapter 9, pages 80–82.

Solid ammonium carbamate decomposes directly into the two gases CO_2 and NH_3.

$$NH_2CO_2NH_4(s) \rightleftharpoons 2\,NH_3(g) + CO_2(g)$$

In this experiment, the equilibrium constant for the heterogeneous equilibrium is calculated directly from the decomposition pressure, at each of several temperatures. From the variation of the equilibrium constant with temperature, the thermodynamic functions $\Delta G°$, $\Delta H°$, and $\Delta S°$ are calculated.

EQUIPMENT AND MATERIALS

The mercury isoteniscope and the associated apparatus and a vacuum pump and trap.
    Solid ammonium carbamate.

DIRECTIONS

Make equilibrium pressure–temperature measurements from 30 to 60°C, using about 5 g of powdered solid.
    Repeat the measurements with about 10 g of solid.

CALCULATIONS AND GRAPHS

Plot the pressure–temperature graph.

Calculate the equilibrium constant at several temperatures from the observed pressures.

Plot ln K (or log K) versus $1/T$ and estimate $\Delta H°$ of the reaction.

Calculate $\Delta G°_{25°C}$ and $\Delta S°_{25°C}$.

NOTES / Powder the solid so that equilibrium is attained rapidly. On evacuating the air, lower the pressure gently so as not to lose any of the powder.

Make sure that no air is present, so that the only gases are the reaction products.

Heat the reaction flask to 60°C in a large beaker of water. Then allow the beaker to cool down to room temperature. Read the pressure and temperature at intervals of about 5 cm Hg.

Ammonium carbamate is sold under the name "ammonium carbonate" because when added to water it forms ammonium carbonate.

EXERCISES

1. Explain why the equilibrium constant should be independent of the amount of solid present. Is it, experimentally?

2. The results are more precise for the descending temperature curve than for the ascending temperature curve. Explain.

3. For what quantity of substance is your computed $\Delta H°$ of reaction? Why?

4. What error(s) might be introduced by the presence of moisture?

Experiment 26

Dimerization of Acetic Acid Vapor

Mercury Isoteniscope Method

OBJECTIVES / To measure the vapor pressure of acetic acid over a range of temperatures and to calculate the thermodynamic quantities for the dissociation of acetic acid dimer.

READING ASSIGNMENT / Chapter 9, pages 80–82.

Carboxylic acids dimerize in the vapor phase and in nonpolar solutions, due to hydrogen bonding. In this experiment, the reaction is studied by measuring the pressure developed by acetic acid vapor at a series of temperatures.

A weighed amount of acetic acid is introduced and vaporized and is assumed to dimerize. The degree of dissociation of the dimer is then calculated from the difference between the observed pressure and that expected from the dimer (assuming the ideal gas law). From the degree of dissociation at each of several temperatures, the dissociation constant is calculated, and from the variation of the dissociation constant with temperature, $\Delta H°$ is obtained. $\Delta G°$ and $\Delta S°$ are then calculated for the dissociation reaction.

EQUIPMENT AND MATERIALS

The mercury isoteniscope and associated apparatus, a vacuum pump and trap, and an oil or silicone bath.
    Glacial acetic acid.

DIRECTIONS

Calculate the amount of acetic acid that, if monomeric, would develop a pressure in the apparatus of slightly less than 1 atm at 200°C. Use this quantity for the experiment.

Measure the pressure at several temperatures in the range 130–200°C, working with both increasing and decreasing temperatures.

Determine the volume of the gas system.

CALCULATIONS AND GRAPHS

At each temperature subtract the vapor pressure of mercury from the observed pressure.

Graph the pressure as a function of temperature. On the same set of coordinates graph the theoretical pressure for the vapor assuming it to be an ideal dimeric gas.

At several arbitrary temperatures calculate the degree of dissociation of the dimer.

Calculate the dissociation constant from each calculated value of the degree of dissociation.

Plot the logarithm of the dissociation constant against the reciprocal of the Kelvin temperature and estimate the enthalpy of dissociation.

Use the van't Hoff equation to estimate the dissociation constant at 25°C.

Calculate $\Delta G°$ and $\Delta S°$ of dissociation for the reaction at 25°C from the $\Delta H°$ of dissociation and the dissociation constant.

NOTES / Heat the bulb in an oil bath. Allow sufficient time for the vapor to reach the temperature of the bath. Be sure the acetic acid is completely vaporized.

Determine the volume of the reaction flask from the weight of water it can contain.

Calculate the volume of the manifold from the measured length and bore. A more accurate, although more difficult, determination of manifold volume is from the amount of mercury needed to fill the manifold between the valves and the bulb.

Either introduce the desired amount of glacial acetic acid with a 1 ml calibrated pipet or else weigh the acetic acid into a 1 ml glass-stoppered vial and insert the vial into the reaction flask.

Freeze the acetic acid with an ice–salt mixture or with an acetone–solid carbon dioxide bath before evacuating the air from the apparatus; otherwise, you may lose some acetic acid vapor. The removal of air without loss of acetic acid is the most difficult part of the procedure, and accounts for most of the experimental error.

EXERCISES

1. Both the monomer and the dimer are assumed to be ideal gases. Comment on the validity of this assumption in light of the fact that acetic acid vapor dimerizes.

2. Assume that there is an experimental error of 1 mm in leveling the mercury in the U tube at an overall pressure of 20 cm. Estimate the effect on the calculated equilibrium constant.

3. Assume an error of 1 mm in the measurement of the bore of the manifold tube. Calculate the error in the volume and the degree of dissociation of the dimer at any particular temperature.

4. Compare the $\Delta H°$ of dissociation with the energies to be expected from two hydrogen bonds per molecule of dimer.

5. If a vapor of a volatile liquid is partially associated or partially dissociated, the heat of vaporization computed from the Clausius–Clapeyron equation is the heat of vaporization per mole (Avogadro's number) of particles in the vapor phase rather than per mole (formula weight) of liquid vaporized. Explain.

Experiment 27

Kinetics of the Iodide-Catalyzed Decomposition of H_2O_2

Gas Volume Measurement

OBJECTIVES / To measure the rates at which gas is evolved from solutions of H_2O_2, KI, and KCl over a range of temperatures and to calculate the rate constants, activation energy, and the order with respect to the catalyst.

In this experiment, the progress of a reaction is studied by measuring the quantity of gas produced as a function of time. The reaction is

$$2\,H_2O_2 \;\rightarrow\; 2\,H_2O + O_2 \uparrow$$

At any given time, the volume of oxygen evolved is directly proportional to the number of moles of peroxide that have decomposed, provided, of course, that the volume of oxygen is measured at constant pressure and temperature. The rate of the reaction is therefore proportional to the rate of oxygen evolution.

The apparatus used in this experiment is shown in Diagram D. The reacting solution is attached to the gas buret. The incoming gas depresses the liquid level. To measure the volume of gas at any time, the leveling bulb is raised or lowered until the liquid is at the same level in both the buret and the bulb. At that moment, the buret reading and the time are both recorded. The pressure of the oxygen in the buret is atmospheric pressure minus the vapor pressure of water in the buret. Since the vapor pressure of water is constant during the run, the oxygen pressure is also constant and the buret reading is directly proportional to the amount of oxygen produced and therefore to the quantity of peroxide decomposed.

The reaction, which is first order with respect to peroxide, is catalyzed by iodide ion. The iodide concentration should therefore be kept constant in each and all runs while the rate constant is being determined. The ionic strength must be kept constant.

Experiment 27. Kinetics of the Iodide-Catalyzed Decomposition of H_2O_2

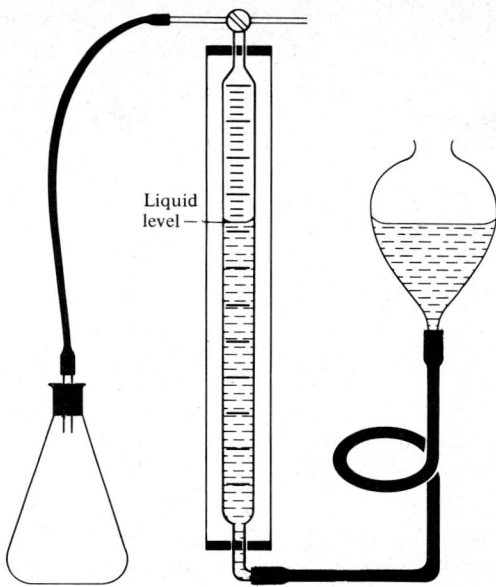

Liquid level —

DIAGRAM D / Gas buret.

EQUIPMENT AND MATERIALS

A 100 ml gas buret, an adjustable thermostat, a timer, and a shaker.
    Approximately 3% H_2O_2 solution, 0.100 M KI, and 0.100 M KCl.

DIRECTIONS

Prepare 10 ml of reactant solution by mixing 5 ml of the KI solution with 5 ml of the peroxide solution.
    Put the reaction mixture into a 25 ml Erlenmeyer flask with a few glass beads and bring it to 30°C in a thermostat. Allow 3–5 min for thermal equilibrium, while shaking it gently.
    Connect the flask to the gas buret and take periodic readings of gas volume as a function of time while shaking the flask in the thermostat at a slow regular rate. Continue readings until less than 0.5 ml of gas is generated per 2 min interval.
    Without detaching the flask from the buret, heat the reaction mixture in a beaker of boiling water for about 1 min in order to decompose the unreacted peroxide. Cool the flask down to 30°C before reading the final volume.
    Repeat at 30°C and make additional replicate runs at several other temperatures ranging from 20 to 50°C.

Repeat the 30°C runs with as many different concentrations of KI as time permits. Maintain a constant ionic strength by adding KCl. The total volume of KI and KCl solution should be kept at 5 ml.

CALCULATIONS AND GRAPHS
Plot

$$\log\left(\frac{v_f - v_0}{v_f - v_t}\right) \text{ vs. time}$$

v_f = final gas volume
v_0 = initial gas volume
v_t = volume at time t

The slope of the graph is $k/2.303$, where k is the specific rate constant.

Plot the logarithm of k, at constant iodide concentration, vs. $1/T$ to obtain an estimate of activation energy.

Plot the graph of the logarithm of the observed rate constant vs. the logarithm of the iodide concentration. From the graph, estimate the order of the reaction with respect to iodide ion.

NOTES / Whenever a reading is taken, the leveling bulb must be adjusted.

In making duplicate runs and in determining temperature coefficients keep the concentration of I⁻ constant.

The time intervals need not be uniform as long as sufficient data are obtained to plot the graph.

Time can be saved and better results obtained by suspending flasks of the reagent solutions in the thermostat and bringing the liquids to the reaction temperature before mixing them.

EXERCISES

1. Explain why $\log[(v_f - v_0)/(v_f - v_t)]$ versus time is plotted rather than the usual $\log[c_0/(c_0 - x)]$ versus time, where c_0 is the initial peroxide concentration and x is the concentration that has reacted.

2. Explain why it is not necessary to know either the initial peroxide concentration or the actual concentration at any time during the reaction.

3. Explain why the flask is cooled down after heating before the final reading is made.

4. Explain the function of the glass beads and of the shaking.

5. Suggest a reaction mechanism.

6. Explain why the ionic strength is kept constant.

# Experiment 28

Rate of the Acid-Catalyzed Hydrolysis of Sucrose

Polarimetry

OBJECTIVES / To measure the optical activity of sucrose solutions during the course of acid-catalyzed hydrolysis. From the data, rate constants of the catalyzed and uncatalyzed reactions are calculated and the order of reaction with respect to H_3O^+ is determined.

READING ASSIGNMENT / Chapter 17, pages 217–219.

In this experiment, the rate of a chemical reaction is followed by observing the rate of change in a physical property, the rotation of polarized light.

Each sucrose molecule hydrolyzes to a molecule of glucose and a molecule of fructose. Each of these substances has its own optical rotation. The rate of the reaction is therefore followed not by the optical rotation itself but by the *changes* in optical rotation. The change in rotation that takes place *after* time t is directly proportional to the quantity of sucrose hydrolyzing *after* time t (i.e., to the amount that has still not reacted at that time).

$$\alpha_t - \alpha_\infty = kc_{\text{suc}}, \text{ at time } t$$

α_t = rotation at time t
α_∞ = rotation at infinite time, i.e., when the reaction is over
c_{suc} = concentration of sucrose

The reaction is really second order, but since there is such a huge excess of water, it appears to be first order (i.e., it is pseudo first order). The rate equation is

$$-\frac{dc_{\text{suc}}}{dt} = kc_{\text{suc}}$$

EXPERIMENTS

The rate constant, k, is compound, since there is an acid-catalyzed rate and an uncatalyzed rate.

$$k = k_0 + k_{H^+} c_{H^+}^n$$

$n = $ order with respect to H_3O^+

EQUIPMENT AND MATERIALS

A polarimeter and a sodium vapor lamp.
    Crystalline sucrose, concentrated HCl solution, and glacial acetic acid.

DIRECTIONS

Prepare a sample of solution 2 M with respect to HCl and containing 10 g of sucrose per 100 ml. Measure the angle of rotation at 1 min intervals for the first 10 min and then at 2 min intervals for the next 20 min. Record each time and reading.
    Heat another sample of the solution for 15 min with a water bath at 60°C. Cool this down to room temperature and determine its optical rotation.
    Repeat the procedure with several different concentrations of acid until sufficient data have been obtained to calculate the rate constants for both the catalyzed and uncatalyzed reactions.
    Repeat with 2 M acetic acid.

CALCULATIONS AND GRAPHS

Arbitrarily consider the time of the first reading as t_0 and the first reading to be α_0, the initial angle of rotation. Plot the log of $(\alpha_0 - \alpha_\infty)/(\alpha_t - \alpha_\infty)$ against time and evaluate the apparent rate constant. α_∞ is the final value of the rotation after heating at 60°C and cooling back down to room temperature.
    Plot a graph of apparent rate constant versus concentration of HCl. Extrapolate to zero to decide if there is an appreciable uncatalyzed rate. If there is, estimate the rate constant for the uncatalyzed reaction and subtract it from the observed rate constants to obtain the corrected rate constants for the catalyzed reaction at each pH.
    Plot a graph of the logarithm of the corrected (if necessary) rate constant versus the logarithm of the HCl concentration to determine the order with respect to H_3O^+. The value at the intercept, $\log[H_3O^+] = 0$, is the logarithm of the specific rate constant for the catalyzed reaction. The slope of the line is the order with respect to H_3O^+.
    From the results, decide whether the reaction is an example of specific H_3O^+ catalysis or of general acid catalysis.
    Write the complete rate equation, inserting all known values.

Experiment 28. Rate of the Acid-Catalyzed Hydrolysis of Sucrose

NOTES / Do not use HCl of concentration higher than 2 M. At high acidities, another, competing, reaction takes place.

Make up the solutions by mixing equal volumes of double strength sucrose solutions and HCl solutions.

Filter the sucrose solution if necessary.

Each time the polarimeter tube is used, rinse it with a portion of the reaction mixture.

It is not necessary to measure out the sucrose to better than 0.5 g/100 ml. Therefore, the trip scale may be used instead of the analytical balance.

Determine the concentration of stock HCl solution by titrating a diluted aliquot portion against a standard base.

If the polarimeter tube has a water jacket, maintain the reaction temperature at 25°C. If not, run the reaction at room temperature.

The polarimeter should be balanced at *minimum* intensity of light (i.e. maximum darkness).

EXERCISES

1. Suggest a mechanism for the catalyzed reaction.
2. Explain why it is not necessary to weigh the sucrose accurately.
3. Explain why the second portion of the reaction mixture is heated.
4. Explain why the rotation, at any time, is not directly proportional to the concentration.

Experiment 29

Effect of Ionic Strength on the Rate of a Reaction

Spectrophotometry

OBJECTIVES / By a spectrophotometric measurement, to determine the time for a predetermined quantity of reagent to be consumed in a chemical reaction. The results are used to determine the charge on the activated complex.

In this reaction, the kinetics are not studied by observing concentration changes as they occur. Instead, the student determines the time for a definite quantity of reagent to react. The reaction is the oxidation of iodide ion to iodine by perdisulfate ion in the presence of starch and thiosulfate ion.

(1) $$2\,I^- + S_2O_8^{2-} \rightleftharpoons \underset{\substack{\text{activated} \\ \text{complex}}}{C^*} \rightarrow I_2 + 2\,SO_4^{2-}$$

The reaction is first order with respect to both reactants.
    The iodine that is produced would be expected to react with the starch to form a blue color, but before it can, the thiosulfate reduces it back to iodine.

(2) $$2\,S_2O_3^{2-} + I_2 \rightarrow 2\,I^- + S_4O_6^{2-}$$

This keeps the iodide concentration constant, with virtually zero iodine present until all the thiosulfate has been used up. At that point the solution turns a deep blue. (This is an example of a "clock" reaction.)
    For each equivalent of $S_2O_8^{2-}$ used up in oxidizing I^-, an equivalent of $S_2O_3^{2-}$ is used up in reducing the iodine back to iodide. Therefore,

(3) $$\frac{\Delta[S_2O_3^{2-}]}{\Delta t} = \frac{\Delta[S_2O_8^{2-}]}{\Delta t}$$

Experiment 29. Effect of Ionic Strength on the Rate of a Reaction

The reaction should be very sensitive to changes in ionic strength. Combining accepted rate theory with the Debye–Hückel limiting law for activity coefficients results in the *Brønsted equation*:

(4) $$\log k = \log k_0 + 2Az_1 z_2 \sqrt{\mu}$$

k = rate constant
μ = ionic strength
A = Debye–Hückel constant
   = 0.509 in water at 25°C
z_1, z_2 = respective charges on the ions forming the activated complex

(Actually, for some reason, in this experiment the equation is followed at concentrations far above those at which Debye–Hückel theory would be expected to hold.)

EQUIPMENT AND MATERIALS

A timer, a thermostat, and a spectrophotometer with a thermostatted sample holder.
Stock solutions of 0.100 M potassium iodide, 0.100 M potassium perdisulfate ($K_2S_2O_8$), 0.100 M sodium thiosulfate ($Na_2S_2O_3$), and freshly prepared 1% starch. Anhydrous magnesium sulfate.

DIRECTIONS

Prepare reaction mixtures containing KI, $K_2S_2O_8$, $Na_2S_2O_3$, and fresh starch solution. Record the time it takes for the blue starch iodine color to appear. Repeat two or three times with fresh solution. To prepare the reaction mixtures, mix 50 ml of each of two reagent solutions. One contains the reductant, 4×10^{-2} M KI, and the other contains the oxidant, $K_2S_2O_8$, at a concentration of 5×10^{-3} M, $Na_2S_2O_3$ at a concentration of 2×10^{-4} M, and 1 ml of starch indicator.
Repeat the procedure with a reaction mixture containing sufficient $MgSO_4$ in addition to the other constituents to be 0.01–0.10 M with respect to $MgSO_4$. Repeat at several different $MgSO_4$ concentrations, each time recording the time for the blue color to appear.

CALCULATIONS AND GRAPHS

Write the equations for all the successive steps in the reaction, indicating relative velocities.
Use the differential form of the rate equation to calculate the rate constant for the removal of persulfate. The reaction is first order with respect to

both reactants. Assume that over the range of experimental concentration changes,

$$-\frac{\Delta[S_2O_8^{2-}]}{\Delta t} = \frac{d[S_2O_8^{2-}]}{dt}$$

Calculate a separate rate constant for each run and average the data for all runs at the same $MgSO_4$ concentration.

Assume that the presence of the added $MgSO_4$ does not change the mechanism of the reaction or its order. Plot a graph of log k versus the square root of ionic strength. Compare the slope of the graph with that predicted by the Brønsted equation, and calculate the charge on the activated complex.

NOTES / The appearance of the blue color is best detected with a spectrophotometer. However, if the temperature in the sample holder is not controlled, the detection must be done visually using a blank and a white background.

Run the reaction at 25° C, bringing the stock solutions to reaction temperature before mixing.

The thiosulfate concentration must be small enough so that only a few per cent of the reactants will have been consumed when the blue color appears.

To improve precision, repeat the measurements at each concentration of $MgSO_4$ once or twice. However, do not continue repeating indefinitely. It is better to get five points on a curve, each the result of two measurements, than to get two points, each the result of five measurements.

In preparing the reaction mixtures, the stock 0.1 M solution of $Na_2S_2O_3$ is too concentrated to use directly. Prepare a 0.005 M $Na_2S_2O_3$ solution from the 0.1 M stock and use this in making up the reaction mixture. The $Na_2S_2O_3$ is kept at 0.1 M because dilute thiosulfate solutions deteriorate on standing.

If the reaction takes more than 15 min, replace the starch solutions. To decrease the reaction time, decrease the thiosulfate concentration.

EXERCISES

1. Explain the observed variation of rate constant with ionic strength on the basis of either the collision theory or the transition state theory of reaction rates.

2. Suggest a method of showing that the thiosulfate ion does not itself take part in the reduction of the persulfate ion.

3. Estimate the importance of the errors introduced by assuming that $dx/dt = \Delta x/\Delta t$ and that the volumes of solution are additive.

Experiment 30

Kinetics of a Fast Reaction, by a Relaxation Technique

Concentration-Jump Method

OBJECTIVES / To disturb an equilibrium system by a concentration = jump method and to follow the relaxation reaction by measuring absorbances with a spectrophotometer. From the data, relaxation times and rate constants are calculated.

READING ASSIGNMENT / Chapter 24, pages 319–320.

The equilibrium reaction between chromate and dichromate ion in water

$$2\,H_3O^+ + 2\,CrO_4^{2-} \rightleftharpoons Cr_2O_7^{2-} + 3\,H_2O$$

is quite rapid and the kinetics cannot conveniently be studied by normal methods. In this experiment, the C-jump relaxation method is used. The reactants are mixed and allowed to reach equilibrium. Then, the concentrations of the reactants and products are suddenly (in about 4 sec) changed and the reaction is monitored during the time taken to return to equilibrium. At t_0, the start of the observations, c_i, the concentration of the species being observed, is not the equilibrium concentration, at t_∞. The difference, at any time, between the concentration, c_{i_t} and c_{i_∞} is Δc_i. This difference, Δc_i, approaches zero as an exponential function of time and at any time

$$\frac{d\,\Delta c_i}{dt} = -\frac{\Delta c_i}{\tau}$$

τ = relaxation time

The relaxation time, τ, is the characteristic constant for the exponential change. It is the time needed for Δc_i to fall from any value to the fraction $1/e$ of that value. It can be obtained from the graph of Δc_i versus time. One simply selects a value of Δc_i at a desired t_1 and then finds from the curve the time t_2 at which

EXPERIMENTS

Δc_i is $1/e$ that of its value at t_1. τ is $t_2 - t_1$. Another way to find τ is to plot the natural logarithm of Δc_i versus time. The slope of the line is $-1/\tau$.

The relaxation time for any reaction is a complicated function of rate constants, concentrations, and equilibrium constants, the form of the function depending on the mechanisms of the reaction. For a simple first order reaction,

$$\frac{1}{\tau} = k_f + k_r$$

k_f, k_r = rate constants for forward, reverse reactions

In this case, the mechanism is

(1) $\qquad H_3O^+ + CrO_4^{2-} \rightleftharpoons HCrO_4^- + H_2O \qquad$ fast

(2) $\qquad 2\ HCrO_4^- \underset{k_r}{\overset{k_f}{\rightleftharpoons}} Cr_2O_7^{2-} + H_2O \qquad$ slow

To follow the H_3O^+ concentration, an indicator is added to the solution, giving rise to a third equilibrium,

(3) $\qquad H_3O^+ + In^- \rightleftharpoons HIn + H_2O \qquad$ fast

The equilibrium constants for the first two reactions are $K_1 = 1.3 \times 10^6$ and $K_2 = 5 \times 10^1$. K_3 depends on the indicator chosen.

Under the conditions of this experiment,

(4) $\qquad \dfrac{1}{\tau} = 4k_f[HCrO_4^-] + k_r[H_2O]$

Therefore, k_f and k_r are easily obtained from a graph of $1/\tau$ against $[HCrO_4^-]$.

In any case, during any one run $[HCrO_4^-]$ is nearly constant (although it differs from run to run) and may be taken to be equal to $[HCrO_4^-]_\infty$ at the end of the run. $[HCrO_4^-]_\infty$ may be calculated from the material balance

(5) $\qquad [Cr_{total}] = [CrO_4^{2-}] + [HCrO_4^-] + 2[Cr_2O_7^{2-}]$

and the equilibrium constants for reactions (1) and (2). The relation is

(6) $\qquad [HCrO_4^-]_\infty$

$$= \frac{1}{2K_2}\left(\sqrt{\frac{K_1[H_3O^+]+1}{K_1[H_3O^+]} + 4K_2[Cr^{VI}]} - \frac{1+K_1[H_3O^+]}{K_1[H_3O^+]} \right)$$

In this experiment, the experimental conditions permitting the use of equation (4) are $[Cr_{total}] \approx 10^{-2}$ and $[H_3O^+] \approx 10^{-7}$.

To run the experiment, two solutions are made which when mixed will be in the desired pH range. Each has a known concentration of Cr^{VI}. The reactants are mixed rapidly and efficiently and the reaction is followed by measuring the absorbance of the indicator at 620 nm on a recording spectrophotometer, such as a Cary 14 or 15, or a Beckman DBG. The concentration of $HCrO_4^-$ is

Experiment 30. Kinetics of a Fast Reaction, by a Relaxation Technique

calculated from the total chromium concentration and the final pH of the solution. The relaxation time is obtained either directly from the curve of absorbance versus time or with a plot of ln(absorbance) versus time, taking the logarithms of points on the experimental curve. $1/\tau$ is then plotted against $[HCrO_4^-]$, and the slope and intercept are determined. The ionic strength must be maintained constant because the rate is very sensitive to secondary salt effects, since all the reactants and intermediate complexes are charged.

EQUIPMENT AND MATERIALS

A recording double beam spectrophotometer, a pH meter, a 3 ml syringe, and a microsyringe.

Stock solutions of 0.1 M KNO_3, 2 M KOH, and 10^{-3} M bromophenol blue (BPB) in methanol. Crystalline $K_2Cr_2O_7$.

DIRECTIONS

Preparation of Solutions. Make up two solutions of $K_2Cr_2O_7$ in 0.1 M KNO_3. For solution A, the more concentrated, put about 1.5 g of $K_2Cr_2O_7$ (weighed to the milligram) in a 50 ml volumetric flask and fill to the mark with 0.1 M KNO_3 solution. Solution A is about 0.1 M in dichromate. For solution B put 10 ml of stock BPB (10^{-3} in methanol) in a 1 liter volumetric flask, add about 2.9 g of $K_2Cr_2O_7$ (weighed to the milligram), and fill with 0.1 M KNO_3, either by adding stock solution or by putting 10.1 g of solid KNO_3 in the flask and filling to the mark with water.

Take 100 ml of solution B and add a few drops of 2 M KOH to bring the pH to the 6.4–7.2 range, as measured with a pH meter.

Mixing the Reactants and Monitoring the Reaction. Put 3 ml of B each in the sample and the reference cell. Start the recorder and set the baseline at 620 nm. With the microsyringe quickly inject a known volume of A (between 0.05 and 0.2 ml) into the solution in the sample cell and, as quickly as possible, with the 3 ml syringe, aspirate and reinject the solution into the cell, to ensure thorough mixing.

After the absorbance stops changing, stop the recorder and measure the final pH of the solution. Wash out the cells and syringes and repeat the procedure, varying the volume of A, the pH of B, and the dichromate concentrations of A and B as desired.

GRAPHS AND CALCULATIONS

For each run, from the absorbance–time graph, either calculate $1/\tau$ directly or plot a graph of ln absorbance versus time and calculate $1/\tau$ from that.

For each run, calculate $[HCrO_4^-]$ from the final pH and the total $[Cr_{total}]$ concentration, using equation (6).

Plot $1/\tau$ versus $[HCrO_4^-]$. Use equation (4) to calculate k_f and k_r. Assume that the concentration of water is 55.5 moles/liter.

From k_r and k_f, calculate K_2 and compare it with the accepted value of 5×10^1.

NOTES / The recorder and spectrophotometer must have a response time of about 1 sec or less.

The molarity of the KNO_3 solution should be known to two significant figures.

Calculate $1/\tau$ from several points on each curve. If the data are good, consecutive results on the same curve should agree within 10%.

Practice the injection-mixing technique with water until you can perform the operation in 4 sec or less.

Be very careful in making up your solution. An error here propagates itself throughout the entire experiment.

EXERCISE
1. Derive equation (6).

Experiment 31

Kinetics of the Formation of Peroxochromic Acid

Stopped-Flow Method

OBJECTIVES / To mix the solutions for a rapid reaction, using the stopped-flow technique and to follow the progress of the reaction, using a spectrophotometer connected to a storage oscilloscope. From the data, half-lives times and rate constants are calculated.

READING ASSIGNMENT / Chapter 24, pages 317–319.

In acid solution, hydrogen peroxide oxidizes dichromate ions to peroxochromic acid in accord with

(1) $$HCrO_4^- + 2\,H_2O_2 + H_3O^+ \rightleftharpoons CrO_5 + 4\,H_2O$$

This reaction goes rapidly to virtual completion, the equilibrium constant being about 10^8. The peroxochromic acid, however, is unstable and rapidly decomposes to Cr^{III}, with the liberation of oxygen. The rate equation for the formation of peroxochromic acid is

(2) $$\frac{d(CrO_5)}{dt} = k(H^+)^m(H_2O_2)^n(HCrO_4^-)^p$$

In the presence of a large excess of H_3O^+ and H_2O_2, the rate law becomes

(3) $$\frac{d(CrO_5)}{dt} = k_{obs}(HCrO_4^-)^p$$

where $k_{obs} = k(H_3O^+)^m(H_2O_2)^n$.

EQUIPMENT AND MATERIALS

The equipment consists of a stopped-flow apparatus (Diagram E), a spectrophotometer equipped with terminals for connection to an oscilloscope, the oscilloscope, and an oscilloscope camera.

Stock solutions of 1 M KNO_3, 1 M $K_2Cr_2O_7$, 1 M HNO_3, and H_2O_2.

EXPERIMENTS

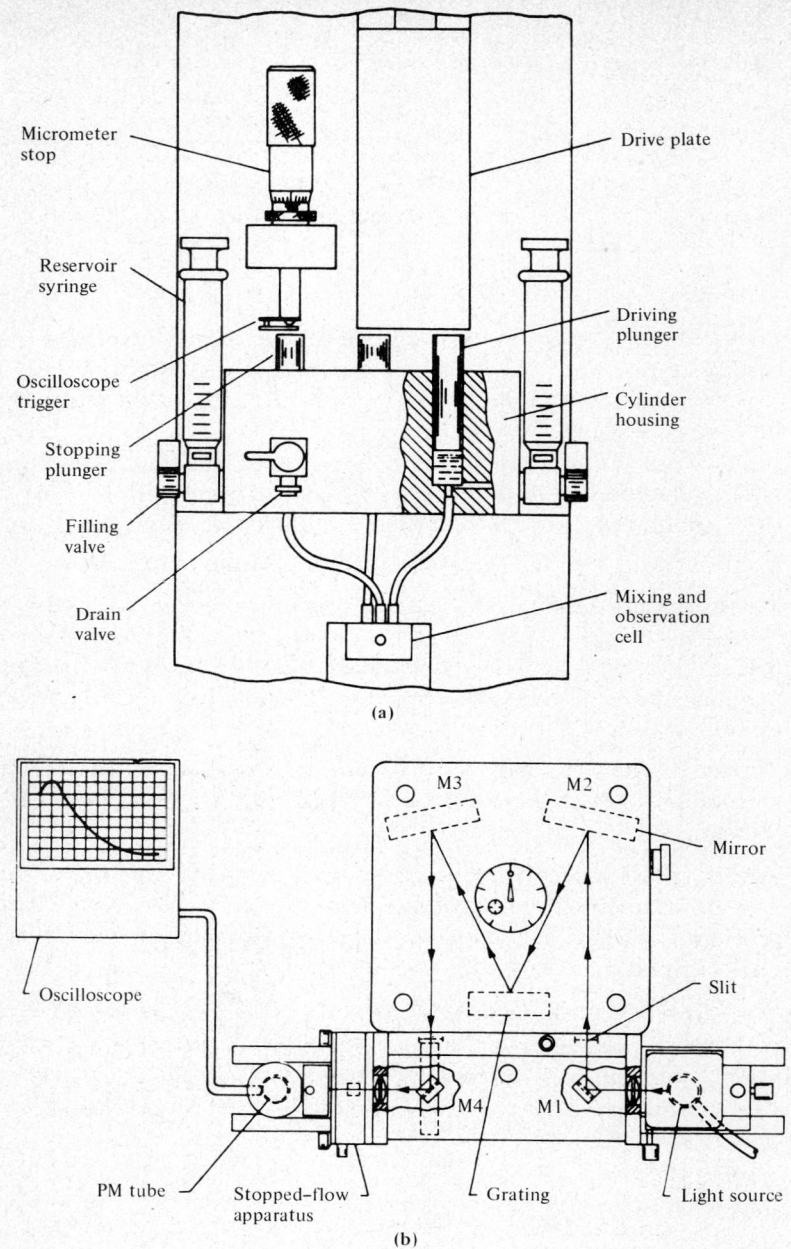

DIAGRAM E / Aminco–Morrow stopped-flow apparatus: (a) mixing mechanisms; (b) overall schematic (courtesy of American Instrument Company).

Experiment 31. Kinetics of the Formation of Peroxochromic Acid

DIRECTIONS

Calibrate the oscilloscope sweep rate.

Follow the manufacturer's instructions for using the stopped-flow apparatus.

Run a series of reactions to determine the order with respect to each reactant. In each reaction the H_3O^+ and the H_2O_2 should be present in good excess. The ionic strength of the reaction mixture should be kept at 0.1.

CALCULATIONS AND GRAPHS

Use the half-life method to determine the order with respect to the reactant whose concentration is varied and the apparent rate constant k_{obs}.

From each value of k_{obs} and the concentrations of the reactants, calculate the overall rate constant, k.

NOTES / The reactants should be kept at 25°C to within 0.1°C.

The reactant solutions should all be made up at the same ionic strength so that there is no change on mixing. KNO_3 is the neutral salt.

The molarities of the reacting mixtures in the cell at zero time should be 0.2–0.5 × 10^{-3} for $Cr_2O_7^{2-}$, 2–5 × 10^{-2} for H_2O_2, and 2–10 × 10^{-2} for H_3O^+. Since each syringe of the apparatus has the same volume, the reactant solution should have twice the $Cr_2O_7^{2-}$ and H_2O_2 concentrations as the desired values of the reaction mixture.

Since reactant solutions must be acidic, for ease in calculation each reactant solution should have the same pH, and thus dilution need not be taken into account.

The reaction product decomposes relatively slowly within the narrow concentration limits given above. Outside these concentration ranges the product forms, but decomposes so rapidly that accurate analysis becomes very difficult [Moore, Kettle, and Wilkins].

The major difficulty in operating the stopped-flow apparatus is the formation of bubbles. These cause irreproducibility. If the syringes are manually operated instead of automatically driven, the student should practice to develop a consistent manner of pushing the driving mechanism.

EXERCISES

1. How would you determine the rapidity with which the solutions are mixed?
2. How would you determine the linear velocity of the fluid moving through the mixing chamber?
3. How would you determine the time required to fill the cell?

References

Caldin, E. F., *Fast Reactions in Solution.* Wiley, New York, 1964.

Chance, B., The accelerated flow method for rapid reactions, *J. Franklin Inst.*, **229**, 455–76, 613–40, 737–66 (1940).

Moore, P., S. F. A. Kettle, and R. G. Wilkins, Kinetics of formation of blue peroxychromic acid in aqueous solution, *Inorg. Chem.* 3, 466 (1966).

Experiment 32

Cross Sectional Area of Molecules in a Soluble Monolayer

Surface Tension Measurement

OBJECTIVES / To measure, accurately, the surface tension of water and of aqueous solutions of an alcohol and to calculate the surface area of an adsorbed alcohol molecule in a monolayer.

READING ASSIGNMENT / Chapter 13, pages 124–128 and 130–132.

Solute molecules that lower surface free energy tend to be adsorbed at the surface. The extent of adsorption is given by the *Gibbs equation*,

$$(13\text{-}8) \qquad \Gamma = -\frac{1}{RT} \frac{d\gamma}{d \ln c}$$

Γ = excess surface concentration, moles/cm^2
R = gas constant, ergs/mole-K
γ = surface tension, dynes/cm
c = concentration in solution

Since the reciprocal of Γ is an area per mole, a graph of $1/\Gamma$ vs. π (the surface pressure) is a two-dimensional analog of a P–V isotherm. From such an isotherm, the point of complete coverage can be determined, and therefore the area per mole and the area per molecule calculated.

In this experiment, surface tension measurements are made on solutions of alcohols. From the results, adsorption and surface pressure calculations are made and used to estimate molecular cross sectional area.

EQUIPMENT AND MATERIALS

The du Noüy tensiometer or the Rosano tensiometer.
    Distilled water and *n*-propyl, *n*-butyl, isobutyl, *tert*-butyl, or *n*-amyl alcohol.

EXPERIMENTS

DIRECTIONS

Prepare a 1 M stock solution of the alcohol (a 0.1 M stock solution if n-amyl is used).

Prepare successive $\frac{1}{10}$ dilutions of the alcohol in water and determine the surface tension of each. Continue diluting until the surface tension of the solution approaches that of pure water.

CALCULATIONS AND GRAPHS

Draw the graph of γ vs. log c and determine the slope at each experimental value of c.

Use the Gibbs adsorption isotherm to calculate Γ, the excess surface concentration, at each solution concentration used.

The difference between γ_0, the surface tension of the solvent, and γ, the surface tension of the solution, is π, the surface pressure. $1/\Gamma$ is the area per mole. Plot a graph of π vs. $1/\Gamma$. This is the two-dimensional isotherm analogous to the three-dimensional p vs. v graph.

From the π vs. $1/\Gamma$ graph, determine the surface pressure and the surface area per mole of adsorbed solute at the point of complete surface coverage (i.e., monolayer formation).

Calculate the surface area per adsorbed molecule at the point of monolayer formation.

The cross sectional area of a hydrocarbon chain is about 22 $Å^2$. Estimate the orientation of the solute molecules relative to the surface and comment on this, if your experimental results warrant a comment, in terms of solubilities and molecular attractions.

From the calculated cross sectional area of a molecule and the density of the alcohol, estimate the thickness of a monolayer.

NOTES / Clean the glassware thoroughly with detergent solution.

In making up the stock solution the alcohol may dissolve slowly. If so, heat the mixture until solution is complete and then cool it down to 25°C.

Allow time for the surface to reach equilibrium by letting the solutions stand in wide-mouth dishes for at least 15–20 min before making measurements. Cover the dishes to minimize contamination.

If necessary for interpolation, make up additional dilutions at intermediate concentrations.

Experiment 33

Insoluble Monolayers on a Liquid Surface

Surface Pressure Measurement

OBJECTIVES / To measure, accurately, the surface tension of acidified water on which a film of an insoluble organic acid has been deposited. From the surface tension measurements at different surface areas, the molecular weight of the acid is estimated and the molecular cross sectional area is calculated.

READING ASSIGNMENT / Chapter 13, pages 124–128 and 130–132.

Just as the molecular weight of a gas, within a few per cent, can be determined from the ideal gas law, so the molecular weight of an adsorbed "two-dimensional gas" can be determined from the two-dimensional analog of the ideal gas law, that is, the *surface isotherm*,

$$\pi A = \frac{gRT}{M}$$

π = surface pressure
A = surface area
g = mass of the surface "gas"
M = molecular weight of the "gas"

In addition to the molecular weight, the cross sectional area of a molecule can be calculated from the surface isotherm. The cross sectional area of the surface (and therefore of the surface film) is decreased until the adsorbed molecules form a complete monolayer, touching each other on all sides. From the area of the surface at this point, and the number of molecules in the film (obtained from the mass, g, and the molecular weight) the cross sectional area of a molecule can easily be calculated.

In this experiment, an insoluble film of a known mass of solute is spread on a water surface whose area can be varied. The surface tension is measured at each area and the surface pressure calculated. An isotherm is graphed and from the minimum area to which the film can be compressed, without destroying it, the area of a monolayer is calculated. From the surface pressures at large areas, where the molecules are far apart, the molecular weight of the solute is calculated. From the molecular weight of the solute, the mass of the surface layer and the area of the monolayer, the cross sectional area of the molecule is obtained.

EQUIPMENT AND MATERIALS

Rosano tensiometer, a paraffin-coated rectangular glass tray, a calibrated syringe or a 1 ml calibrated pipet, a ruler, and two waxed glass rods or flat glass strips.

Solutions of accurately known concentration of myristic acid and of stearic acid in petroleum ether containing about 100 mg of acid per 100 ml of solution.

0.01 M hydrochloric acid and calcined talcum powder.

DIRECTIONS

Fill the tray with 0.01 M hydrochloric acid solution and clean the surface. Place a glass strip or rod at one end of the tray to act as a movable barrier.

Set the tensiometer in position so that the blade is in contact with the liquid at the end of the tray opposite to the movable strip.

INSOLUBLE MONOLAYERS ON A LIQUID

Measure and record the force exerted by the liquid on the blade.

Spread 0.2–0.3 ml of the acid petroleum ether solution, measured accurately, onto the surface and allow the solvent to evaporate. Measure and record the force exerted on the blade. Measure and record the surface area.

Slide the glass strip a few centimeters toward the other end of the tray, decreasing the area. Measure and record the area and the restoring force.

Repeat the procedure, decreasing the area in successive steps until either the movable strip reaches the other end of the tray or the restoring force passes through a maximum and then decreases.

Repeat the procedure with the other acid.

CALCULATIONS AND GRAPHS

Calculate the surface pressure at each area for which there is a surface tension measurement.

Plot the graph of π vs. A, the surface area.

At several positions on the isotherm, corresponding to low surface pressures and large areas, calculate the molecular weight of the solute, as a check on the equations.

Extrapolate the linear high pressure portion of the graph down to the area axis. From the intercept on this axis and the number of molecules in the film, estimate the cross sectional area of the molecule.

Use the calculated value of the cross sectional area and the accepted value of the density of the solute to calculate the length of the molecule. From the difference between the calculated chain length of stearic acid (C_{18}) and myristic acid (C_{14}) and the accepted value of 109° for the carbon–carbon single bond angle, estimate the carbon–carbon single bond distance and compare this to the accepted value.

NOTES / The top edge of the tray must have a paraffin coating.

Clean the surface by dusting a small quantity of calcined talcum powder over it and then sweeping away the talcum film along with any surface dust or oil. Use either waxed glass rods or glass strips to sweep the surface, or gently blow filtered air over it, collecting the talcum and dirt at one corner. Then remove the film of dirt and talcum with a pipet connected to an aspirator. The talcum shows any surface dirt present, since dust particles or oil droplets that are not visible to the eye produce large visible gaps in the talcum film.

As the area gets smaller, decrease the distance through which the rod is moved between measurements.

Use the n-hexane fraction of the petroleum ether for best results.

EXERCISE

1. Explain the function of the hydrochloric acid.

Experiment 34

Surface Area of a Powder

BET Gas Adsorption Method

OBJECTIVE / To measure the pressure of nitrogen in a system containing a sample of powder kept at liquid air temperature. From the measured pressure, the volume of adsorbed nitrogen (at STP) is calculated. From the results of measurements at several different pressures, the surface area of the powder is estimated.

READING ASSIGNMENT / Chapter 13, pages 132–134.

In this experiment, the surface area of a sample of powder is calculated from the volumes of nitrogen adsorbed at a series of pressures. The Brunauer–Emmett–Teller equation is used for the calculations.

In principle, the experiment is simple. A sample of powder is put into a chamber and cooled down to the temperature at which nitrogen liquefies at atmospheric pressure. A known quantity of nitrogen gas, at a pressure of less than 1 atm, is admitted to the sample chamber, where some is adsorbed on the powder. The pressure is measured. From the pressure and temperature and the volume of the chamber, the quantity of nitrogen *not adsorbed* is calculated. The difference between the initial amount of nitrogen gas and the quantity left after adsorption is the quantity of adsorbed nitrogen, which is a variable in the BET equation. Many manipulations are necessary and great care is required.

The apparatus (Diagram F) consists of a series of bulbs of known volume connected to a central space, also of known volume. This central space is in turn connected to the sample chamber, to a manometer, and to a manifold for adding or removing gas. The manifold is connected to a vacuum pump and to a nitrogen tank. All volumes must be known, including the volume between the central space and the manometric liquid, so that, from the pressure and temperature, the quantity of nitrogen in each portion of the apparatus can be calculated. In addition, the free volume or dead space in the sample chamber must

Experiment 34. Surface Area of a Powder

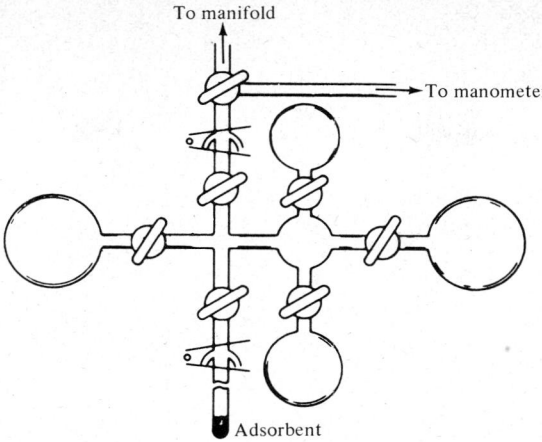

DIAGRAM F / Gas buret for BET apparatus.

be known. This is the volume in the sample chamber that is not filled with solid and so is open to the gas. It includes not only the space above the sample but the spaces between the grains of the solid.

To measure the volumes of the bulbs and the other spaces in the buret and manometer, a calibrated bulb or flask is used. This is a bulb whose volume has been obtained from the weight of water it can contain. The calibrated bulb containing air at atmospheric pressure is fastened to the buret in place of the sample chamber and its stopcock closed. The rest of the system is then evacuated and all the stopcocks closed. The stopcocks between the calibrated volume and the central space are opened and the pressure measured. Using Boyle's law, from the original pressure and volume and the new pressure, the new volume is calculated, giving the volume of the central space and manometer lines. Each bulb is opened, in turn, and each time the new pressure determined, enabling the bulb volume to be calculated. After all the system volumes have been measured, the calibrating volume is replaced with the sample holder and sample.

To measure the dead space, the sample chamber is evacuated and the center space is filled with a known pressure of nitrogen. The stopcock is opened to the sample chamber, which for this measurement is kept at room temperature, so that there is no adsorption. From the new pressure, the total volume and therefore the dead space may be calculated.

After the various volumes are measured, the same technique is used to determine the adsorption of nitrogen. The entire system is evacuated, the sample chamber is closed off, and nitrogen is admitted to the rest of the system until the pressure is about 10–20 mm, measured accurately. Each portion of the system now contains a definite amount of nitrogen that can easily be calculated from the ideal gas law. All stopcocks are now closed, isolating the buret from

EXPERIMENTS

the manifold and shutting off each bulb. The sample chamber is immersed in a Dewar of liquid nitrogen, cooling the sample down to the temperature at which nitrogen can be adsorbed. The stopcock to the sample chamber is opened and nitrogen enters from the central space. Some nitrogen is slowly adsorbed, reducing the pressure below that which would have been observed without adsorption. After about 15–30 min, adsorption is complete and the pressure is steady. The pressure is recorded and one of the bulbs is opened, admitting an additional increment of nitrogen while increasing the volume of the gas space. After equilibrium is reached (in about 5 min) another bulb is opened, and so on. The equilibrium pressure is recorded each time. When all the bulbs have been opened, there should be sufficient data for the calculation of the isotherm.

DIRECTIONS

First calibrate all volumes of the system, including the dead space. Once the volumes are known, evacuate the sample chamber, cool it with liquid nitrogen, and adsorb successive known increments of nitrogen from the gas buret. Each time, record the equilibrium pressure and volume.

CALCULATIONS AND GRAPHS

Calibration of the Volumes. Obtain the volume of the capillary manifold from the several values of the initial pressure in the calibrated bulb and the final pressure in the bulb and capillary manifold. Use Boyle's law in the form

$$p_i v = p_f (v + v_x)$$

p_i = initial pressure
v = calibrated volume
p_f = final pressure
v_x = volume of the capillary manifold

The manifold volume v_x may be calculated from the results of the different runs and the values then averaged. Alternatively, a graph of $v(p_i - p_f)$ vs. p_f may be plotted, giving a straight line whose slope is v_x.

For the volume of the bulbs of the buret, use the calculated value of the initial volume, that is, that of the calibrated bulb, the capillary manifold, and any buret bulbs previously measured. The same calculations are performed as for the volume of the manifold.

For the volume of the dead space, repeat the calculation, using as the initial volume the calculated value of v_x and the various buret bulbs.

Calculation of the Volume of Adsorbed Nitrogen. From the initial volume, pressure, and temperature of the nitrogen, calculate the number of moles of gas in the system.

From the equilibrium pressure and the volume and temperature in the capillary manifold and in the sample chamber, calculate the number of moles of gas present at equilibrium.

From the difference, calculate the number of moles of adsorbed nitrogen and its volume at STP.

Plot the BET graph of $p/v(p_0 - p)$ vs. p/p_0. Use the barometric pressure for p_0. From the graph calculate v_m, the volume of a monolayer.

The surface area of an adsorbed nitrogen molecule is 12 Å^2. Calculate the surface area of the sample.

From the known weight of the sample and the calculated surface area, calculate the real area (as opposed to the apparent or geometrical area) per gram.

Plot the v vs. p isotherm.

NOTES / **The temperature of the gas in the system is that of the buret, since most of the gas is in the buret. In measuring the system temperature, keep the thermometer as close to the buret as possible.**

The longer the system is degassed before use, the more precise and accurate the measurements will be. If necessary, come into the laboratory early and start the vacuum pump so that the system will be ready when the laboratory period begins.

Test the system for leaks. If there are any, call the instructor. Do not attempt to seal the apparatus by yourself.

Use two hands to turn the stopcocks and turn them with a slight inward pressure to prevent leakage of air into the system. Be gentle with the stopcocks.

Either shield the apparatus or wear safety gear, in case of implosion.

Protect the pumping system with adequate traps. If a diffusion pump is used, do not start it until the pressure has been reduced to the proper range by the forepump. In addition, do not bring the pressure back up until the diffusion pump has been turned off and its contents allowed to cool.

The sample chamber should be made of quartz or vycor glass (why?). The sample should be between 0.25 and 0.30 g.

The measured volumes should be precise to 5%.

Every time a new volume is determined by pressure change, the smaller the initial volume, the greater the pressure change and the more precise the calculated result will be.

EXERCISES

1. State the assumptions, explicit and implicit, of the BET equation. Comment on their validity.

2. What error, if any, results from the presence of 0.50% of oxygen in the tank nitrogen? How could the oxygen be removed?

3. On silica gel the results of surface area determinations with nitrogen, neon, helium, and carbon monoxide agree with each other but disagree with the results with butane and pentane, which are considerably lower. Explain.

4. Look up the thermal coefficients of cubical expansion of quartz and of pyrex. Calculate the change in dead space on cooling a pyrex and a quartz sample chamber from room temperature to the nitrogen boiling point.

References

Brunauer, S., *The Adsorption of Gases and Vapors*, Vol. I. Princeton University Press, Princeton, N.J., 1943.

Brunauer, S., P. H. Emmett, and E. Teller, Adsorption of gas in multimolecular layers, *J. Amer. Chem. Soc.* **60**, 309 (1938).

Experiment 35

Dipole Moment of a Polar Liquid

Dielectric Constant Measurement and Refractometry

OBJECTIVES / To determine the dielectric constant and the density of each of a series of solutions of a polar liquid in benzene. From these results and the refractive index of the solute, the dipole moment of the solute is calculated.

READING ASSIGNMENT / Chapter 14, pages 165–167; Chapter 8, pages 75–77; Chapter 18, pages 221–225.

The dipole moment of a molecule may be obtained in several different ways, but usually it is calculated from the dielectric constant. In measuring the dielectric constant, the material is inserted between the plates of a capacitor. If the molecules of the substance have dipoles, they will tend to line up in the electric field, decreasing the field strength and increasing the capacitance.

$$\varepsilon = \frac{C}{C_0}$$

C_0 = capacitance with vacuum between plates
C = observed capacitance with substance between plates

The ratio is called the *dielectric constant*. The larger the dipole moment, the greater the interaction between the molecules and the field and the larger the dielectric constant. The proportionality is not direct, however, because the field also distorts or "polarizes" the molecules, inducing an additional dipole moment in each. The observed dielectric constant therefore depends on two factors, the permanent dipole moment and the polarizability of the molecules. There is, of course, also a density factor. The more molecules there are between the plates (provided that the plate area and interplate distance are constant), the greater the interactions and the greater the dielectric constant. The more dilute the solution, the lower the dielectric constant. Gases will have lower dielectric constants than liquids. Air has a dielectric constant so close to that of a vacuum that for most measurements, the capacitance in air may be substituted for C_o, the capacitance in a vacuum.

To obtain the molecular dipole moment from the observed dielectric constant, a quantity called the *molar polarization* is used. This is defined as

(1) $$P = \frac{\varepsilon - 1}{\varepsilon + 2}\left(\frac{M}{\rho}\right)$$

P = molar polarization
M = molecular weight
ρ = density

It has the units of a molar volume, since the dielectric constant is dimensionless. For gases and for nonpolar liquids, the connection between the dipole moment of the molecule and the molar polarization is given by

(2) $$P = \frac{4\pi N \mu^2}{9kT} + \frac{4\pi N \alpha}{3}$$

μ = dipole moment
k = Boltzmann's constant
N = Avogadro's number
T = Kelvin temperature
α = polarizability of the molecule

The first term of the right-hand side is the orientation polarization, due to the permanent dipole moment. The second term is the induction polarization, due to an additional dipole moment induced by the electric field used to measure the dielectric constant. Any field strong enough to be used to orient molecules will also distort them, inducing a dipole whether or not there was one originally.

To calculate the dipole moment, the induction (or distortion) polarization must be taken into account. One way to do this is to obtain the molar polarization at a series of temperatures and then to plot a graph of P vs. $1/T$. The slope of the line is $4\pi N \mu^2 / 9k$.

Another way to eliminate the induced polarization is to calculate it and then subtract it from P. For this calculation, light is considered as an electromagnetic radiation whose frequency is so high that although it induces dipoles, it does not orient molecules (i.e., before the molecule can be oriented by the field of the light, the sign of the field reverses). From the interaction of radiation with the molecules, the *molar refraction* is calculated. This is defined as

(3) $$R = \frac{\eta^2 - 1}{\eta^2 + 2}\left(\frac{M}{\rho}\right)$$

η = refractive index

The resemblance between equations (3) and (1) is obvious. For light of "infinite" wavelength, the molar refraction equals the induced polarization term of equation (2):

$$R_\infty = \tfrac{4}{3}\pi N\alpha$$

Unless there is an absorption band in the visible spectrum, light at the wavelength of the sodium D line may be substituted for light of infinite wavelength. To obtain the dipole moment, therefore,

$$P - R_D = \frac{4\pi N\mu^2}{9kT}$$

R_D = molar refraction at wavelength of the sodium D line

However, one important difficulty arises. Equation (2) cannot be used for polar liquids because of the interactions between molecules. The dipoles distort the adjacent molecules, producing additional induced dipoles that increase the measured polarization. In this experiment, instead of calculating and correcting algebraically for this perturbation, it is minimized by an experimental procedure.

The polar liquid is dissolved in a nonpolar solvent, separating the polar molecules and decreasing the dipole-induced dipole interactions. Measurements are made at successive dilutions and the results are extrapolated to zero concentration, or infinite dilution, where there is presumably no interaction between the molecules, and equation (2) can then be applied.

The molar polarizations of a series of solutions of a polar liquid in a nonpolar solvent, such as benzene, are determined. For each solution

$$P_{12} = \frac{\varepsilon - 1}{\varepsilon + 2}\left(\frac{M_1 X_1 + M_2 X_2}{\rho}\right)$$

P = molar polarization
M = molecular weight
X = mole fraction
1 refers to solvent
2 refers to solute
ρ = density of solution

The term in the numerator of the second fraction is the number average molecular weight,

$$\frac{\sum n_i M_i}{\sum n_i} = \sum X_i M_i$$

The molar polarization of the solution is related to that of the solvent by

$$P_{12} = P_1 X_1 + P_2 X_2$$

For dilute solutions of component 2 in component 1, P_1 is assumed to equal P_1°, the molar polarization of the pure solvent (component 1). Therefore,

$$P_2 = \frac{P_{12} - P_1^\circ X_1}{X_2}$$

Values of P_2 at each concentration can be calculated from the molar polarizations of solution and solvent. A graph of P_2 versus X_2 is plotted and extrapolated to infinite dilution. The extrapolated value of solute polarization is used with the molar refraction of the pure liquid to calculate the dipole moment.

EQUIPMENT AND MATERIALS

An impedance bridge, a capacitance cell, an Abbe refractometer, and a pycnometer.

Benzene as the solvent and chlorobenzene, chloroform, or nitrobenzene as the polar liquid.

DIRECTIONS

Determine the density and refractive index of the solute at 25°C.

Determine the density and dielectric constant of the solvent and of solutions of polar liquid in benzene at mole fractions of solute ranging from 0.003 to 0.1. Use at least four or five different concentrations.

CALCULATIONS AND GRAPHS

Calculate the molar polarization of the solvent and of each solution.

Calculate the molar polarization of the solute in each solution.

Plot a graph of P_2 vs. X_2 and extrapolate to zero mole fraction. Make additional measurements if necessary.

Calculate the dipole moment of the solute, using the molar refraction and the molar polarization at infinite dilution.

NOTES / Measure the dielectric constants to at least three significant figures.

Obtain densities and mole fractions to at least three significant figures.

In graphing select a scale appropriate to the precision of the calculated values of P_2.

Experiment 35. Dipole Moment of a Polar Liquid

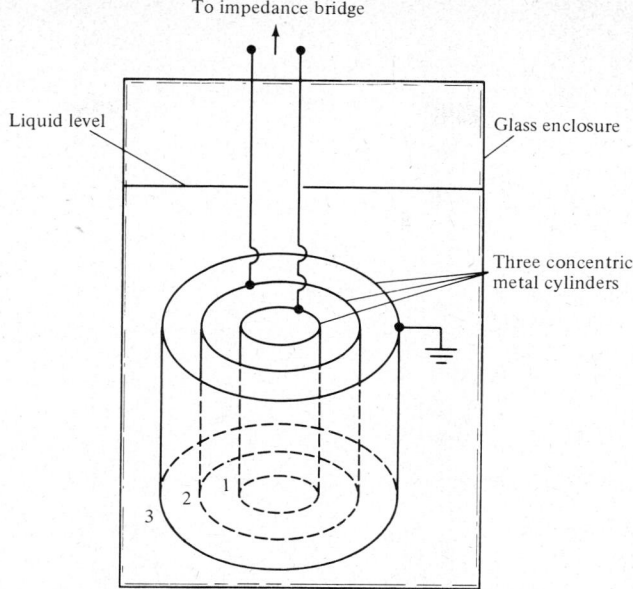

DIAGRAM G / Cell for measuring dielectric constants of liquids. Cylinders 1 and 2 go to bridge; cylinder 3 goes to ground.

The best type of capacitance cell consists of two concentric cylinders, arranged as shown in Diagram G and enclosed within a third cylinder that serves as an electrical shield.

Chlorinated solvents are very toxic, and nitrobenzene and benzene are worse. Work in the hood and absolutely *do not* spill solvent on the skin.

EXERCISES

1. Explain the terms induction polarization, orientation polarization, and dipole–induced dipole polarization.

2. Would the environment of the solute molecule at infinite dilution be similar to that in the gas phase? Would its dipole moment be the same as that in the gas phase? Explain, with reference to dispersion forces between nonpolar molecules.

3. Your procedure is based on the implicit assumption that the dielectric constant of a material without a permanent dipole is equal to the square of the refractive index. Explain the limits of the validity of this assumption and its effect on your calculated results.

EXPERIMENTS

4. The dipole moment of chlorobenzene is 1.70 D (debyes). Predict the moments of ortho-, meta-, and para-dichlorobenzenes.

5. Carefully consider and discuss whether or not a permanent dipole moment exists in *p*-diaminobenzene, *p*-dihydroxybenzene, and other compounds in which the axis of symmetry of the substituent does not pass through the center of the benzene ring, or in which the substituent does not have an axis of symmetry.

Chapter 36

Bond Moments, Bond Angles, and Dipole Moments

Dielectric Constant Measurement and Refractometry

OBJECTIVES / To measure the capacitance and refractive index and determine the density and dielectric constant of a monosubstituted benzene and its ortho- and meta-disubstituted derivatives. From the results, the dipole moments, bond moments, and bond angles are calculated.

READING ASSIGNMENT / Chapter 14, pages 165–167; Chapter 18, pages 221–225.

No completely satisfactory relationship between dipole moments and dielectric constants of polar liquids has yet been developed. One common method of determining dipole moments of polar liquids is the one employing several dilutions in nonpolar solvents and extrapolating to zero solute concentration (Experiment 35). Even with this technique, however, there is sometimes a discrepancy between dipole moments from measurements on gases and those from solutions. There are some semiempirical relationships that give reasonably good results. The *Böttcher* [1943] *equation*,

$$\mu^2 = \frac{9kT(\varepsilon - \eta^2)(2\varepsilon + \eta^2)M}{4\pi N\varepsilon(\eta^2 + 2)\rho}$$

μ = dipole moment, debyes
k = Boltzmann's constant
ε = dielectric constant
η = refractive index
M = molecular weight
N = Avogadro's number
ρ = density

is based upon the Onsager [1936] theory of liquid dielectrics. It must be regarded as an approximation, but it is a useful approximation.

EQUIPMENT AND MATERIALS

An impedance bridge, a capacitor suitable for dielectric constant measurements on liquids, a pycnometer, and a refractometer.

Toluene, nitrobenzene, and meta- and ortho-nitrotoluenes; or chlorobenzene and ortho- and meta-dichlorobenzenes.

DIRECTIONS

Measure the refractive index and density of each liquid at 25°C.
Determine the dielectric constant of each liquid at 25°C.

CALCULATIONS

Calculate the dipole moment of each liquid.

From the moments of the mono- and disubstituted compounds, calculate the angles between the substituent groups ortho and meta to each other.

NOTES / The best type of capacitor for use in this experiment is shown on page 449.

All substances used here are toxic. Work in the hood and do not spill any on the skin.

EXERCISE

1. Compare the calculated angles between ortho substituents with one half of the calculated bond angle between meta substituents and with the 60° to be expected from geometry. Explain any major discrepancy.

References

Böttcher, C. F. J., Eine neue Method zur Berechnung von Dipolmomentum, *Rec. Trav. Chim.* **62**, 119 (1943).

Onsager, L., Electric moments of molecules in liquids, *J. Amer. Chem. Soc.* **58**, 1486 (1936).

Experiment 37

Dissociation Energy of Halogen Gases

Ultraviolet Spectrophotometry

OBJECTIVES / To obtain the UV spectrum of a halogen gas and to calculate the bond dissociation energy.

READING ASSIGNMENT / Chapter 19, pages 234–237.

In this experiment, halogen gases at the ground electronic and vibrational levels are excited, by ultraviolet radiation, to the first excited electronic state. Since the electronic transition is accompanied by a change in vibrational level (from the zero vibrational level in the ground state to other quantum levels in the excited state), an entire vibrational spectrum appears and is recorded.

The difference in energy between adjacent spectral lines represents the difference in energy between two consecutive vibrational quantum levels of the excited state. At higher and higher vibrational quantum levels, this energy difference gets smaller and becomes zero at the convergence limit, where the molecule dissociates. The sum of the bond dissociation energy plus the energy of ionization of one of the dissociated atoms is calculated from the wave number of the convergence limit.

EQUIPMENT AND MATERIALS

A dual beam recording uv–visible spectrophotometer with quartz sample cells.
    Crystalline iodine and liquid bromine.

DIRECTIONS

Calibrate the recorder chart in terms of wavelength, assuming that both the chart speed and the scan speed are constant.
    Put a few crystals of I_2 in a glass-stoppered sample cell and record the spectrum in the range 400–650 nm using an empty sample cell as reference.

In a hood, pour some dry bromine vapor into a glass-stoppered cell and record the spectrum in the range 450–650 nm.

CALCULATIONS AND GRAPHS

Select a series of well-defined peaks in the I_2 spectrum and determine the convergence limit with the graphical method discussed in Chapter 19. The convergence limit in the Br_2 spectrum is obtained directly from the recorded spectrum.

The energy required to promote an electron from the ground state to the first electronically excited level is 7599 cm^{-1} for iodine and 3860 cm^{-1} for bromine. Calculate the bond dissociation energies and compare them to accepted values.

NOTES / **Do not touch the optical surfaces of the cells. Fingerprints leave etch marks.**

Protect the spectrometer against corrosive vapors by covering the top of the cell with a vapor trap, such as a pipet bulb, if necessary.

Also protect yourself. Halogen vapors are toxic and cause burns.

Since the wavelength rather than the absorbance is of interest, any fixed reference may be used provided that it does not absorb in the region being investigated.

Because iodine has such a low vapor pressure, the small changes due to absorbance at the spectral peaks will be too small to show up on the chart unless the recorder sensitivity is greatly increased. If the detector output is proportional to transmittance, however, increasing the recorder sensitivity will produce a signal that goes off scale. For example, if the transmittance is in the 95% range, the output signal might be 95 mV and the deflections due to absorbance at spectral peaks might be 1–2 mV. Increasing the recorder sensitivity to 10 mV full-scale deflections would amplify the deflections tenfold, but the 95 mV signal would be far off scale. This problem is solved by using a potentiometer to buck out most of the signal. In the example given above, a constant 90 mV bucking potential would result in a recorder signal in the center of the chart on the 10 mV scale moving up and down by 10–20% of the scale.

The bromine vapor has a much higher pressure than the iodine, but as its convergence limit is reached, the changes in transmittance become smaller and smaller until they are no longer visible. In order to follow these changes, the recorder sensitivity must be increased and the bucking potential occasionally varied to keep the recorder pen on scale as absorbance changes.

The greater the chart speed, the more precise the calibration, but the more difficult it is to locate the tip of the peak.

Experiment 37. Dissociation Energy of Halogen Gases

EXERCISES

1. How would increased temperature affect the spectrum?

2. In the case of the iodine the observed spectrum consists of two series of lines. One series, starting from the first excited vibrational state, is in the red to green region of the spectrum. The other, starting from the ground vibrational state, is in the green and violet region. Explain why the series starting from the ground state is in a higher frequency region than the other.

3. Explain why the peaks come closer to each other near the convergence limit.

4. Comment on the assumption that the molecules are harmonic oscillators. Base your statements on your observations or calculations.

5. Draw a Morse function of the ground level and the first excited electronic level for both bromine and iodine. Indicate on this graph the various quantities involved in this experiment.

6. Are the dimensions of the electronically excited iodine molecules the same as those of the ground state molecules? Explain. Answer the same question for bromine.

Experiment 38

Molecular Weight and Shape of Dissolved Polymer Molecules

Viscometry

OBJECTIVES / To determine, with a Ubbelohde viscometer, the viscosity of solutions of a polymer in alcohol and in water. The results are used to calculate specific and intrinsic viscosities at each concentration and to estimate the molecular weight of the polymer.

READING ASSIGNMENT / Chapter 12, pages 113–115 and 118–119.

In solutions of high polymers the relationship between the shape of the solute particles and the viscosity of the solution is given by equation (12-4). In this experiment the intrinsic viscosities of solutions of polyethylene glycol in water and in ethanol are determined. From the constant the shape of the polymer molecules is predicted and the molecular weight is estimated.

EQUIPMENT AND MATERIALS

Ubbelohde viscometer, a thermostat, a timer, a 10 ml pipet, and a 50 ml volumetric flask.

Carbowax 20M is the solute, and water and ethanol are the solvents.

DIRECTIONS

Start with 20 ml of 2% solution of the polymer in alcohol and use the viscometer to determine the flow time at 25°C. Repeat until 0.5 sec precision is obtained.

Repeat with three successive dilutions of the starting solution and with the pure solvent.

Repeat the entire procedure with water as the solvent.

Experiment 38. Molecular Weight and Shape of Dissolved Polymer Molecules

CALCULATIONS AND GRAPHS

Calculate the specific viscosity of the polymer solution at each concentration in each solvent.

Plot η_{sp}/c against c, and from the graph obtain the intrinsic viscosity.

Plot $\ln(\eta_c/\eta_0)/c$ against c and from the graph obtain the intrinsic viscosity. Compare this with the result previously calculated.

From the intrinsic viscosity, calculate the characteristic constant in equation (12-4).

From the value of the constant, predict the shape of the polymer molecules. If there are any differences between the polymer in water and in alcohol, explain these. From equation (12-7) and the data of Table 12-1, calculate the molecular weight of the polymer.

NOTES / **If there is curvature or excessive scatter in the graph, repeat the measurements.**

It is necessary to agitate the solutions to obtain the 2% starting solution. The agitation might cause foaming, which could make it difficult to time the rate of liquid fall. If there is foam, be sure that the timer starts and stops when the liquid level is at the mark, rather than when the foam is at the mark.

Experiment 39

Electronic Structure of an Ion

Magnetic Susceptibility Method

OBJECTIVES / To obtain measurements of magnetic force on solutions of transition metal nitrates from which magnetic susceptibilities can be calculated and the number of unpaired spins calculated for each of the metal ions.

READING ASSIGNMENT / Chapter 21, pages 271-275.

Although in most substances, paramagnetic susceptibility has both spin and orbital components, in the ions of the first transition series, the orbital component is negligible. The magnetic moment is given by

(21-7) $\quad \mu = [n(n-2)]^{1/2} B$

μ = magnetic dipole moment
n = number of unpaired spins
B = Bohr magneton

If the induced magnetic dipole moment (produced by the field) is considered negligible, equations (21-5) and (21-7) gives a simple formula for the number of unpaired spins,

(1) $\quad n(n+1) = 0.80 \chi_m T$

χ_m = molar magnetic susceptibility
T = Kelvin temperature

This equation applies to the transition metal ions.

In the apparatus used here, the magnet is a large permanent magnet of about 0.36 T (3600 G). The exact field strength is not important provided that it is constant, because a relative measurement is employed. A sample tube filled to a given height with water is suspended in the field and the apparent mass determined. The field is then removed by rolling away the magnet (with an electromagnet, the current would be turned off). The weight is then taken again. The tube is then filled with solution to the same height and the procedure repeated. The results must be corrected for the magnetic susceptibility of the glass of the tube and the water in the solution.

$$(2) \quad \frac{\chi_{sp, salt}}{\chi_{sp, water}} = \frac{m_1}{m_2} \left(\frac{(w_2 - w_0) - (w_1 - w_0)(m_3/m_1)}{w_1 - w_0} \right)$$

χ_{sp} = specific magnetic susceptibility
m_1 = mass of water used for calibration
m_2 = weight of salt in the solution
m_3 = weight of water in the solution
w_0 = weight change (in the field and out) of the glass tube
w_1 = weight change of the tube and water
w_2 = weight change of the tube and solution

At 25°C, the specific susceptibility per gram of water is accepted as 0.72×10^{-12} T. The molar susceptibility is obtained by multiplying the specific susceptibility by the molecular weight.

EQUIPMENT AND MATERIALS

Solutions of cobalt, manganous, ferric, nickel, and copper nitrate, anywhere from 0.2 M to 1 M. If the solutions are not available, make them up, weighing out the salts to three significant figures. To prevent hydrolysis, add about 1 ml of concentrated nitric acid per 100 ml of the nitrate solution.

A permanent magnet, mounted on a dolly or other movable support, or an electromagnet.

An ordinary test tube, with glass hooks on the sides so that it may be suspended from a balance.

NOTES / The tube is suspended by a thread from the balance. Make sure that the thread does not stretch or touch anything that might produce friction.

In weighing the empty tube, be sure that it is clean and dry.

Do not let the tube touch the side of the magnet during the weighing.

Use the same volume of liquid and make sure that the sample is only part way into the field and that the magnet is always in the same position when readings are taken.

Do not let the solution wet the supporting thread.

CALCULATIONS

In the calculation of the weight change due to the salt solution, the correction for the effect of the water in the salt solution is the right-hand term in the brackets in equation (2). To get the mass of the water in the salt solution, you need the density, but this experiment is imprecise enough for you to get the density from the mass of the total solution and the volume. The volume is the same as the volume of water, which you get from the mass of the water, in the calibration measurement.

Calculate the number of unpaired spins for each salt and comment on the applicability of Hund's rule and the Aufbau principle.

Experiment 40

Isotopic Composition of an Element

Mass Spectrometry

OBJECTIVES / To obtain the mass spectrum of CCl_4 and to calculate the isotopic composition of elemental chlorine.

READING ASSIGNMENT / Chapter 22, pages 289–297.

Since there are two isotopes of chlorine, ^{35}Cl and ^{37}Cl, the molecular weight of a chlorine compound such as CCl_4 is really an average of the molecular weights of the molecules containing different ratios of ^{35}Cl and ^{37}Cl. In most methods of determining molecular weights, only this average is measured. In mass spectrometry, however, each of the species produce a peak and the magnitude of the peak gives the relative abundance of the particular particle. From the relative abundances, the isotopic composition of the chlorine can be determined and the average molecular weight of CCl_4 calculated.

In addition to the chlorine isotopes, carbon has two stable isotopes. If the mass spectrometer is sufficiently sensitive, there will be two spectra, one corresponding to ^{12}C and the other to ^{13}C. The ^{13}C spectrum is not usually observed with the relatively insensitive spectrometers used as teaching tools.

EQUIPMENT AND MATERIALS

The mass spectrometer and an inlet system. The sample pressure may be reduced either in a multipurpose high-vacuum rack or with a special leak valve designed for the particular model mass spectrometer and supplied by the manufacturer.

DIRECTIONS

Obtain a background scan through mass 150. Then introduce the carbon tetrachloride sample and scan through the same range. Identify each peak present in terms of mass number and intensity.

CALCULATIONS AND GRAPHS

From the mass number of each peak, determine the composition of the fragment.

From the intensities of the peaks of the CCl_3^+ and CCl_2^+ fragments, calculate the isotope distribution ratio.

Using the experimentally determined isotope ratio, calculate the atomic weight of chlorine.

NOTES / The sample must be introduced as a vapor of pressure between 10^{-1} and 10^1 torr. This is best accomplished by connecting a tube of CCl_4 to a vacuum rack, pumping away the air and some of the vapor, allowing the vapor to fill the rack at the vapor pressure, and then pumping the vapor down to the desired pressure range.

It may be necessary to scan manually over part of the range for precise location and determination of peak heights. It is necessary to obtain a steady base line for accurate determination of peak height.

The peak for the CCl_4^+ molecular ion will be extremely small and may not be observed at all.

EXERCISES

1. Explain the presence of peaks at masses 36 and 38.
2. Explain the low intensity of the CCl_4^+ peak, and the high intensity of the CCl_3^+ peak.
3. There is a slight difference between the energy of the $C-{}^{35}Cl$ bond and that of the $C-{}^{37}Cl$ bond. Explain the reason for this energy difference and state what effect you would expect it to have on your observed isotope ratio.
4. Compare the cracking pattern you obtained with that found in the literature and comment on any differences.

Experiment 41

Appearance Potentials of Gaseous Ions

Mass Spectrometry

OBJECTIVES / To measure the ionization potentials at which the mass 29 fragment of ethane and propane appear, using a mass spectrometer. From these appearance potentials, the dissociation energy of the CH_3-H bond is estimated.

READING ASSIGNMENT / Chapter 22, pages 289–294 and 298–301.

In this experiment, appearance potentials of gaseous ions are determined and used to calculate bond dissociation energies.

The compound investigated is CH_4 so that ionization potentials are determined for C_2H_5H and $C_2H_5CH_3$ (ethane and propane).

EQUIPMENT

The mass spectrometer and an inlet system. The inlet may be either a multi-purpose high-vacuum rack or a special inlet system.

DIRECTIONS

Make a background scan to be certain that the peaks at mass 29 and 30 are negligibly small.

Introduce a sample of ethane and find the $C_2H_5^+$, mass 29 peak. Keeping the ion acceleration potential constant at the value for mass 29, measure the detector current at a series of electron beam potentials between 0 and 30 eV.

Replace the ethane with propane and repeat the measurements of detector current versus electron accelerator potential without changing the ion acceleration potential.

CALCULATIONS AND GRAPHS

Plot detector current versus electron beam potential for both gases on the same graph.

At a series of detector currents, measure the difference in electron beam potentials for the two gases (i.e., at a series of ordinates determine the difference in abscissae).

Plot the potential differences as a function of detector current and extrapolate to zero detector current.

From the difference in appearance potential between ethane and propane, the dissociation energy of the hydrogen molecule, and the heats of formation of methane, ethane, and propane calculate the energy of the CH_3-H bond.

NOTES / Both gases should be introduced to the mass spectrometer at about the same pressure. If the inlet system has a dosing volume (i.e. if a fixed volume of sample is introduced and expanded to another, larger fixed volume), this requirement will be automatically fulfilled.

To prevent the ethane peak from contributing to the propane values, the inlet should be closed for about 15 min and the residual value of the mass-29 peak should be checked before admitting the propane.

Zero readings should be made from time to time. This is done most conveniently by switching the ion accelerating voltage to mass 3. Since there is almost zero tritium present, the detector current should read zero. More measurements should be taken at the lower current potential values, since this is the region of interest.

EXERCISES

1. Explain the rationale of the extrapolation procedure.

2. Why is it *not* necessary to calibrate the electron beam scale so as to obtain absolute values?

3. The two gases not only have different sensitivities toward ionization by electron impact, but the mass 29 fragment represents a different fraction of each gas. Does this materially affect your calculations? If not, explain why not. If it does, suggest a method of calculating a correction factor or an experimental method for avoiding this complication.

4. Why is it that the excess ethane cannot be flushed out with a neutral gas, such as nitrogen or helium?

Experiment 42

The Hydrogen Emission Spectrum

Emission Spectrography

OBJECTIVES / To photograph the visible emission spectrum of hydrogen gas and to calculate the Rydberg constant and the series term for the Balmer series.

READING ASSIGNMENT / Chapter 19, pages 229–230 and 244–246.

In this experiment, hydrogen gas is subjected to a high energy electric discharge, producing atoms in each of the electronically excited quantum states. These atoms drop back to lower energy states, finally reaching the ground state, emitting radiation in the process. The radiation emitted in each transition is a photon whose energy is equal to the energy difference between the initial and final energy levels. Since the energy levels are narrowly defined, the emission spectrum consists of several series of bright lines, following the general formula given in Chapter 19:

(19–2a) $$\bar{\nu} = \frac{1}{\lambda} = R_H \left(\frac{1}{n_1^2} - \frac{1}{n_2^2} \right)$$

The spectrum is photographed, and the wavelength of each line of the observed spectrum is determined. To determine the wavelength of each line, the apparatus is calibrated using light of known wavelengths. In this experiment, the mercury vapor spectrum and the sodium vapor spectrum are both photographed on the same film as the hydrogen spectrum to aid in indexing the hydrogen lines. Because the wavelengths of the mercury lines are accurately known, these lines are used as calibration points. However, there are so many in the spectrum that it is difficult to identify the individual lines. The D line of the sodium spectrum, an intense line at 589 nm, is therefore used as a calibration point to identify the mercury lines, which in turn are used to index the hydrogen lines.

For each hydrogen line, there are three unknown quantities, R_H, n_1, and n_2. To calculate R_H and to obtain the series constant, n_1, and the term constant, n_2, in the equation, the energy differences, in wavenumbers, are determined for several successive lines. From equation (19–2a) the difference in wavenumber between two successive lines e.g., line a and line b, is

$$\Delta \bar{v}_{ab} = \bar{v}_b - \bar{v}_a = R_H \left(\frac{1}{n_a^2} - \frac{1}{n_b^2} \right) = R_H \left[\frac{1}{n_a^2} - \frac{1}{(n_a + 1)^2} \right]$$

$$= R_H \frac{2n_a + 1}{n_a^2 (n_a + 1)^2}$$

where n_a is n_2 for line a. The difference in wavenumber between line a and the third line in the sequence, line c, would be

$$\Delta \bar{v}_{ac} = \bar{v}_c - \bar{v}_a = R_H \frac{4n_a + 4}{n_a^2 (n_a + 2)^2}$$

The ratio

$$\frac{\Delta \bar{v}_{ac}}{\Delta \bar{v}_{ab}} = \frac{4n_a + 4}{2n_a + 1} \left(\frac{n_a + 1}{n_a + 2} \right)^2$$

can be solved for n_a.

As soon as n_a is determined, it is rounded off to the exact integer, and either $\Delta \bar{v}_{ab}$ or $\Delta \bar{v}_{ac}$ is used to obtain a close value of R_H. This and n_a are inserted into equation (19–2a) and n_1 determined. Once n_1 is known, it, too, is rounded off to an exact integer. Finally, R_H is recalculated for each of the spectral lines, using the exact values of n_1 and n_2.

EQUIPMENT AND MATERIALS

A grating spectrograph equipped with a Hartmann diaphragm; sodium, mercury, and hydrogen vapor lamps; and a filmstrip reader.

Either 35 mm film or photoplates, depending on the particular spectrograph being used.

DIRECTIONS

Load the camera in a darkroom, use the Hartmann diaphragm to take several exposures of each spectrum—the hydrogen, the mercury, and the sodium—on one film.

Mount the developed and dry photograph on the film reader and record the positions of the sodium D line, the hydrogen lines, and two mercury lines bracketing each hydrogen line.

CALCULATIONS

From the known wavelengths of sodium and mercury lines in Table 19-2 plot a graph of wavelength versus position. Compare this chart with your film to identify the various mercury lines in the photograph.

Calculate the wavelength and wavenumber of each hydrogen line from the wavelengths of the two mercury lines bracketing it, assuming a linear interpolation over the interval measured.

Calculate the Rydberg constant and the series term for the Balmer series.

NOTES / Try to have the hydrogen exposure adjacent to mercury exposures.

Do not take two identical spectra on adjacent portions of the film.

The mercury lamp is very intense, and 5 sec of exposure time through a 10 μ slit should be sufficient. The sodium lamp is much less intense, and would take 5 min or so if it passed through the same 10μ slit. For the hydrogen lamp, a wider slit, of 50 μ, is recommended. Several exposure times should be tried. While 3, 10, and 20 min are suggested initial exposure times, shorter or longer times may be needed for a particular lamp.

Each lamp requires 10-15 min to warm up.

Shield the mercury and hydrogen light from your eyes. Ultraviolet light can cause severe eye burns.

EXERCISES

1. From your data, estimate the ionization potential of the hydrogen atom.

2. In what region of the spectrum would you look for the spectral series of hydrogen, closest to the Balmer series?

3. State how the wavelengths or frequencies of a line spectrum could be measured precisely and accurately without using a known series for a reference.

Experiment 43

Vibrational–Rotational Spectra of Gases

Infrared Spectrophotometry

OBJECTIVES / To obtain the ir spectrum of a heteropolar diatomic gas and to calculate the vibrational force constant, vibration partition function, vibrational heat capacity, intermolecular distance, and moment of inertia.

READING ASSIGNMENT / Chapter 19, pages 228–229 and 232–234.

In this experiment, a diatomic gas is excited, by infrared radiation, from the vibrational ground state to the first excited vibrational state. The excitation corresponds to a change of +1 in the vibrational quantum number. Associated with this are changes of ±1 in the rotational quantum number, so the spectrum consists of a band of sharp peaks, similar to that of Figure 19-3. There are two sets of peaks separated from each other by a narrow gap. One set corresponds to a decrease in rotational quantum number, and therefore a lower energy change and a longer wavelength. The other corresponds to an increase in rotational quantum number, an increased energy change, and therefore a shorter wavelength for the exciting radiation. The gap corresponds to the forbidden transition, with no change in rotational quantum number.

EQUIPMENT AND MATERIALS

An infrared recording spectrophotometer equipped with a gas sample cell. Cylinders of dry HCl, HBr, and CO.

DIRECTIONS

Follow the manufacturer's instructions for operating the spectrophotometer, and obtain the spectra of the gases.

CALCULATIONS

Determine the wavenumbers of the fundamental vibration and of any overtones. The wavenumber is the reciprocal of the wavelength. It is the number of waves per unit length, with dimensions of reciprocal centimeters (cm^{-1}).

Calculate K, the force constant for the fundamental vibration, using the formula

$$K = 4\pi^2 \bar{v}^2 c^2 \mu$$

$\bar{v}$ = wavenumber
c = speed of light
μ = reduced mass, $m_1 m_2 / (m_1 + m_2)$
m = mass of the atom

Calculate Z_{vib}, the vibrational partition function,

$$Z_{vib} = \frac{e^{-hc\bar{v}/2kT}}{1 - e^{-hc\bar{v}/kT}}$$

h = Planck's constant
k = Boltzmann's constant, R/N
T = Kelvin temperature

Calculate the vibrational energy per mole from the equation

$$E_{vib} = \frac{Nhc\bar{v}}{e^{hc\bar{v}/kT} - 1} + \frac{Nhc\bar{v}}{2}$$

N = Avogadro's number

The first term on the right is the vibrational energy in excess of the zero-point energy, per mole. The second term is the zero-point vibrational energy.

Calculate the vibrational contribution to the heat capacity at 300°K using the relationship

$$C_{vib} = \frac{Re^{hc\bar{v}/kT}}{(e^{hc\bar{v}/kT} - 1)^2} \left(\frac{hc\bar{v}}{kT}\right)^2$$

Compare the calculated vibrational heat capacity with that expected from the equipartition principle. Explain any discrepancy.

Assume that the equipartition principle is valid for translational and rotational energy. Subtract the contributions for translation and rotation from the accepted value of the heat capacity to obtain the vibrational heat capacity. Compare the result with your calculated value of the vibrational contribution to the heat capacity.

Determine the wavenumbers or wavelengths at the peaks corresponding to changes in rotational quantum number.

The difference, in wavenumbers, between adjacent lines (except at the origin) in the rotation–vibration spectrum is $2B$:

$$B = \frac{h}{8\pi^2 Ic} = \tfrac{1}{2}\Delta\bar{\nu}$$

$$I = \frac{m_1 m_2 r^2}{m_1 + m_2}$$

$r =$ internuclear distance

Calculate I, the moment of inertia, for HCl and HBr and the interatomic distances.

NOTE / Dry the gases before filling the cell and fill the cell in a hood.

EXERCISES

1. In the equation for calculating the vibrational energy, the second term is the zero-point energy. Explain what is meant by zero-point energy.

References

Karplus, M., and R. N. Porter, *Atoms and Molecules*. Benjamin, New York, 1970.

Moelwyn-Hughes, E. A., *Physical Chemistry*, 2nd ed. Pergamon Press, Oxford, 1964.

Moore, W. J., *Physical Chemistry*, 5th ed. Prentice-Hall, Englewood Cliffs, N.J., 1973.

Experiment 44

Effect of Isotopic Mass Change on Molecular Vibration

Infrared Spectrophotometry

OBJECTIVES / To obtain the ir spectra of DCl and of HCl and to determine the shift in the fundamental vibration frequency and force constant due to the difference in isotopic mass.

READING ASSIGNMENT / Chapter 19, pages 228–229 and 232–234.

Since the fundamental vibration frequency and the moment of inertia depend on reduced mass, the spectrum of DCl should be similar to that of HCl, but shifted to longer wavelengths.

EQUIPMENT AND MATERIALS

An infrared recording spectrophotometer equipped with a gas sample cell.
Cylinders of dry HCl and dry DCl.

DIRECTIONS

Follow the manufacturer's instructions for operating the spectrophotometer, and obtain ir spectra of samples of DCl and of HCl in the wavelength region 2.5–8.0 μm (microns).

CALCULATIONS

Determine the fundamental vibration frequency of each gas and compare the ratio of the experimentally determined frequencies with the theoretical relationship

$$\frac{\nu_{DCl}}{\nu_{HCl}} = \sqrt{\frac{\mu_{HCl}}{\mu_{DCl}}}$$

ν = vibration frequency
μ = reduced mass, $m_1 m_2/(m_1 + m_2)$

For each gas calculate the force constant (in newtons per meter) for the fundamental vibration from the relationship

$$K = 4\pi^2 \nu^2 \mu$$

From the wavenumbers of the peaks corresponding to the changes in rotational quantum number and from the difference in wavenumbers between the rotational peaks, calculate the moment of inertia and the internuclear distance for both HCl and DCl. The difference between adjacent lines (except at the origin) in the rotation–vibration spectrum is $2B$:

$$B = \frac{h}{8\pi^2 I c} = \tfrac{1}{2} \Delta \bar{\nu}$$
$$I = r^2 \mu$$

$\bar{\nu}$ = wavenumber
I = moment of inertia
r = internuclear distance

NOTES / Fill the cell in a hood.

Dry the gases before passing them into the cell.

Wavenumbers are not SI units. Be careful.

EXERCISES

1. Theoretically, the force constants for HCl and for DCl are identical. Explain, on the basis of classical theory, why this should be so (i.e., why the isotopic mass should have no effect). If your calculated force constants did, in fact, differ, analyze the data and decide if your deviations from theory are significant or if they fall within the range of experimental error.

2. Compute the difference in zero-point vibrational energy between DCl and HCl. From this difference, compute the expected isotope effect at room temperature on the rate of a hypothetical reaction with an overall activation energy of 80 kJ per mole for HCl.

Experiment 45

The ESR Spectrum of a Free Radical

OBJECTIVES / To obtain the esr spectrum of a radical and to determine the number of equivalent protons and the charge density on each carbon.

READING ASSIGNMENT / Chapter 21, pages 269–271 and 275–287.

In this experiment the *p*-benzosemiquinone radical anion (BSQ$^-$) is produced from hydroquinone and its esr spectrum obtained in order to decide its electronic structure.

The reaction producing the radical is

$$\text{hydroquinone} + 2\text{OH}^- \longrightarrow \text{dianion} \xrightarrow{\text{oxygen}} \text{BSQ}^-$$

From the number of peaks in the spectrum the number of equivalent protons is obtained. From the peak separations the odd-electron charge density is calculated, using the McConnell relation.

Diphenylpicrylhydrazyl (DPPH) may be used to check the calibration of the spectrometer and to establish the conditions for obtaining an optimum spectrum.

EQUIPMENT AND MATERIALS

An esr spectrometer and samples of hydroquinone and, if desired, DPPH.

DIRECTIONS

Follow the manufacturer's directions for operating the spectrometer.

Prepare the BSQ$^-$ by mixing about 0.1 g of hydroquinone with a few milliliters of alcohol and 2 or 3 drops of 1 M NaOH. If the signal becomes attenuated, shake the container to dissolve more solid. The mixture need not be filtered before being added to the sample tube and may, in fact, be prepared in the sample tube.

CALCULATIONS

From the number of peaks in the BSQ$^-$ spectrum, and their intensities, determine the number of equivalent protons, and therefore carbons.

Determine the hyperfine splitting constant a (Tesla) from the separation of the peaks. To calculate the odd-electron charge density on each of the equivalent carbon atoms, use the McConnell relationship

$$a = Q\rho$$

a = hyperfine splitting constant
$Q = 2.25 \times 10^{-4}$ (empirical value)
ρ = odd-electron charge density on equivalent carbon atoms

Experiment 46

Quantum Efficiency of a Photochemical Reaction

Chemical Actinometry

OBJECTIVES / To irradiate a reaction mixture with monochromatic radiation from an actinometrically calibrated source and from the measured absorbances of the irradiated solution to calculate the quantum yield of the reaction.

READING ASSIGNMENT / Chapter 19, pages 237–241.

Molecular bromine reacts with cinnamic acid, adding across the double bond. The reaction takes place slowly in the dark, but when the reaction mixture is exposed to radiation, a rapid chain reaction occurs (provided that the protons have sufficient energy). In this experiment, the length of the reaction chain is determined. The reaction chain is the number of molecules reacting per photon of absorbed radiation. Another term for this is *quantum efficiency* or *quantum yield*.

To calculate quantum yield, the reaction mixture is irradiated with a known number of photons at 436 nm, and the amount of Br_2 used is then determined spectrophotometrically.

Irradiation with a known number of photons is not as easy as it sounds. The radiation source must be calibrated. In this case, the calibration is done actinometrically. A solution of known quantum efficiency is irradiated for a measured time period and from the extent of the reaction, the amount of radiation per unit time is calculated.

To compensate for the dark (unirradiated) reaction, a blank is determined in both the actinometric and the photobromination runs. Each time a run is made, the result of the blank reaction is subtracted from the result of the irradiated reaction. Note that each run is different because the composition of the reactant changes with time. It is therefore not possible to average the blanks and the runs and to subtract one average from the other. Instead, what is averaged is the difference between each blank and the corresponding irradiated result.

EQUIPMENT AND MATERIALS

For the irradiation a mercury lamp, a power supply, and a monochromator are necessary. The sample cell should be contained in an opaque housing equipped with a window. The sample cell should either be of silica or of pyrex with silica windows.

The actinometer solution is 0.125 M potassium ferrioxalate trihydrate in 0.1 M sulfuric acid. If necessary, the ferrioxalate should be purified by recrystallization from warm water. The developer for the actinometer solution contains 1 g of 1,10-phenanthroline monohydrate per liter of water. The buffer used to dilute the complex is 0.1 M sodium acetate in 0.018 M sulfuric acid.

Both the bromine and the cinnamic acid stock solutions are in carbon tetrachloride at a concentration of 0.05 M.

DIRECTIONS

Calibration of the Light Source—Actinometry. Irradiate a sample of actinometer solution at 436 nm for about 5 min, measuring the time accurately. Then develop the color by adding a 3 ml aliquot to 2 ml of the complexing phenanthroline solution, bringing the volume to 10 ml with buffer, and letting it stand for $\frac{1}{2}$ hr. Determine the absorbance at 510 nm. Repeat the procedure with additional samples of actinometer solution to obtain data for calculating precision.

Determine the absorbance of another sample of the actinometer solution at 436 nm, to find the fraction of the light that is adsorbed.

Photobromination Reaction. Irradiate a sample of a mixture of equal parts of bromine and cinnamic acid stock solutions for an accurately measured time. Measure its absorbance at 436 nm.

Determine the absorbance of an aliquot of the unirradiated reaction mixture at 436 nm.

Repeat the procedure until sufficient data have been obtained for the calculation of precision.

CALCULATIONS

Actinometry. The difference between the absorbance of the irradiated solution and an irradiated blank is due to the Fe^{2+} produced by irradiation. Calculate the Fe^{2+} concentration in the ferroin solution using Beer's law. The absorptivity of the ferroin complex is 1.11×10^4 liters/mole-cm.

The quantum efficiency of the actinometer reaction is 1.11

molecules/photon or moles/einstein. Calculate the number of einsteins absorbed.

From the absorbance of the unirradiated solution calculate the fraction of the incident radiation absorbed by the solution. From this fraction and the number of einsteins absorbed calculate the number of einsteins incident upon the solution per minute.

Photobromination Reaction. From the absorbance of the unirradiated solution and the previously determined radiation intensity, calculate the number of einsteins absorbed during the irradiation period.

The absorptivity of bromine in carbon tetrachloride solution at room temperature is 448 liters/mole-cm. Calculate the extent of the reaction, using a blank to compensate for the simultaneous slow dark reaction (the reaction not catalyzed by light).

Calculate the quantum yield.

NOTES / Allow the lamp to warm up and stabilize for 1–2 hr before irradiating a sample. Do not flip the lamp on and off during that period. If the light is shut off, do not start it again for at least 10 min or it may burn out.

Make sure that the ultraviolet light does not shine into or reflect into anyone's eyes. Ultraviolet light can cause severe eye damage.

In all cases the irradiated sample must be compared with a blank. This is a sample of solution that is treated exactly like the irradiated sample except that it is wrapped in an opaque material and not exposed to light. Both the sample and the blank should be shielded from any fluorescent light emitted and from sunlight.

If the absorbances are too high for accurate measurement, assume that Beer's law holds and dilute the solutions appropriately.

In calculating the extent of the photochemical reactions, all dilutions and aliquot samplings must be taken into account.

Keep the distance between the light and the cell constant.

Compute the concentration of the bromine from its absorbance. Do not rely on the label of the stock bottle.

Do not run the irradiation so long that you get a clear solution. If you do, you will have irradiated nonabsorbing solution, and your results will not be valid.

Either use the same cuvet throughout or correct the result for different optical path lengths, using Beer's law.

Do not wash the cell with water or acetone unless you dry it and rinse it with CCl_4 or with your reaction mixture. Otherwise, the radicals formed will react with the water or the acetone, invalidating your results.

The differences in results can be traced to differences in the surfaces of the cell, different amounts of dissolved oxygen or other impurities, and power fluctuations. In any case, you are taking an average value over a sufficiently long time period for the power fluctuations to cancel out and with a sufficient concentration change so that impurity effects may become unimportant.

Poor dilution technique is your greatest source of error.

EXERCISES

1. Write the equations for the photobromination reactions.
2. The bromine stock solution deteriorates on standing. If the bromine concentration is too low, the apparent quantum yield is less. Explain.

References

Bauer, W. H., and F. Daniels, Separation of photochemical and thermal action in the photobromination of cinnamic acid, *J. Amer. Chem. Soc.* **56**, 378 (1934).

Bauer, W. H., and F. Daniels, Oxygen a factor in the bromination of cinnamic acid, *J. Amer. Chem. Soc.* **56**, 2014 (1934).

Experiment 47

Equilibrium Constant of a System in an Excited State

Fluorimetry

OBJECTIVE / To determine the absorption and emission spectra of β-naphthol and β-naphthoxide ion and from measurement of emission at controlled pH's to calculate the pK_a^*.

READING ASSIGNMENT / Chapter 19, pages 226–228 and 230–231.

β-Naphthol is a very weak acid, ionizing in basic solution according to reaction (1):

(1) $\qquad C_{10}H_7OH \rightleftharpoons H^+ + C_{10}H_7O^-$

Both the naphthol and the naphthoxide ion may be excited by ultraviolet radiation and then emit the excitation energy as fluorescence. Reaction (2) shows the excitation of the molecule:

(2) $\qquad C_{10}H_7OH + h\nu \rightarrow C_{10}H_7OH^*$

(The asterisk refers to the excited state.) Electronically excited β-naphthol is a much stronger acid than ground state β-naphthol. It ionizes according to reaction (3):

(3) $\qquad C_{10}H_7OH^* \rightleftharpoons H^+ + C_{10}H_7O^{-*}$

The ionization constant, K_a^*, is therefore greater than K_a.

In strongly acid solution, the β-naphthol system is almost completely in the acid form. At higher pH's, the ratio of the ion to the molecule increases. From this ratio, the pK_a and pK_a^* can be calculated, since pK = pH when the concentration of naphthoxide equals that of the nonionized naphthol. This 50% point can be identified in two different ways. As in all weak acid titrations, the graph of naphthoxide ion versus pH shows an inflection point at 50% ionization, because here the buffering effect is greatest. Also, at the 50% point, the naphthoxide concentration is half of the total concentration of β-naphthol.

At very low pH's there is almost zero naphthoxide. In this region, the only species that can be excited by radiation is the naphthol molecule. Any fluorescence due to naphthoxide ion is therefore the result of an excited naphthol molecule having ionized and the excited ion then emitting radiation. The overall reaction scheme is shown in reactions (2)–(5):

(2) $\qquad C_{10}H_7OH + h\nu \rightarrow C_{10}H_7OH^*$
(3) $\qquad C_{10}H_7OH^* \rightleftharpoons H^+ + C_{10}H_7O^{-*}$
(4) $\qquad C_{10}H_7OH^* \rightarrow h\nu' + C_{10}H_7OH$
(5) $\qquad C_{10}H_7O^{-*} \rightarrow h\nu'' + C_{10}H_7O^-$

As shown in (4) and (5), both excited species emit radiation, but if the radiation detector is set for a wavelength that is emitted only by the ion and not by the molecule, the emitted radiation can be used as a quantitative measure of the amount of the excited naphthoxide present.

In this experiment, solutions of β-naphthol at different pH's are irradiated by ultraviolet radiation and the relative concentration of naphthoxide ion determined by measuring the emitted radiation. From the graphs of naphthoxide fluorescence versus pH, the pK_a^* is obtained.

EQUIPMENT AND MATERIALS

A spectrophotofluorimeter and a pH meter.
 A stock solution of saturated β-naphthol in water.
 1.0 M and 0.10 M HCl and 10^{-4} M NaOH.

DIRECTIONS

Irradiate a solution of β-naphthol in acid at pH 1.0 and determine the wavelength at which the emitted radiation is at a maximum. Irradiate at 313 nm. Use a solution of 1.0 M HCl diluted with 9 volumes of naphthol stock solution. At this pH, the naphthol is completely in the un-ionized form.

At the same wavelength, irradiate a solution of naphthol at pH 9, in which the naphthol is almost entirely ionized. Again determine the wavelength of maximum emission. Use a solution of 10^{-4} M NaOH diluted with 9 volumes of naphthol stock solution.

Prepare solutions of naphthol in dilute acid that will have the same amount of naphthol as those above but with pH's at 2.0, 2.2, 2.4, 2.6, 2.8, 3.0, 3.2, 3.4, 3.6, 3.8, and 4.0. In each case irradiate at 313 and measure the fluorescence at the emission wavelength of the naphthoxide ion.

To prepare a naphthol solution of pH 2.0, dilute 0.10 M HCl with nine times its volume of stock solution, in effect, diluting the acid 1 to 10. For pH 2.2 solution, dilute acid of pH 1.2 with 9 times its volume of stock solution.

Experiment 47. Equilibrium Constant of a System in an Excited State

Prepare all other solutions the same way, in each case using an acid whose pH is 1.0 pH units less than that of the desired naphthalene solution and diluting it with 9 times its volume of stock solution.

To prepare the acids at the different pH's, there are two methods. The first is to add HCl to distilled water, observing the pH of the resulting solution with a pH meter. Use either 1.0 or 0.10 M HCl, adding it with a buret or drop by drop with a pipet until the meter indicates the desired pH. The second way is to put a calculated volume of 0.10 M HCl into a 100 ml flask using a buret, and then to add water to the mark. The formula for calculating the correct amount of 0.10 M to be added is

$$x \text{ ml } 0.10 \ M \text{ acid} \times 0.10 = 100 \text{ ml} \times \text{molarity of desired acid}$$

For example, for pH 1.2, the molarity of acid is 0.063. Therefore, 63 ml of 0.10 M acid is needed.

CALCULATIONS AND GRAPHS

Plot naphthoxide fluorescence as a dependent function of pH. Note that pK_a^* cannot be estimated accurately from these graphs because the slope is nearly constant over a wide range on both sides of the inflection point, where $pH = pK_a^*$.

Calculate Δ(fluorescence)$/\Delta$pH for each increment of 0.2 pH unit. Plot these values as a function of the midpoint of the pH interval. Draw a smooth curve through the points. The maximum occurs where $pH = pK_a^*$. Estimate the pH at the maximum by eye or, if the data warrant, as described in Chapter 4.

NOTES / Wear rubber gloves. Do not get the β-naphthol on your skin. It is toxic.

Naphthol oxidizes rapidly in air, especially in basic solutions. Do not make up solutions in advance and let them stand around. Make up each one as needed and use it directly.

Keep the cells scrupulously clean. Never touch a cell face with your fingers.

Use distilled water as the standard for zeroing the spectrophotometer. Use the 10^{-4} M NaOH as the 100% standard, if your instrument requires one. Recheck the 100% standard from time to time to be sure that it has not oxidized in the air.

To make up the stock solution, shake up a few milligrams of naphthol in 100 ml of distilled water and then filter off the excess solid.

Although it is time consuming to make up the solutions of the desired pH's, this procedure is preferable to employing buffers of known pH. The buffers often absorb radiation, either that of excitation or of emission.

It is not necessary to know the concentration of either form of the naphthol or even the total naphthol concentration, provided that the latter is kept constant in all solutions. (Explain.)

References

Weller, A., Fast Reactions in Excited Molecules, in *Progress in Reaction Kinetics*, G. Porter (ed.), Vol. 1. Pergamon Press, New York, 1961.

Experiment 48

Electrokinetic Phenomena

OBJECTIVES / From measurements of voltage, current, pressures, and flow rates, to determine osmotic pressure, streaming current, streaming potential, and electroosmosis across a porous plug filled with dilute electrolyte.

READING ASSIGNMENT / Chapter 23, pages 312–314.

This experiment has a twofold objective. Measurements are made from which electroosmosis, electroosmotic pressure, streaming current, and streaming potential are calculated. The results of these calculations are used to check the validity of the Onsager relations.

Quantities measured are potential, current, hydrostatic pressure, and rate of liquid flow. The cell used is shown in Diagram H. It is difficult to reproduce data because the measurements are sensitive to impurities and often vary from time to time. The equipment must be kept scrupulously clean. If possible, use conductance water and recrystallized reagents. As with many irreversible systems, each apparatus has its own characteristics, and each set of results applies only to the particular diaphragm and solution used. (The ratios, however, are meaningful.)

EQUIPMENT AND MATERIALS

The cell shown in Diagram H, a digital voltmeter, a DC power supply capable of reaching at least 100 V, a microammeter that can be read to at least 0.1 μA, and a cell.

The solutions used are 10^{-3}–10^{-4} M KCl, or NaCl.

The electrodes are Ag|AgCl(s), reversible to the anion.

A timer or a stopwatch.

EXPERIMENTS

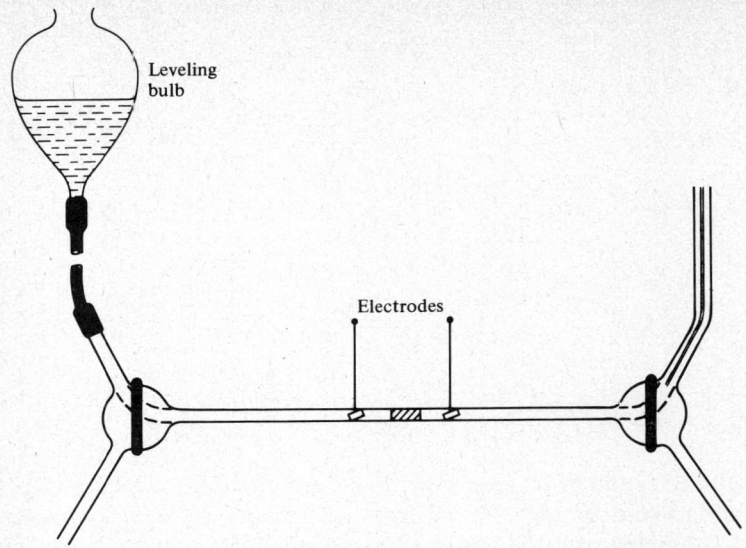

DIAGRAM H / Apparatus for measuring electrokinetic phenomena.

INSTRUCTIONS FOR THE APPARATUS

Fill the cell with the solution, using suction. Air bubbles must be eliminated. Do not grease the stopcocks. Make sure that the electrodes have a good coating of silver chloride, to prevent bubbles being formed by the current.

When the cell is filled, mount it with the leveling bulb and vertical capillary in front of the meter stick. Make sure that it is horizontal.

The conductance cell is calibrated with a standard solution, such as 0.100 M KCl. The capillaries must be of known cross sectional area.

DIRECTIONS

Electroosmotic Pressure

$$\text{EOP} = \left(\frac{\Delta P}{E}\right)_{J_m = 0}$$

Open the stopcocks so that liquid can flow from the leveling bulb through the vertical capillary.

Connect the electrodes to the power supply. Then raise the leveling bulb to a height of several feet, and, adjusting the position to cause the liquid to rise in the vertical capillary, obtain a flow rate of 1 cm every few seconds. Record the flow rate and pressure head. Repeat several times.

Apply a DC potential in the direction that will oppose the pressure head, causing the liquid to fall in the vertical capillary. Adjust the pressure head and the potential until the liquid level is stationary. Record both the pressure head and the potential. Measure the distance across the diaphragm. Repeat for several combinations of bulb height and potential.

Streaming Current

$$\text{SC} = \left(\frac{J_{el}}{J_m}\right)_{E=0}$$

Connect the electrodes to a microammeter. By raising or lowering the leveling bulb, force the liquid alternately up and down in the vertical capillary, noting the current. Select the best combination of easily controlled flow rates and currents large enough to be observed accurately. Record the flow rate and the current for rising and for falling levels, repeating several times to obtain data for calculating precision. The voltmeter must read zero.

Streaming Potential

$$\text{SP} = \left(\frac{E}{\Delta P}\right)_{J_{el}=0}$$

Connect the electrodes to the digital voltmeter and microammeter.
Alternately raise and lower the leveling bulb, so as to keep a constant pressure head, while measuring the pressure head and the potential between the electrodes. Repeat several times. The microammeter must read zero.

Electroosmosis

$$\text{EO} = \left(\frac{J_m}{J_{el}}\right)_{\Delta P=0}$$

Connect the microammeter in series with the cell and connect the power supply to the circuit.

Adjust the right-hand stopcock so that liquid can flow out of the cell through the lower capillary. Then lower the leveling bulb to the level of the lower end of the capillary so that there is no pressure difference and liquid does not flow out of the cell.

Apply a potential sufficient to produce a measurable current and liquid flow. Let the liquid flow into a weighted container for a convenient time interval and record the current and the mass of liquid.

Second Electroosmosis

$$\text{2nd EO} = \left(\frac{J_m}{E}\right)_{\Delta P=0}$$

Connect the power supply and the digital voltmeter to the electrodes. Adjust the leveling bulb so that there is no pressure forcing liquid out of the cell and open the stopcock to the lower capillary.

Apply a sufficient potential for liquid to be forced out of the cell into a weighed container, for a convenient time interval. Record the potential and the mass of the solution.

Connect the impedance bridge and measure the resistance between the electrodes.

CALCULATIONS

Convert all voltage and current measurements to SI units, using volts, amperes, and ohms. The pressure head, in cm of H_2O, must be multiplied by the factor 98.3 to convert it to newtons/meter^2.

Calculate the specific conductivity and equivalent conductivity of the solution. If the conductance cell is not already calibrated, calibrate it with a standard solution such as 0.0100 M KCl.

Calculate

$$EOP = \left(\frac{\Delta P}{E}\right)_{J_m=0}$$

$$SC = \left(\frac{J_{el}}{J_m}\right)_{E=0}$$

$$SP = \left(\frac{E}{\Delta P}\right)_{J_{el}=0} = \left(\frac{d\phi}{dP}\right)_{J_{el}=0}$$

$$EO = \left(\frac{J_m}{J_{el}}\right)_{\Delta P=0}$$

$$\text{Second EO} = \left(\frac{J_m}{E}\right)_{\Delta P=0}$$

ΔP = pressure head, newtons per meter^2
J_{el} = current, amperes
J_m = milliliters of liquid flowing per second
E = potential, volts applied per meter of distance
    = $d\phi/dx$

Compare your results with those predicted by the Onsager relations. Within the rather large experimental error, these should be

$$SC = -EOP$$
$$SP = EO = r \text{ (2nd EO)}$$

r = resistance

Experiment 48. Electrokinetic Phenomena

NOTES / If the electrodes are platinum, it is necessary to plate them with silver and then to anodize the silver in a chloride solution. The cell should be thoroughly washed with distilled water.

Once you have used one electrolyte for the cell, do not change to another electrolyte.

If possible, use Teflon stopcocks.

The area of the electrodes should be large enough so that no bubbles are formed during the electrolysis. Wire mesh electrodes are preferred.

Do not allow the electrodes to stand for extended periods of time in a room with fluorescent lights or in strong sunlight.

Use transparent tubing to connect the leveling bulb and cell, so as to detect the presence of any air bubbles.

Experiment 49

Correlation of Polarographic Reduction Potentials and Electron Affinity

Polarography

OBJECTIVES / To obtain polarograms of aromatic hydrocarbon reduction at the dropping mercury electrode. From the half-wave potentials, the value of the Hückel β is obtained.

READING ASSIGNMENT / Chapter 16, pages 213-215.

A number of workers [Maccol; Hoijtink and Van Schooten; Bergman] have established a linear correlation between the half-wave reduction potentials of a series of aromatic hydrocarbons and their electron affinities, based on the idea that the electron affinity is the energy of the lowest vacant molecular orbital [Streitweiser]. The relationship is

$$E_{1/2} = -m\beta + c$$

m = appropriate root of the secular equation
β = Hückel "beta"
c = compound constant

The object of this experiment is to duplicate some of the experimental half-wave potentials and to see if the relationship does in fact hold.

DIRECTIONS

Determine the half-wave potentials of three or more aromatic hydrocarbons at the dropping mercury electrode.

The solvent is dimethylformamide. The electrolyte is 0.1 M tetraethylammonium bromide. The resistance is sufficiently low that even with a two-electrode technique it will not be necessary to determine the resistance of the solution. The recommended hydrocarbons are naphthalene, biphenyl, anthracene, and chrysene.

Experiment 49. Correlation of Polarographic Reduction Potentials/Electron Affinity

CALCULATIONS

For each hydrocarbon, calculate the energy of the lowest unoccupied electronic level, using the Hückel molecular orbital approximation given by Streitwieser.

Plot the half-wave potentials against the calculated electronic energies and determine the value of β. Compare these with accepted data.

NOTES / Dimethylformamide is toxic. Work in a hood. Chrysene may be *carcinogenic.* **Wear rubber gloves.** *Be careful.*

References

Bergman, I., Polarography of polycyclic aromatic hydrocarbons and the relation between the half-wave potentials and absorption spectra, *Trans. Faraday Soc.* **50**, 829 (1954).

Hoijtink, G. J., and J. van Schooten, Polarographic reduction of conjugated hydrocarbons. I. A theoretical discussion of the polarographic reduction of aromatic hydrocarbons, *Rec. Trav. Chim.* **71**, 1089 (1952); II. Reduction of 1,n-diphenylpolyenes, **72**, 691 (1953); III. Comparison of molecular orbital approximation with half-wave potentials of alternant hydrocarbons, **72**, 903 (1953).

Maccol, A., Reduction potentials of conjugated systems, *Nature* **163**, 178 (1949).

Streitweiser, A., *Molecular Orbital Theory for Organic Chemists.* Wiley, New York, 1961.

Appendix 1

Atomic Weights of the Elements

Element	Symbol	Atomic Number	Atomic Weight[a]
Actinium	Ac	89	[227]
Aluminum	Al	13	26.9815
Americium	Am	95	[243]
Antimony	Sb	51	121.75
Argon	Ar	18	39.948
Arsenic	As	33	74.9216
Astatine	At	85	[210]
Barium	Ba	56	137.34
Berkelium	Bk	97	[249]
Beryllium	Be	4	9.01218
Bismuth	Bi	83	208.9806
Boron	B	5	10.81
Bromine	Br	35	79.904
Cadmium	Cd	48	112.40
Calcium	Ca	20	40.08
Californium	Cf	98	[251]
Carbon	C	6	12.011

[a] Based on atomic weight of carbon-12 = 12.0000. To convert to old atomic weights based upon O = 16 multiply by 1.0003203.

The atomic weight of some elements varies because of natural variations in isotope composition. The variations are: B, ± 0.003; C, ± 0.00005; H, ± 0.00001; O, ± 0.0001; Si, ± 0.001; S, ± 0.003.

Square brackets indicate mass number of most common or most stable isotope.

APPENDIX 1

Element	Symbol	Atomic Number	Atomic Weight[a]
Cerium	Ce	58	140.12
Cesium	Cs	55	132.9055
Chlorine	Cl	17	35.453
Chromium	Cr	24	51.996
Cobalt	Co	27	58.9332
Copper	Cu	29	63.546
Curium	Cm	96	[247]
Dysprosium	Dy	66	162.50
Einsteinium	Es	99	[254]
Erbium	Er	68	167.26
Europium	Eu	63	151.96
Fermium	Fm	100	[253]
Fluorine	F	9	18.9984
Francium	Fr	87	[223]
Gadolinium	Gd	64	157.25
Gallium	Ga	31	69.72
Germanium	Ge	32	72.59
Gold	Au	79	196.9665
Hafnium	Hf	72	178.49
Helium	He	2	4.00260
Holmium	Ho	67	164.9303
Hydrogen	H	1	1.0080
Indium	In	49	114.82
Iodine	I	53	126.9045
Iridium	Ir	77	192.22
Iron	Fe	26	55.847
Krypton	Kr	36	83.80
Lanthanum	La	57	138.9055
Lawrencium	Lr	103	[257]
Lead	Pb	82	207.2
Lithium	Li	3	6.941
Lutetium	Lu	71	174.97
Magnesium	Mg	12	24.305
Manganese	Mn	25	54.9380
Mendelevium	Md	101	[256]
Mercury	Hg	80	200.59
Molybdenum	Mo	42	95.94
Neodymium	Nd	60	144.24
Neon	Ne	10	20.179
Neptunium	Np	93	237.0482
Nickel	Ni	28	58.71
Niobium	Nb	41	92.9064
Nitrogen	N	7	14.0067
Nobelium	No	102	[254]

Atomic Weights of the Elements

Element	Symbol	Atomic Number	Atomic Weight[a]
Osmium	Os	76	190.2
Oxygen	O	8	15.9994
Palladium	Pd	46	106.4
Phosphorus	P	15	30.9738
Platinum	Pt	78	195.09
Plutonium	Pu	94	[242]
Polonium	Po	84	[210]
Potassium	K	19	39.102
Praseodymium	Pr	59	140.9077
Promethium	Pm	61	[145]
Protactinium	Pa	91	231.0359
Radium	Ra	88	226.0254
Radon	Rn	86	[222]
Rhenium	Re	75	186.2
Rhodium	Rh	45	102.9055
Rubidium	Rb	37	85.4678
Ruthenium	Ru	44	101.07
Samarium	Sm	62	150.4
Scandium	Sc	21	44.9559
Selenium	Se	34	78.96
Silicon	Si	14	28.086
Silver	Ag	47	107.868
Sodium	Na	11	22.9898
Strontium	Sr	38	87.62
Sulfur	S	16	32.06
Tantalum	Ta	73	180.9479
Technetium	Tc	43	98.9062
Tellurium	Te	52	127.60
Terbium	Tb	65	158.9254
Thallium	Tl	81	204.37
Thorium	Th	90	232.0381
Thulium	Tm	69	168.9342
Tin	Sn	50	118.69
Titanium	Ti	22	47.90
Tungsten	W	74	183.85
Uranium	U	92	238.029
Vanadium	V	23	50.9414
Xenon	Xe	54	131.30
Ytterbium	Yb	70	173.04
Yttrium	Y	39	88.9059
Zinc	Zn	30	65.37
Zirconium	Zr	40	91.22

Appendix 2

Units and Important Physical Constants

We have attempted in this book to use SI (Système International) units, except in a few instances where the pull of traditional usage is still overwhelming. In general, any data that enter into a calculation in a scientific paper or laboratory report should be expressed in SI units. In this appendix we also offer a few conversion factors to cgs and to traditional units still in use to make older literature accessible.

The SI units are based on seven fundamental quantities:

Quantity	Unit (Abbreviation)	Definition
length, l	meter (m)	1650763.73 wavelengths in vacuum of the orange-red line of the spectrum of ^{86}Kr
mass, m	kilogram (kg)	mass of a Pt–Ir bar kept by the International Bureau of Weights & Measures
time, t	second (s)	duration of 9,192,631,770 periods of the radiation corresponding to the transition between the two hyperfine levels of the ground state ^{133}Cs
current, I	ampere (A)	the constant current that, if maintained in two straight parallel conductors of infinite length and negligible cross section and separated from each other by a distance of 1 meter in vacuum, will produce between these conductors a force of 2×10^{-7} Newton per meter of length
temperature, T	Kelvin (K)	1/273.16 of the thermodynamic temperature of the triple point of water

Units and Important Physical Constants

Quantity	Unit (Abbreviation)	Definition
quantity of substance	mole (mol)	the quantity of substance equal to the molecular weight of that substance in grams[a]
luminous intensity	candela (cd)	the luminous intensity of 1/600,000 m^2 of a [black body] radiating cavity at the temperature of freezing Pt (2042 K)

[a] The use of gram-mole is one concession to older usage, since the SI unit of mass is the kilogram.

All other units are derived; some of the most important derived units are given below. In the SI system *all* dimensions are expressed with superscripts. Thus, for example, density is ml^{-3} (kg·m^{-3}) rather than m/l^3 (kg/m^3).

Quantity	Dimensions	Unit (Abbreviation)	Fundamental Units
density	ml^{-3}	—	kg·m^{-3}
velocity	mt^{-1}	—	m·s^{-1}
force	mlt^{-2}	Newton (N)	kg·m·s^{-2}
pressure	$ml^{-1}t^{-2}$	Pascal (Pa)	kg·m^{-1}·s^{-2} = N·m^{-2}
energy	ml^2t^{-2}	joule (J)	kg·m^2·s^{-2}
power	ml^2t^{-3}	watt (W)	kg·m^2·s^{-3} = J·s^{-1}
frequency	t^{-1}	Hertz (Hz)	s^{-1}
electric charge	IT	coulomb (C)	A·s
electric potential, emf	$ml^2t^{-3}I^{-1}$	volt (V)	kg·m^2·s^{-3}·A^{-1}
electrical resistance	$ml^2t^{-3}I^{-2}$	Ohm (Ω)	kg·m^2·s^{-3}·A^{-2}
electrical resistivity	$ml^3t^{-3}I^{-2}$	ohm meter (Ω·m)	kg·m^3·s^{-3}·A^{-2}
capacitance	$ml^{-2}t^4I^2$	farad (F)	kg·m^{-2}·s^4·A^2
inductance	$ml^2t^{-2}I^{-2}$	Henry (H)	kg·m^2·s^{-2}·A^{-2}
magnetic flux	$ml^2t^{-2}I^{-1}$	Weber (Wb)	kg·m^2·s^{-2}·A^{-1}
magnetic induction	$mt^{-2}I^{-1}$	Tesla (T)	kg·s^{-2}·A^{-1} = Wb·m^{-2}
thermal conductivity	$mlt^{-3}K^{-1}$	—	kg·m·s^{-3}·K^{-1} = W·m^{-1}·K^{-1}

Angular measure (SI): 1 radian (rad) = $1/2\pi$ of one revolution = angle subtended at center of circle by arc length equal to the radius of the circle.

1 degree = $2\pi/360$ radian = 1/360 revolution

1 minute = 1/60 degree

1 second = 1/60 minute = 1/3600 degree.

Solid angle (SI):

1 steradian (sr) = $1/4\pi$ sphere

A solid angle, in steradians, is the fraction of the surface area of the sphere intersected by the angle, multiplied by 4π.

In addition to these, some units that may still be used, pending adoption of replacements consistent with the SI system, include:

1. Electron volt (eV)—a unit of energy equal to that of an electron accelerated through a 1 V potential difference = 1.602×10^{-19} J/particle, or 96.48 kJ mol^{-1} (this is an acceptable "natural unit").
2. Kayser (K; formerly known as "wavenumber"). 1 K = 1 cm^{-1} = 100 m^{-1}. Used in spectroscopy as a unit of energy given in terms of reciprocal wavelength. 8066 K = 1 eV. (This corresponds to a photon of wavelength 1/8066 cm, or 1239.85 nm). The abbreviation is the same as that for Kelvin. Beware of confusion.
3. Poise (P), Stokes (St); the cgs units for viscosity, and kinematic viscosity are both still used (especially the centipoise, (cp).
4. Debye (D), a unit of electric dipole strength = 1×10^{-18} statcoulomb cm = 3.33564×10^{-30} C·m.

Radioactivity: The Curie (abbreviation Ci) is the quantity of a radioactive nuclide that produces 3.7×10^{10} disintegration per second. It is acceptable for use with the SI system.

Certain logarithmic units are acceptable.

1. pH = $-\log_{10}[a_{H+}]$, where mol, not kmol, is used for concentration.
2. Bel (B) = $\log_{10}(P_1/P_2)$, where P_1/P_2 is the ratio of two values of power. The decibel (dB) is more commonly used; 1 B = 10 dB.
3. Neper (Np) = $\ln(x_1/x_2)$, where x_1/x_2 is the ratio of two values of current, voltage, or power; 1 Np = 8.686 dB.

The calorie, as a unit of energy (= 4.184 J) is obsolete, although it is still used.

Below are some cgs units, with conversion to SI.

Quantity	cgs Unit	Multiplied by	Gives SI Unit
energy	erg	10^7	J
force	dyne	10^5	N
electric potential	statvolt (esu^a)	1/300	V
electric charge	statcoulomb (esu)	3×10^9	C
magnetic induction	gauss (emu^b)	10^4	T
pressure	torr	1.332×10^2	Pa
	atm	1.013×10^5	Pa

^{a} Electrostatic unit.
^{b} Electromagnetic unit.

Because many units are inconveniently large or small, or their physical values may cover an extremely wide range, the following accepted prefixes, indicating magnitudes as shown, are used.

Prefix	Symbol	Multiple
atto	a	10^{-18}
femto	f	10^{-15}
pico	p	10^{-12}
nano	n	10^{-9}
micro	μ	10^{-6}
milli	m	10^{-3}
centi	c	10^{-2}
deci	d	10^{-1}
deka	da	10^{1}
hecto	h	10^{2}
kilo	k	10^{3}
mega	M	10^{6}
giga	G	10^{9}
tera	T	10^{12}

Finally, certain fundamental physical constants are consistently used. The following table is based on the compilation of B. N. Taylor, W. H. Parker, and D. N. Laugenberg, *Revs. Mod. Phys.* **41**, 375 (1969).

Constant	Symbol	Value[Uncertainty] and Unit
Avogadro's number	N	$6.022169[40] \times 10^{23}$ mol^{-1}
gas constant	R	$8.31434[35]$ J $\cdot$ mol$^{-1} \cdot$ K^{-1}
Boltzmann's constant (R/N)	k	$1.380622[59] \times 10^{-23}$ J $\cdot$ K^{-1}
Planck's constant	h	$6.626196[50] \times 10^{-34}$ J $\cdot$ s
Faraday constant (Ne)	F	$9.646670[54] \times 10^{4}$ C $\cdot$ mol^{-1}
proton rest mass	m_p	$1.672614[11] \times 10^{-27}$ kg
	$m_p{}^*$	$1.00727661[08]$ amu
atomic mass unit	amu	$1.660531[11] \times 10^{-31}$ kg
electron rest mass	m_e	$9.109558[54] \times 10^{-31}$ kg
	$m_e{}^*$	$5.485930[34] \times 10^{-4}$ amu
elementary charge	e	$1.6021917[70] \times 10^{-19}$ C
charge to mass ratio, electron	e/m_e	$1.7588028[54] \times 10^{11}$ C $\cdot$ kg^{-1}
speed of light in vacuum	c	$2.9979250[10] \times 10^{8}$ m $\cdot$ s^{-1}
gyromagnetic ratio of protons in water	γ_p	$2.6751965[82] \times 10^{8}$ rad s$^{-1} \cdot$ T^{-1}
corrected for diamagnetism of water	$\gamma_p/2\pi$	$4.257597[13] \times 10^{7}$ Hz $\cdot$ T^{-1}
Bohr magneton	μ_B	$9.274096[65] \times 10^{-24}$ J $\cdot$ T^{-1}
Rydberg constant	R_∞	$1.09737312[11] \times 10^{7}$ m^{-1}
Bohr radius	a_0	$5.2917715[81] \times 10^{-11}$ m
Stefan–Boltzmann constant	σ	$5.66961[96] \times 10^{-8}$ W $\cdot$ m$^{-2} \cdot$ K^{4}
gravitational constant	G	$6.6732[31] \times 10^{-11}$ N $\cdot$ m$^{2} \cdot$ kg^{-2}
permittivity of free space	ε_0	8.855×10^{-12} F $\cdot$ m^{-1}
permittivity of free space	μ_0	$4\pi \times 10^{-7}$ Wb $\cdot$ A$^{-1} \cdot$ m^{-1}
standard volume of ideal gas		2.24136×10^{-2} m$^{3} \cdot$ mol^{-1}

Appendix 3

Relative Density and Volume of Water

Relative Density of Water[a]

Temp. (°C)	0	1	2	3	4	5	6	7	8	9
−10	0.99815	0.99843	0.99869	0.99892	0.99912	0.99930	0.99945	0.99958	0.99970	0.99979
0	0.99987	0.99993	0.99997	0.99999	1.00000	0.99999	0.99997	0.99993	0.99988	0.99981
10	0.99973	0.99963	0.99952	0.99940	0.99927	0.99913	0.99897	0.99880	0.99862	0.99843
20	0.99823	0.99802	0.99780	0.99756	0.99732	0.99707	0.99681	0.99654	0.99626	0.99597
30	0.99567	0.99537	0.99505	0.99473	0.99440	0.99406	0.99371	0.99336	0.99299	0.99262
40	0.99224	0.99186	0.99147	0.99107	0.99066	0.99025	0.98982	0.98940	0.98896	0.98852
50	0.98807	0.98762	0.98715	0.98669	0.98621	0.98573	—	—	—	—
60	0.98324	—	—	—	—	0.98059	—	—	—	—
70	0.97781	—	—	—	—	0.97489	—	—	—	—
80	0.97183	—	—	—	—	0.96865	—	—	—	—
90	0.96534	—	—	—	—	0.96192	—	—	—	—
100	0.95838	—	—	—	—	—	—	—	—	—
110	0.9510	—	—	—	—	—	—	—	—	—

[a] From Smithsonian Tables. Values are in grams per milliliter.

Relative Volume of Water[a]

Temp. (°C)	0	1	2	3	4	5	6	7	8	9
−10	1.00186	1.00157	1.00131	1.00108	1.00088	1.00070	1.00055	1.00042	1.00031	1.00021
0	1.00013	1.00007	1.00003	1.00001	1.00000	1.00001	1.00003	1.00007	1.00012	1.00019
10	1.00027	1.00037	1.00048	1.00060	1.00073	1.00087	1.00103	1.00120	1.00138	1.00157
20	1.00177	1.00198	1.00221	1.00244	1.00268	1.00294	1.00320	1.00347	1.00375	1.00405
30	1.00435	1.00466	1.00497	1.00530	1.00563	1.00598	1.00633	1.00669	1.00706	1.00743
40	1.00782	1.00821	1.00861	1.00901	1.00943	1.00985	1.01028	1.01072	1.01116	1.01162
50	1.01207	1.01254	1.01301	1.01349	1.01398	1.01448	—	—	—	—
60	1.01705	—	—	—	—	1.01979	—	—	—	—
70	1.02270	—	—	—	—	1.02576	—	—	—	—
80	1.02899	—	—	—	—	1.03237	—	—	—	—
90	1.03590	—	—	—	—	1.03959	—	—	—	—
100	1.04343	—	—	—	—	—	—	—	—	—
110	1.0515	—	—	—	—	—	—	—	—	—

[a] From Smithsonian Tables. Values are in milliliters per gram.

Appendix 4

Temperature Correction for Barometric Readings

Temperature Correction for Barometric Readings[a]

Temp. (°C)	Observed height (mm)									
	700	710	720	730	740	750	760	770	780	790
10	1.14	1.16	1.17	1.19	1.21	1.22	1.24	1.26	1.27	1.29
11	1.26	1.27	1.29	1.31	1.33	1.35	1.36	1.38	1.40	1.42
12	1.37	1.39	1.41	1.43	1.45	1.47	1.49	1.51	1.53	1.55
13	1.48	1.50	1.53	1.55	1.57	1.59	1.61	1.63	1.65	1.67
14	1.60	1.62	1.64	1.67	1.69	1.71	1.73	1.76	1.78	1.80
15	1.71	1.74	1.76	1.78	1.81	1.83	1.86	1.88	1.91	1.93
16	1.82	1.85	1.88	1.90	1.93	1.96	1.98	2.01	2.03	2.06
17	1.94	1.97	1.99	2.02	2.05	2.08	2.10	2.13	2.16	2.19
18	2.05	2.08	2.11	2.14	2.17	2.20	2.23	2.26	2.29	2.32
19	2.17	2.20	2.23	2.26	2.29	2.32	2.35	2.38	2.41	2.44
20	2.28	2.31	2.34	2.38	2.41	2.44	2.47	2.51	2.54	2.57
21	2.39	2.43	2.46	2.50	2.53	2.56	2.60	2.63	2.67	2.70
22	2.51	2.54	2.58	2.61	2.65	2.69	2.72	2.76	2.79	2.83
23	2.62	2.66	2.69	2.73	2.77	2.81	2.84	2.88	2.92	2.96
24	2.73	2.77	2.81	2.85	2.89	2.93	2.97	3.01	3.05	3.08
25	2.85	2.89	2.93	2.97	3.01	3.05	3.09	3.13	3.17	3.21
26	2.96	3.00	3.04	3.09	3.13	3.17	3.21	3.26	3.30	3.34
27	3.07	3.12	3.16	3.20	3.25	3.29	3.34	3.38	3.42	3.47
28	3.19	3.23	3.28	3.32	3.37	3.41	3.46	3.51	3.55	3.60
29	3.30	3.35	3.39	3.44	3.49	3.54	3.58	3.63	3.68	3.72
30	3.41	3.46	3.51	3.56	3.61	3.66	3.71	3.75	3.80	3.85
31	3.53	3.58	3.63	3.68	3.73	3.78	3.83	3.88	3.93	3.98
32	3.64	3.69	3.74	3.79	3.85	3.90	3.95	4.00	4.05	4.11
33	3.75	3.81	3.86	3.91	3.97	4.02	4.07	4.13	4.18	4.23
34	3.87	3.92	3.98	4.03	4.09	4.14	4.20	4.25	4.31	4.36
35	3.98	4.03	4.09	4.15	4.21	4.26	4.32	4.38	4.43	4.49

[a] To reduce readings of a mercury barometer with a brass scale to 0°C, subtract the tabulated quantity. Corrections are in torr.

Appendix 5

Density of Mercury

Density of Mercury[a]

Temp. (°C)	0	1	2	3	4	5	6	7	8	9
−10	13.6202	13.6177	13.6152	13.6128	13.6103	13.6078	13.6053	13.6029	13.6004	13.5979
0	13.5955	13.5930	13.5906	13.5881	13.5856	13.5832	13.5807	13.5782	13.5758	13.5733
10	13.5708	13.5684	13.5659	13.5634	13.5610	13.5585	13.5561	13.5536	13.5512	13.5487
20	13.5462	13.5438	13.5413	13.5389	13.5364	13.5340	13.5315	13.5291	13.5266	13.5242
30	13.5217	13.5193	13.5168	13.5144	13.5119	13.5095	13.5070	13.5046	13.5021	13.4997
40	13.4973	—	—	—	—	—	—	—	—	—

[a] From Smithsonian Tables. Values are in grams per milliliter.

Appendix 6

Glassworking

In principle, all that need be done to work glass is to soften the glass by heating, shape it while it is soft, and then allow it to cool. In practice, there are some difficulties, but with care and common sense almost anyone can become proficient enough to make repairs and to construct simple laboratory setups.

The basic operations are cutting, bending, sealing, blowing, annealing, healing cracks, and fire-polishing. Whatever the operations to be performed, wash the glass beforehand to remove grease, dust, and any contaminants, and then dry it. Use goggles for protection against glass splinters and to prevent eye damage and headaches from the glare of the intense sodium flame. Didymium glass or goggles are most satisfactory. Heated glass cannot be worked until after the yellow sodium flame appears and softening of the glass is noticed. (The sole criterion of when to stop heating and to start working the glass is the good judgment acquired from previous mistakes.) When heating glass in a stationary flame, rotate the glass slowly (unless it is fastened in place or too awkward to move) to prevent the softened glass from sagging out of shape. Be careful to rotate it with both hands at the same rate.

In sealing glass to glass, never seal together two pieces of very different softening point and coefficient of thermal expansion. If it is necessary to connect two dissimilar glasses, a graded seal should be used. This consists of several pieces of glass of intermediate composition, sealed one to the other, with properties varying by small steps over the entire range. Graded seals of various types and sizes are commercially available from a number of sources, for soft glass to quartz, pyrex to quartz, and so on.

CUTTING

The best way to cut glass (except for small diameter tubing or rods) is with a glass saw, a thin circular disk containing an abrasive material. If no saw is

Glassworking

available, use a file or a knife or cut the glass with a flame. To cut tubing or rods less than 15 mm in diameter, first scratch the glass with a sharp file or a glass knife, then grasp the glass with the thumbs directly opposed to the scratch and gently bend the glass away from the scratch until it breaks. If scratched properly, the glass breaks cleanly along the scratch line. If the scratch is not sharp, the glass will shatter instead of breaking cleanly. If it does not break easily, deepen the scratch and try again to break the glass. Wetting the scratch with a drop of water before breaking the glass improves the results and should be adopted as routine procedure in all glass-cutting methods. Wrap a cloth or towel around the glass when breaking it to avoid cuts by flying glass.

To cut close to an end or to cut wide tubing or rods, scratch the glass and wet the scratch. Then press the molten end of a 2–3 mm rod against the end of the scratch. The rapid expansion of the heated scratched area breaks the glass at the scratch. Do not remove the hot glass rod until the end is no longer soft, or the cut will not be sharp and clean.

If for some reason the glass cannot be scratched and broken, melt the glass, sealing off the tube at the point to be cut. Blow the molten glass out into a bubble. Scrape off the thin edge of the bubble with a wire gauze and fire-polish the rough edge.

BENDING

Soften the glass by heating over the area where the bend is to be made. Then, very gently and gradually, apply force to produce the bend (gravity works well). For a smooth bend the softened area should be large enough so that the outer bend is not stretched thin while the inner bend is crumpled. As a rule of thumb, heat and soften a length of about six or seven times the diameter of the tubing. Be sure to rotate the glass while heating. If the glass cannot be rotated, move the flame around the glass, heating it at the line where the bend is to start. As the glass softens, move the flame slowly in the direction of extension of the bend. The flame should be at least as wide as the tube. Figure A–1 shows correct and incorrect bends.

FIGURE A–1 / Correct and incorrect bends.

SEALING TUBES CLOSED

Soften the glass at the point to be sealed and then slowly pull the ends of the softened glass apart while rotating the hot part of the glass in the flame (this procedure is called "pulling a point"). Melt the tip shut in a small flame. The closed end of the sealed tube should be hemispherical and the glass wall should have the same thickness at the end as at the sides. If necessary, thicken the wall by heating it. If the end is too thick, soften it and then gently blow air into the tube. The air will push the soft glass out toward the end, enlarging the tube slightly and thinning the wall. Another way to thin out the wall is to touch the molten glass with a bit of "cane" (glass rod used for adding or removing glass or forming handles). The soft hot glass will stick to the cane and can then be pulled off the end of the tube.

If the seal is so close to the end that there is no room to hold the glass, make a handle by sealing a piece of glass to the end to be removed.

GLASS BLOWING

To get the desired thickness of glass, simply force air into the hot glass to push the soft walls out. This is best done by series of gentle puffs rather than by a single blast. Remove the hot glass from the flame, hold it straight up in the air, and rotate it to minimize sag. Allow about 3 sec to elapse and then puff. The 3 sec delay allows the thin glass to cool while the thick glass is still soft. The blowing then stretches the thick glass more than the thin glass, giving a uniform wall thickness. There are other methods, but this is the simplest for beginners. Remember to close all the air outlets before blowing.

JOINING TUBES

The surfaces should be smooth and should fit together without air gaps. Square the ends off, either by fire-polishing or by blowing out the surfaces to be joined. If the glass is not squared off, the seal will be lumpy, strained, and fragile and may leak. The tubes must have the same wall thickness and the same inner diameters at the point of contact. If the tubes are of slightly different diameters, enlarge the smaller tube by heating and flaring it out with a carbon rod until it looks like Figure A-2a. If the tubes are originally of widely differing diameters, seal the larger tube, then blow an opening of the desired diameter, as in Figure A-2b and c.

After preparing the tubes, heat both ends to be joined until they are soft. Then bring the two ends squarely together with a slight pressure. The two pieces will adhere. To complete the seal, heat the junction until the two pieces of glass melt and flow into each other. Blow into the tube, when necessary, to prevent collapse of the wall during heating (one of the two tubes must be shut or plugged beforehand). If the glass wall at the joint is too

Glassworking

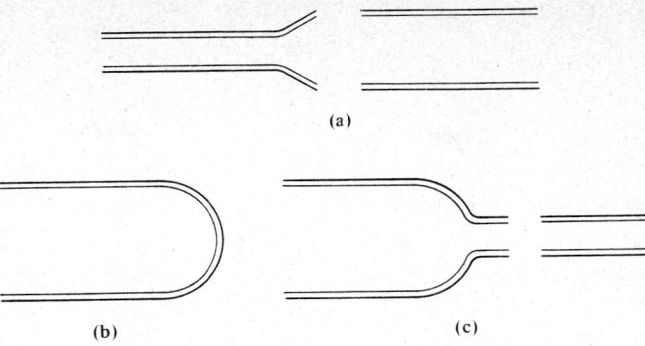

FIGURE A-2 / End-to-end seals.

thick, stretch the glass by pulling gently at the two ends to thin out the wall. Several heatings, stretchings, and blowings may be necessary to make a smooth uniform joint.

If the glass cannot be rotated, use the "spotting" method. Keep the joint warm with a large flame and use a pinpoint flame to melt a small part of the circumference at a time.

To make a T tube (Figure A-3) form a hole in the wall of one tube by heating a spot and then blowing it out. Flare the hole with a carbon rod or an iron file to form a flanged opening with the same diameter as the other tube. Seal the second tube to the flanged end of the first tube.

RING SEALS

Method 1. First flare the end of the smaller tube into a flange. Then insert the smaller tube into the larger one, as shown in Figure A-4a. Heat the contact areas until the tubes melt together at the contact surfaces. Remove any excess glass by blowing it out, if necessary.

Method 2. First make a hole in the larger outer tube. Then blow a slight bulge in the inner tube at the point where it is to be sealed, and, finally,

FIGURE A-3 / T-seals.

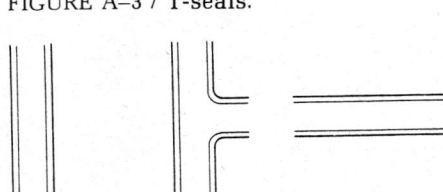

APPENDIX 6

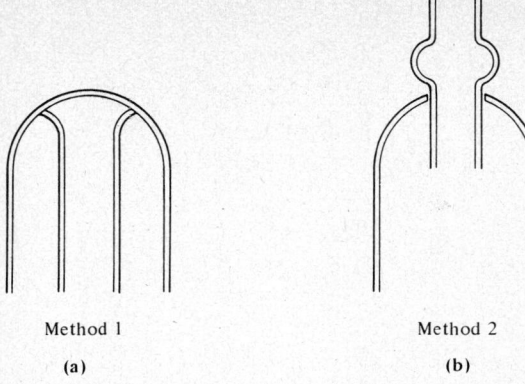

Method 1   Method 2
(a)   (b)

FIGURE A–4 / Ring seals.

insert the inner tube into the outer. The bulge will be in contact with the edge of the hole in the outer tube. Seal the glass at the contact area (Figure A–4b).

ANNEALING

This is probably the most important single step in glassworking. Glass is a poor conductor of heat. If it is allowed to cool rapidly, the outside will cool at a faster rate than the inside, setting up thermal strains that weaken and may shatter the glass. To prevent the development of these stresses, cool the glass slowly at a controlled rate in an oven. If a piece of glass is already stressed, anneal it by heating it slowly almost to the softening point and then cooling it slowly in an oven.

If a suitable oven is not available, let the hot glass cool in a luminous flame. The deposit of carbon that forms on it as it cools will act as a thermal insulator. This is not as satisfactory as using an oven, but with care it is adequate.

HEALING CRACKS

When cracks develop, heat the glass gently with a wide, luminous flame at points on either side of the crack. Then play the flame into the crack and allow the surfaces in contact to sinter, *not melt*, into each other, healing the crack (if you heat rapidly, the glass will melt and surface tension will pull the surfaces apart before they can sinter together). As the crack starts to disappear, gradually make the flame hotter, until the glass softens, eliminating the line where the crack had been. If a hole develops during the process, seal it with cane.

FIRE-POLISHING

Heat the edges in a flame until the glass softens and surface tension smoothes out any roughness. All rough and jagged edges should be fire-polished.

SEALING WIRE TO GLASS

The glass used must have about the same coefficient of thermal expansion as the wire, or on cooling the glass will either crack or pull away from the wire, producing a leak. Soft glass is used with platinum, and pyrex or Vycor is used with tungsten.

One simple method of sealing wire to glass is to insert the wire into a thick-walled capillary and then to slowly melt the capillary shut. Another method is to melt a bead of glass around the wire at the point where the seal is to be made. The bead is then sealed to glass.

Appendix 7

Electrical Work

SPLICING

Figure A-5 illustrates the various kinds of splices.

To connect two wires permanently, use the *Western Union splice*. When made on single strand conductor, this splice is strong, gives good electrical contact, and need not be soldered, although (like all splices) it is improved by soldering. If multistrand conductor is used, the Western Union splice must be soldered because the fine wires are too flexible to hold.

The *pigtail splice* is the simplest and easiest to make of all splices. However, it is bulky, hard to tape, and weak. For use in a permanent

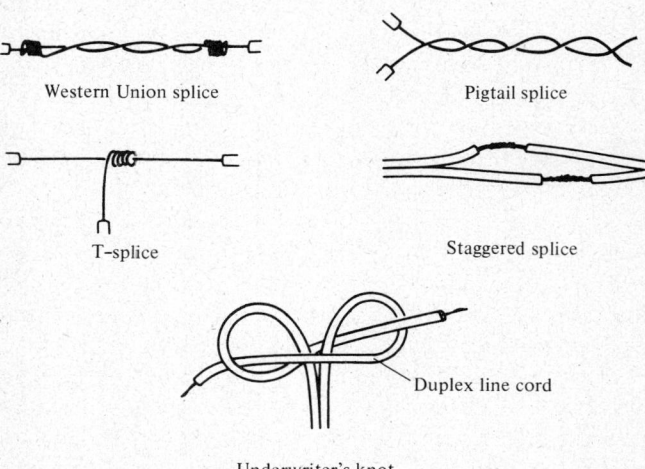

FIGURE A-5 / Slices and knots.

connection it must be fastened by soldering or by capping with a wire nut (also called a solderless connector).

For joining three leads, make a *T-splice*, either as shown in Figure A–5 or by twisting three wires into a pigtail. Always solder the T-splice unless it is made with stiff single strand conductor that will not unwrap under load.

When joining lengths of duplex insulated wire (lamp cord), use the *staggered splice* to make a neater joint with less danger of a short circuit.

When wiring an ordinary two-prong plug to duplex wire, make an *underwriter's knot* and loop the wires around the prongs to avoid pull on the binding screws if the plug is inadvertently pulled out by the cord.

Insulate all splices by wrapping with either vinyl electrical tape or friction tape. For a neater, more compact, and better-insulated splice, split the common $\frac{3}{4}$ in. friction tape into strips $\frac{3}{8}$ in. wide.

FUSING AND GROUNDING

Common 110 V AC is supplied to outlets through one grounded wire and one ungrounded wire. The accepted (although not always followed) color code is *black* insulation on the "hot" or ungrounded line and *white* insulation on the grounded line. The color code for DC is *red* for positive and *black* for negative. As a rule, the positive is at ground level. Unfortunately, it is not safe to rely on the color code, especially with old wiring. Always test the grounding of the lines. One convenient method is to place a test lamp across one wire and a known ground. If the lamp lights, the wire is hot.

For safety all circuits have fuses. The fuse should be in the ungrounded line between the power source and the outlet. It is wrong and often dangerous to put a fuse in the grounded line or to fuse both lines. If a fuse blows in the grounded line, everything between the fuse and the power source is still at the potential of the hot line, 110 V above ground, and cannot be handled without danger to the experimenter and/or the apparatus and supply lines. If it is necessary to handle live apparatus, the experimenter should himself be insulated from ground. Furthermore, he should work with only one hand, keeping the other either in his pocket or behind his back. Any switches in the apparatus should be placed in the hot line. Correct and incorrect placement of fuses and switches are shown in Figure A–6.

Proper fusing presents a problem for apparatus that is plugged into wall outlets with a two-prong plug. Unless the polarity of both the plug and the wall outlet are known, half the time the plug will be inserted with the fuse on the wrong line. To avoid this problem, use three-prong or polarized two-prong outlets and plugs which can be connected in only one way.

Instruments, especially those with metal cases or external metal parts, frequently have an independent ground connected to the external metal. This is a third wire, coded green, in the power line, which usually terminates at the ground prong of a three-prong plug. The independent ground keeps all

APPENDIX 7

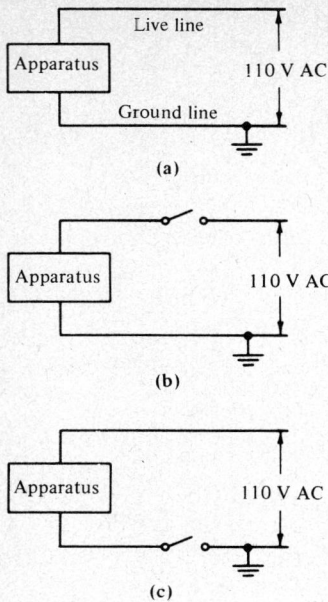

FIGURE A-6 / Types of grounds. (a) Closed circuit; (b) Live line open, apparatus at ground potential. (c) Ground line open, apparatus at line potential.

external metal parts at ground potential. As long as the circuit within the instrument remains insulated from the exposed metal, this ground does nothing except furnish some electrostatic shielding. However, if a short circuit develops between the circuit and the case, the fuse in the ungrounded line will blow, eliminating the possibility of receiving shocks from the instrument case. All instruments, especially those that operate at high voltages, should be properly fused and have grounded cases.

Appendix 8

Color Code for Electronic Components

Color Code for Electronic Components

Color	Significant Figure	Decimal Multiplier	Tolerance (%)	Voltage Rating (capacitors only)
Black	0	1	—	—
Brown	1	10	—	100
Red	2	100	—	200
Orange	3	1000	—	300
Yellow	4	10,000	—	400
Green	5	100,000	—	500
Blue	6	1,000,000	—	600
Violet	7	10,000,000	—	700
Gray	8	100,000,000	—	800
White	9	1,000,000,000	—	900
Gold	—	0.1	±5	1000
Silver	—	0.01	±10	2000

Resistor

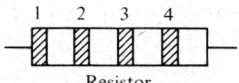
Resistor

1. First significant figure, resistance in ohms.
2. Second significant figure, resistance in ohms.
3. Decimal multiplier.
4. Tolerance (±20% if omitted).

Fixed Mica Capacitor

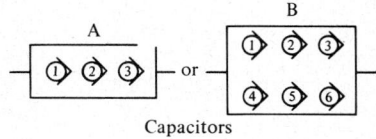
Capacitors

1. First significant figure.
2. Second significant figure.
3. Decimal multiplier (500 V, 20% tolerance only).

1. First significant figure.
2. Second significant figure.
3. Third significant figure.
4. Voltage rating.
5. Tolerance.
6. Decimal multiplier.

Appendix 9

Soldering

Soldering is a method for bonding pieces of metal. An alloy with a melting point lower than that of either metal is melted and allowed to cool and set in the joint. *Soft solders* are low-melting alloys of tin and lead. These have little mechanical strength and are best thought of as glues. *Hard solders* (sometimes known as silver solders or brazing alloys, depending on the composition) fuse at red heat and produce connections with considerable mechanical strength. Hard soldering is sometimes a useful alternative to welding. There are also special solders for specific applications, such as soldering aluminum.

The principles of soft soldering and hard soldering are similar. Differences in technique arise from differences in the melting point and the mechanical properties of the solders. In all cases the surfaces must be freed of dirt and grease and any oxide films must be removed by *fluxes*, materials which either reduce or dissolve the oxides.

Soft soldering is the method of choice because it is easier to do and because soft-soldered joints may be easily reopened by heating. However, soft solder cannot be used if either great mechanical strength or moderate temperature resistance is required, or if the surface oxide is tenacious and difficult to remove (as on stainless steel, aluminum, and magnesium), or if the metal has a high heat capacity and thermal conductivity (as aluminum and magnesium and their alloys).

Before the solder can adhere, the surface oxides must be removed by a suitable flux. The flux may be applied together with the solder (hollow wire solder with a core of flux is very convenient) or the flux may be applied directly to the metal, which is then heated. The heat may be applied with a torch, as in sweat soldering, or with some form of soldering iron (e.g., an electric soldering iron, a soldering copper, or a soldering gun). If a hot iron or a soldering gun is used, the copper tip must also be freed from oxide by first scraping off any scale and then "tinned" by applying a flux and solder. When a metal is properly cleaned and clear of oxide molten solder will wet and flow over the surface, coating it evenly and producing a "tinned" appearance. The most common fluxes for soft solder are *acid fluxes*, such as solid ammonium chloride or zinc chloride, *rosin*, which is a reducing flux, and various proprietary mixtures such as *Nokorode* soldering paste. Any of these is satisfactory for mechanical joints. For electrical and electronic connections, however, rosin or a similar nonelectrolytic flux must be used in order to avoid emf's produced by cells formed by the residual electrolyte from the flux and the metal of the leads or contacts.

To make electrical connections, splice the wires together, heat the joint with a small iron or gun, and apply rosin core solder directly to the heated joint. Continue heating until the solder melts completely and flows into the joint. Applying the solder to the iron just wastes most of the flux, so extra flux is necessary. A similar technique is used for connecting wires to plugs, switches, and terminals. Whenever potentials are important, solder all connections to avoid contact resistance and "noise."

Structural joints, as distinct from electrical connections, are best made by films of solder a few thousandths of an inch thick between the two matching surfaces. Pretinning the pieces produces a sounder joint because it ensures that all parts of both surfaces will be wet by the solder and will therefore bond. However, pretinning is more trouble, and it is not necessary if the soldered area is large enough to make a strong joint even if some regions are not bonded. It cannot be done in cases where the fit of two matched surfaces might be spoiled.

To make pretinned joints, apply flux and solder to each separate surface and heat. Remove the excess solder by quickly wiping the surface with a clean rag while the solder is still liquid. Place the tinned surfaces together or clamp them, if necessary, and heat. Extra solder may be fed in at the joint.

To make a joint without pretinning the metal, put flux on the cleaned surfaces and place them in contact, clamping if necessary. Heat the joint and feed solder in from one or both edges.

Hard soldering, or *brazing*, must be done with a torch, using a nonoxidizing flame, or in a furnace. An iron will not reach the 600°C or more required to melt hard solder. The most common flux is borax, which removes metal oxides by forming a brittle borate glass which is easily cracked off after the joint has cooled. There are also various proprietary fluxes which do not leave an undesirable residue. The solder may either be placed in or on the joint before heating or be fed into the joint during heating.

To hard solder two wires together, crimp a small piece of hard solder around the joint and then paint the joint with a paste of borax in water. Heat with a torch or burner until the solder melts and runs into the joint. To braze a flat surface, make a sandwich of the two metal pieces and thin sheet of hard solder the size and shape of the joint. Paint the solder and the surfaces with flux, clamp the assembly if necessary, and heat until the solder melts and bonds the joint.

Index

Abbe refractometer, 223–25
Absorbance, 231
Absorption spectrophotometry
 ionization constant by, 402–405
 instrumentation for, 241–49
 molecular structure and properties by, 232–37
 in quantitative analysis, 231
Absorptivity, 231
AC, 153–67
 capacitance, 155–57, 165–67
 impedance, 159–61
 inductance, 157
 filter circuits, 161–63
 measurement
  of capacitance, 165–67
  of dielectric constant, 165–67
  of inductance, 165
 power supplies, 172–73
 reactance, 155–61
Accuracy, 9
Actinometry, 239–41, 475–78
Activity
 of electrolyte, 372–75
 by potentiometry, 193
 of solvent by vapor pressure, 369–71
Activity coefficient, of electrolyte, 372–75
Adiabatic calorimeter, 63, 65–66
Adsorption
 Gibbs isotherm, 131–32, 435–36
 on liquid surfaces, 130–32
 on solid surfaces, 132–34, 440–44
  BET apparatus, 441–42
  BET method, 133–34, 440–44
  Freundlich isotherm, 133
  Langmuir isotherm, 132–33
Ammeter, 142–44
Amperometry, 207–16
 see also Polarography
Amplifiers, 174–81
 electron multiplier, 178
 feedback, 179–81
 frequency response, 176
 impedance
  input, 174–75
  output, 175–76
 linearity, 177
 noise, 177–78
 power gain, 174
 power output, 176
 stability, 176–77
Appearance potential, in mass spectrometry, 298–301, 463–64
Arithmetic mean, 11–13

Atomic weights of elements, 491–93
Averages, method of, 26–29

Balmer spectral series, 229, 465–67
Barometer, 82–83
 temperature corrections for, 500
Beckmann cryoscopic apparatus, 107–108, 364–68
Beckmann differential thermometer, 54–56
Beer–Lambert law, 231
BET method, 133–34, 440–44
 apparatus for, 441–42
Bilayer lipid membranes, 136
Bohr magneton, 270–71
Boiling point, 99, 100–104
Bomb (constant volume) calorimeter, 72–74, 379–82
Böttcher equation, 451
Bragg relation, 256–57
Brønsted equation, 425
Bridges
 AC, 164–65
 Wheatstone, 151–52
Brunauer–Emmett–Teller (BET) equation, 133–34
Bubble pressure method, for surface tension, 129–30

Calomel electrode, 190–91
Calorimeter, 63–69
 constant volume (bomb), 72–74, 379–82
 reaction, 67–72
  isothermally jacketed, 69–70
  two-container (Condon), 70–72, 483–85
Capacitance, 155–57, 165–67
 differential, 167
Chemical kinetics, techniques for fast flow methods, 317–19
 continuous flow, 318
 rapid quenching, 317–18
 stopped flow, 318–19, 431–34
 flash photolysis, 320
 photochemical methods, 321
 relaxation methods, 319–20, 427–30
Chemisorption, 132–33
Choppin–Cottrell apparatus, 396–97
Clausius–Clapeyron equation, 99–100
Cold cathode ionization gauge, 87
Cold traps, 93–94
Computers, 324–42
 analog, 324–25

Computers (Cont.)
 components, 329–33
 language, 337–40
 logic, 325–27
 programming, 333–37
 terminology, 329–32
Concentration-jump method, for kinetics of fast reactions, 427–30
Condon calorimeter, 70–72, 483–85
Conductance, 200–204, 376–78, 406–408
 equivalent, 201–204
 ionic mobility by, 376–78
 ionization constant by, 406–408
 measurement of, 203–204
 specific, 201
 of water, 203
Confidence limits, 13–14, 17–18
Cooling curves, 108–12
Cryoscopic constant, 366–68
Cryoscopy, 105–108
 molecular weight by, 105–106, 364–65
Current, electrical, measurement of, 142–44

D'Arsonval galvanometer, 142–43
DC, 140–53
Debye–Hull–Scherrer powder method of X-ray diffraction, 264–68, 356–59
Decomposition, of hydrogen peroxide, 418–20
Decomposition pressure, of ammonium carbamate, 413–14
Definitions, 494
Density
 of liquids, 75–78
  by pycnometer, 75–77
  by Westphal balance, 77–78
 of mercury, 501
 of solids, 77
 of water, 498
Deviations
 average, 14
  of the mean, 16
 probable, 15
  of the mean, 16
 root mean square, 15
  of the mean, 16
 standard, 15
  of the mean, 16
Diamagnetism, 272
Dielectric constant
 by bridge method, 165–67
 cell for, 449
 in dipole moment determination, 445–52

Index

Differential scanning calorimetry, 111–12
Differential thermal analysis, 109–11
Differentiation, graphical, 37–41
   limiting secant method, 38–39
Diffusion pump, 91–92
Digital instruments, 184–85
Dimerization, of acetic acid, 415–17
Dipole moment, 445–52
   Böttcher equation for, 451
   see also Magnetic dipole moment
Dissociation energy
   mass spectrometry, 298–301, 463–64
   by ultraviolet spectrometry, 234–37, 453–54
Dropping mercury electrode, 213–15
Du Noüy tensiometer, 125–26

Electrical components, color code for, 511
Electrical double layer, 186–88
Electrical work, 508
Electrode potential, equilibrium, 186–99
   applications, 193–99
   measurement of, 189–93
Electrodes, 189–98
   calomel, 190–91
   dropping mercury, 213–15
   pH, 196–98
   silver/silver chloride, 190–92
Electrokinetics, 312–14, 483–87
   electroosmosis, 313–14, 485–86
   second electroosmosis, 485–87
   streaming current, 312, 314, 486
   streaming potential, 312, 314, 485–86
Electrometer, 147
Electron affinity, by polarography, 488–89
Electron spin resonance, 277–88
   of *p*-benzosemiquinone radical anion, 473–74
   hyperfine coupling constant, 278–83
   instrumentation for, 283–88
   field modulation, 284–87
   high field spectrometry, 283–87
   low field spectrometry, 287–88
Emf
   activity coefficients from, 194–95, 372–75
   entropy, enthalpy, and free energy from, 195, 386–88
   measurement of, 192–200
   cells for, 198–99
   standard cell for, 192–93
   solubility product from, 195, 411–12
   transference number from, 206–207
Energy of dissociation
   by mass spectrometry, 298–301, 463–64
   by ultraviolet spectrometry, 234–37, 453–54

Enthalpy
   of combustion, 379–84
   of formation, 379–80
   of reaction, 386–88
   of vaporization, 354–55
Entropy, of reaction, 386–88
Eötvos equation, 125
Error
   analysis of, in report, 47
   of calculated result
   maximum, 20
   most probable, 19–20

Feedback, 179–81
Fick's law of diffusion, 208, 303
Filter circuits, 161–63
Fluorimetry, 230–31
   pK^* by, 479–82
Force constant, of harmonic oscillator, 233, 469
Freezing point, 105–108
   determination by Beckmann method, 106–108
Freundlich isotherm, 133
Fuses, 508

g-factor, for electronic and nuclear spin, 271, 274, 277
Galvanometer
   ballistic, 152–53
   d'Arsonval, 142–43
Gas
   flow-rate measurement, 96–97
   valves for, 94–96
   viscosity, 120–23, 345–46
Gas buret, 418–19
Gas thermometer, 51–52
Gibbs absorption isotherm, 131–32, 435–36
Gibbs–Duhem equation, 390
Glassworking, 502
Graphical methods
   differentiation by, 37–39
   integration by, 34–37
   location of maxima and minima by, 39–41

Half wave potential, 209–12
Hartmann diaphragm, 245
Hartmann formula, 246
Hazards, health and safety, 5–7
Hund's rules, 273–74
Hydrolysis, of sucrose, 421–23

Impedance, 159–61, 174–76
   bridges, 164–65
Inductance, 157, 165
Infrared spectrophotometry, 232–34, 248, 468–72
Integration, 34–37
   by counting, 34
   by planimetry, 34
   Simpson's method, 36–37
   trapezoidal method, 35–36
Internal manometer, 80–82, 400–401, 413–17
Intrinsic viscosity, 114
Ion exchange membrane, 134–36
Ionic mobility, 376–78
Ionic strength, effect on rate of reaction, 424–26
Ionization constant
   by conductance, 406–408

Ionization constant (*Cont.*)
   by spectrophotometry, 402–405
Isoteniscope, 100–102
   mercury, 80–83, 400–401, 413–17
Isotope effect, on molecular vibrations, 228–29, 471–72

K-factor, in confidence limit, 13–14
Kirchoff's conservation laws, 140
Kohlrausch's law, 407

Landé g-factor, 271, 274, 277
Langmuir isotherm, 232–33
Lasers, 240–43
Leak testing, of vacuum systems, 89
Least squares, method of, 29–32
Linear forms of common equations, 26
Lissajous figures, 182–83

McCabe–Thiele diagram, 398
McConnell relationship, 474
McLeod gauge, 84–86
Magnetic angular moment, 270
Magnetic dipole moment, 269–74
   susceptibilities and, 271–75, 458–60
   unpaired spins and, 273
Magnetic resonance methods, 275–88
   electron spin resonance, 277–88, 473–74
   hyperfine coupling constant, 278–83
   instrumentation for, 283–88
   Zeeman effect, 275–77
Manometers, 79–82
Mass spectrometry
   applications, 294–301, 461–64
   dissociation energy, 298–301, 463–64
   gas analysis, 294–96, 461–62
   kinetics, 297
   instrumentation for, 290–94
Membranes, 134–37
   bilayer lipid, 136
   ion exchange, 134–36
   plasma, 137
Mercury, density, 501
Mercury-in-glass thermometer, 52–54
Miller indices, 257–59
Molar polarization, 446–48
Molar refraction, 222, 446–47
Molecular weight
   by cryoscopy, 105–106, 364–65
   of polymer by viscometry, 456–57
Moving boundary apparatus, 204–206, 376–78

Nicol prism, 217–18
Notebook, laboratory, 42–44
Nuclear magneton, 271
Nuclear spins, table of, 271

Ohmmeter, 150–51
Ohm's law
   for AC, 156

515

Index

Ohm's law (*Cont.*)
 for DC, 140
Onsager equation for conductance, 201–202
Onsager relation, 309
Oscillating crystal method, of X-ray diffraction, 260–63
Oscilloscopes, 181–83
Ostwald–Sprengel pycnometer, 75–77
Ostwald viscometer, 116–18, 351–53

Paramagnetism, 271–72
Partial molar volume, 389–91
Partition function, vibrational, 233, 469
Pascal's triangle, 281–82
Perdisulfate oxidation of iodide ion, 108–11
pH, measurement of, 196–98
pH electrode, 196–98
Phase diagram
 of liquid–solid system, 392–94
 of liquid–vapor system, 395–99
Photochemistry, 237–41, 475–78
pK
 by fluorimetry, 479–82
 by spectrophotometry, 402–405
Poiseuille's equation for liquid viscosity, 116
Polarimetry, 218–19, 421–23
Polarography, 208–16
 applications, 209–13
 composition of complex ions, 210–12, 409–10
 electron affinity, 488–89
 kinetics, 212–13
 reversibility, 209–10
 techniques, 213–16
Potentiometer, 148–50
Potentiometry, 193–200, 372–75, 411–12
Power sources, 168–73
 AC, 172–73
 DC, 168–72
 batteries, 168–70
 from AC, 169–72
 voltage regulation, 170–72
Precession camera, 264
Precision, 9–11, 18–20
Prefixes for indicating magnitude, 494
Pumps
 diffusion, 91–92
 rotary, 90–91
 Toepler, 89–90
Pycnometer, 75–78

Radiation
 absorption
  actinometry, 239–41, 475–78
  analysis by, 231
  instrumentation for, 241–49
  selection rules for, 228–30
  theory of, 226–28
 emission
  fluorescence, 230–31
  hydrogen spectrum, 229–30, 465–67
  theory of, 226–28

Radiation, emission (*Cont.*)
 wavelength determination by, 244–46
Ramsey–Shields–Eötvos equation, 125
Ramsey–Young vapor pressure method, 102–104, 369–71
Raoult's law, 100
Reactance, 155–61
Recorders, 183–84
Refractometry, 221–25, 445–52
Relaxation techniques, 319–20, 427–30
Resistance thermometers, 56–58
Resistors, color code for, 511
Resonance stabilization energy, by constant volume calorimetry, 381–82
Rosano tensiometer, 127–28
Rotating crystal method of X-ray diffraction, 260–63, 360–61, 362–63

Safety precautions, 4
Selection rules for diatomic spectra, 228–30
Significant figure operations, 10–11
Soldering, 512
Solubility product, by potentiometry, 195, 411–14
Spectroscopic reference lines, 245
Speed of sound, 347–50
Stopped flow method for kinetics of fast reactions, 318–19, 431–34
Straight line
 estimate of errors in parameters, 33
 by eye, 23
 method of averages, 26–28
 method of least squares, 29–32
Sublimation pressure, 400–401
Surface area of powder, 440–44
Surface pressure, 130–32, 437–39
Surface tension, 124–30
 by bubble pressure, 129–30
 by capillary rise, 128–29
 by du Noüy tensiometer, 125–26
 molecular cross sectional area by, 435–36
 by Rosano tensiometer, 127–28

t-factor for confidence limit, 17–18
Temperature
 control of, 60–62
 correction for, of barometer readings, 500
 measurement of, 50–62
  Beckmann thermometer, 54–56
  gas thermometer, 51–52
  mercury-in-glass thermometer, 52–54
  resistance thermometer, 56–58
  thermocouple thermometer, 58–60, 311–12
Tensiometers
 du Noüy, 125–26

Tensiometers (*Cont.*)
 Rosano, 127–28
Thermal analysis, 108–109, 392–94
Thermogravimetry, 112
Thermostats, 60–62
Transducers, 83–84
Transference number, 204–207
 by emf method, 206–207
 by moving boundary method, 204–206, 376–78

Ubbelohde dilution viscometer, 118–19, 456–57
Units, SI, 494

Vacuum gauges, 84–87
Vacuum pumps, 89–92
Vacuum systems, 87–94
Valves
 needle, 94–95
 reducing, 95–96
Vapor pressure
 by internal manometer method, 80–83, 400–401, 413–17
 by isoteniscope method, 100–102
 of liquid, versus temperature, 354–55
 by Ramsey–Young method, 102–104, 369–71
 of solvent, 369–71
 by transpiration method, 104
Velocity of gas molecules, 347–50
Viscometers
 Hoepler, 120
 Ostwald, 116–18, 351–53
 Saybolt, 119
 Ubbelohde, 118–19, 456–57
Viscosity
 of gas, 120–23, 345–46
 of liquid, 113–20, 351–53
 Poiseuille equation, 116
 of polymer solution, 113–14, 118–19, 456–57
Voltage regulation, 170–73
Voltmeter, 144–47

Water
 conductance of, 203
 relative density and volume, 498
Wavelength, by emission spectrometry, 244–46
Weissenberg X-ray camera, 264
Weston cell, 192–93
Westphal balance, 77–78

X-ray diffraction, 256–58
 powder method, 264–68, 356–58
 single crystal methods
  moving film, 263–64
  oscillating-crystal, 263
  rotating-crystal, 260–63, 360–61, 362–63
X rays
 beam production, 251–56
 Bragg relation, 256–57
 detection, 255
 theory, 256–59
 white radiation, 252

Zeeman energy, 275–78

516